Hazardous Materials Chemistry *for* Emergency Responders

Second Edition

Robert Burke

LEWIS PUBLISHERS

A CRC Press Company
Boca Raton London New York Washington, D.C.

Library of Congress Cataloging-in-Publication Data

Burke, Robert
Hazardous materials chemistry for emergency responders / Robert Burke.--2nd ed.
p. cm.
Includes bibliographical references and index.
ISBN 1-56670-580-0
1. Hazardous substance. I. Title.

T55.3.H3 B87 2002
604.7--dc21 2002073195

Direct all inquiries to CRC Press LLC, 2000 N.W. Corporate Blvd., Boca Raton, Florida 33431.

Visit the CRC Press Web site at www.crcpress.com

Lewis Publishers is an imprint of CRC Press LLC

International Standard Book Number 1-56670-580-0
Library of Congress Card Number 2002073195
Printed in the United States of America 2 3 4 5 6 7 8 9 0
Printed on acid-free paper

Preface

There has been much discussion about the validity of emergency response personnel studying chemistry. Obviously, from the title of this book, the author's opinion is clear; however, the chemistry presented here is more "street chemistry" than college chemistry. I have tried to make the chemistry subject matter appropriate for response personnel and to express the information in understandable terms.

The book is organized into the U.S. Department of Transportation's (DOT) nine hazard classes, with which emergency response personnel should be intimately familiar. Each hazard class will be covered in its own chapter.

Almost every hazardous material presents more than one hazard; the DOT's placarding and labeling system only identifies the most severe hazards. Along with the hazard classes, the typical information is provided about each hazard class. Individual chemicals are discussed, along with their hazards and their physical and chemical characteristics, both as distinct chemicals and within chemical families. Furthermore, the multiple dangers of hazardous materials, including "hidden" dangers, are studied throughout the book.

Common industrial chemicals, along with other hazardous materials, are presented throughout the book. Learning about these chemicals will provide responders with an overview of the varying dangers presented by hazardous materials, and will show the similarities and differences among chemical family members, as well as other hazardous materials. Selected reports of various incidents involving hazardous materials are presented to emphasize the effects that chemical and physical characteristics can have on an incident outcome. Responders should become familiar with the chemical terminology that they will encounter when researching reference sources, including books, computer databases, material safety data sheets (MSDS), and shipping papers. Basic chemistry provides emergency responders with a background that will help them not only to understand chemical terminology, but also to talk intelligently with CHEMTREC (Chemical Transportation Emergency Center), shippers, and industry representatives about chemicals involved in hazardous materials incidents.

Robert Burke, BA, CFPS
Hazardous Materials
Specialist/Instructor

Dedication

This book is affectionately dedicated to my daughter Ashley.

The Author

Robert A. Burke, born in Nebraska and raised in Illinois, earned an A.A. in Fire Protection Technology from Catonsville Community College and a B.S. in Fire Science from the University of Maryland. He has also completed graduate work at the University of Baltimore in Public Administration. Mr. Burke has over 25 years experience in the emergency services as a career and volunteer firefighter, and has served as an assistant fire chief for the Verdigris Fire Protection District in Claremore, Oklahoma; deputy state fire marshal in the state of Nebraska; a private fire protection and hazardous materials consultant; an exercise and training officer for the Chemical Stockpile Emergency Preparedness Program (CSEPP) for the Maryland Emergency Management Agency; and is currently the fire marshal for the University of Maryland, Baltimore. He has served on several volunteer fire companies, including West Dundee, Illinois; Carpentersville, Illinois; Ord, Nebraska; and Earleigh Heights Volunteer Fire Company in Severna Park, Maryland, which is a part of the Anne Arundel County, Maryland, Fire Department.

He is an adjunct instructor at the National Fire Academy in Emmitsburg, Maryland, and the Community College of Baltimore County, Catonsville Campus. Mr. Burke is also a contributing editor for *Firehouse Magazine*, with a bimonthly column titled "Hazmat Studies," and he has also had numerous articles published in *Firehouse, Fire Chief*, and *Fire Engineering* magazines. He writes a monthly column for the Firehouse.com Web site titled "The Street Chemist."

Mr. Burke has developed several CD-ROM-based training programs, including the Emergency Response Guide Book, Hazardous Materials and Terrorism Awareness for Dispatchers and 911 Operators, Hazardous Materials and Terrorism Awareness for Law Enforcement, Chemistry of Hazardous Materials Course, and Chemistry of Hazardous Materials Refresher. He has also published a second book titled *Counterterrorism for Emergency Responders*. He can be reached on the Internet at *robert.burke@att.net.*

Acknowledgments

This book would not have been possible if not for the hundreds of emergency response personnel and fellow instructors from various organizations whom I have had the distinct pleasure to know, teach with, and learn from during classes at the National Fire Academy and across the country. It was their inspiration, friendship, and support that gave me the desire and dedication to put together a book of this type. Many of the ideas for material presented here have come from the input of these responders, students, and instructors; for them, and the many lasting friendships that have developed over the years, I am grateful.

Contents

Chapter 1

Introduction

Mention the word *chemistry* to the average firefighter or other emergency responder and it strikes fear in his heart. Maybe that is a bit overstated, but it certainly creates a great deal of anxiety. Most emergency personnel do not want any part of chemistry. Actually, chemistry is all around us. Everything on Earth, including the human body (see Figure 1.1), is made up of one of the chemical elements or a combination of those elements from the Periodic Table.

In 1869, Russian chemist Dmitri Mendeleev put forth the theory of periodicity. Through his work, the first Periodic Table of Elements was developed. His efforts were the beginning of the understanding of the concept of elements, which led to the understanding of compounds, or what we refer to today as hazardous materials.

The modern-day coinage of the term "hazardous material" occurred in the mid-1970s when the United States Department of Transportation (DOT) established a definition of hazardous material. DOT began the first major regulation of hazardous materials in transportation, including a hazard class and placard and label system for the identification of hazardous materials. During this same time period, the first emergency response guidebook was developed.

As other federal agencies began developing regulations dealing with hazardous materials storage and use, different names were also created. The United States Occupational Safety and Health Administration (OSHA) and the United States Environmental Protection Agency (EPA) both refer to hazardous materials as "hazardous substances." The EPA also regulates chemicals that no longer have a commercial value. When chemicals are no longer useful for the purpose they were intended, they become hazardous waste. Hazardous waste is regulated in the workplaces where it is generated, during transportation to a disposal site, and when it is disposed of. For example, gasoline, when transported, is a *hazardous material* regulated by DOT. When a gasoline tanker offloads gasoline into an underground storage tank at a gasoline station, it becomes a *hazardous substance* regulated by OSHA and EPA. If some of the gasoline were spilled on the ground during the offloading, it would become *hazardous waste*, regulated by OSHA, EPA, and DOT. Three different names for the same gasoline, depending on whether it was transported, in fixed storage, or spilled. For the purposes of this book, we will use the term "hazardous material" interchangeably with all other agency terminology.

Chemical Makeup of the Human Body

Oxygen	65%	Sulfur	
Carbon	18%	Iron	
Hydrogen	10%	Sodium	
Nitrogen	0.3%	Zinc	1%
Calcium	1.5%	Magnesium	
Phosphorus	0.1%	Silicon	
		Potassium	

Figure 1.1

On December 3, 1984, Bhopal, India, experienced a release of approximately 40 metric tons of methyl isocyanate (MIC) at the Union Carbide pesticide plant. Over 100,000 were injured and 3000 people were killed, and many more are likely to die from the long-term effects. The accident occurred around 12:40 a.m. local time, when most of the victims were sleeping. The dead included large numbers of infants, children, and older men and women. These age groups are often adversely impacted by toxic exposures.

It was this incident that led to the Emergency Planning and Community Right-To-Know Act of 1986 (EPCRA). This same chemical has been released on several occasions from the Union Carbide plant in Institute, West Virginia, shortly after Bhopal and as recently as 1996. Fortunately, these releases did not affect the surrounding community.

Following the incident in Bhopal, the U.S. Congress was concerned that such an incident could happen here. It was also concerned with the level of preparedness and training available to deal with an incident of the magnitude of Bhopal. In 1986, Congress passed the Emergency Planning and Community Right-to-Know Act (EPCRA), sometimes referred to as the Superfund Amendments and Reauthorization Act (SARA).

With the passage of this important legislation, the federal government for the first time mandated training and competency for emergency responders to hazardous materials releases. Congress mandated that the EPA also create a list of extremely hazardous substances, most of which are poisons, that would require reporting under EPCRA by companies manufacturing, storing, or using them. EPCRA created the State Emergency Response Commission (SERC) and the Local Emergency Planning Committee (LEPC) to facilitate the reporting and oversight process. Also part of the legislation was the requirement that information about extremely hazardous substances be made available to the public so they know what types of hazardous materials may be found in their communities.

According to the EPA, 85 million Americans live, work, and play within a 5-mile radius of 66,000 facilities handling regulated amounts of high-hazard chemicals. Regulations also require that local emergency responders, particularly the fire department, be given access to chemical facilities for the purposes of preplanning for emergencies. The EPCRA also called for the distribution of Material Safety Data Sheets (MSDS) to the local fire department.

The National Fire Protection Association (NFPA) also has developed standards that set forth competencies that emergency responders should be able to display concerning hazardous materials response. Among the competency requirements are certain chemical and physical characteristics of hazardous materials of which responders must have knowledge. From the requirements of EPCRA, and OSHA and NFPA training competencies, has evolved the study of the chemistry of hazardous materials, sometimes called "street chemistry." You do not need to be a chemist to safely and effectively respond to hazardous materials incidents. The information contained in this book is intended to cover the chemistry requirements of all levels of EPCRA, OSHA, EPA, DOT, and NFPA hazardous materials regulation and response concepts and well beyond.

The difficulty of the subject of chemistry is not the material itself; it is in convincing emergency response personnel that they need to study the concepts to make them better hazardous materials responders at all levels.

Firefighters learn about fire behavior as part of their basic training in order to have a better understanding of fire. Fire behavior is part of the chemistry of fire, which is really a chemical chain reaction. Understanding fire and how it behaves helps firefighters extinguish fires safely and effectively.

Emergency medical personnel take courses to learn how to care for sick and injured patients. Sometimes, drugs are used by paramedics as part of the initial treatment. Drugs are chemicals and can be hazardous if they are not handled properly. Emergency medical personnel learn about drugs and treatment techniques in order to effectively treat their patients. Law enforcement personnel take courses in criminal justice to better prepare for their job of enforcing the law. All emergency response personnel receive some type of basic training to better prepare for their jobs. Responders are called upon daily to deal with incidents of all kinds that often involve hazardous chemicals. In order for personnel to better understand the hazardous chemicals they face on a daily basis, an understanding of basic chemistry principles and terminology is as essential as other emergency response training.

This book presents emergency response personnel with a view of chemistry as it applies to the hazardous materials that may be encountered in any emergency response. Some of the concepts presented may bend the rules of chemistry a bit. However, the purpose of this book is not to educate chemists, but rather to teach response personnel about basic chemistry concepts in a format that most responders, regardless of educational background, can understand. Concepts taught will work in the street application of chemistry when dealing with hazardous materials.

This book also presents firefighters, police, and emergency medical services personnel (EMS) with some basic tools to assist them in understanding hazardous materials and their behavior. This information may help keep them from being injured or killed at the scene of a hazardous materials incident.

HAZARDOUS MATERIALS STATISTICS

The Chemical Abstract Service (CAS) lists over 63,000 chemicals used outside the laboratory environment, and the number increases each year. The DOT regulates over 3800 hazardous materials in transportation, as listed in 49 CFR. Chemicals are also listed by the DOT in the *Emergency Response Guide Book* (ERG). While specific

EPA Top 15 Chemical Accidents

1. Polychlorinated Biphenyls
2. Anhydrous Ammonia
3. Sulfuric Acid
4. Chlorine
5. Hydrochloric Acid
6. Ethylene Glycol
7. Sulfur Dioxide
8. Radioactive Material
9. Benzene
10. Hydrogen Sulfide
11. Sodium Hydroxide
12. Vinyl Chloride
13. Toluene
14. Mercury
15. Ethylene Oxide

Figure 1.2

information on hazardous materials in the ERG is limited, the current edition uses chemistry terminology as a heading in each of the orange guide pages. OSHA regulates the occupational exposure of over 600 hazardous substances. Other lists of chemicals are compiled and regulated by various governmental regulatory agencies, such as the EPA. EPA has several listings of hazardous materials, depending on whether they are in storage or hazardous wastes. The Resource Conservation and Recovery Act (RCRA) deals with hazardous wastes exhibiting the characteristics of ignitibility, corrosivity, and reactivity. Regulations are found in 40 CFR 261.33. The Comprehensive Environmental Response, Compensation, and Liability Act of 1980 (CERCLA) lists hazardous substances in 40 CFR 302, Table 302.4. The Clean Air Act (CAA), Section 112r also publishes a list of toxic chemicals for accidental-release prevention. EPA has a listing of the top 15 chemicals released from chemical accidents in the United States (see Figure 1.2). These 15 chemicals account for two-thirds of all chemical releases.

Over 70% of all hazardous materials on the rail are shipped in tank cars.

Top 23 Hazardous Materials Shipped by Rail

1. Liquified Petroleum Gas
2. Sodium Hydroxide
3. Sulfuric Acid
4. Molten Sulfur
5. Anhydrous Ammonia
6. Elevated Temperature Material
7. Chlorine
8. Methyl Alcohol
9. Fuel Oil (Flammable)
10. Vinyl Chloride
11. Phosphoric Acid
12. Stryene Monomer, Inhibited
13. Carbon Dioxide Cryogenic
14. Hydrochloric Acid
15. Fuel Oil (Combustible Liquid)
16. Ammonium Nitrate
17. Sodium Chlorate
18. Gasoline
19. Butadiene
20. Phenol Molten
21. Methyl Tert Butyl Ether
22. Adipic Acid
23. Crude Oil, Petroleum

Figure 1.3

The National Institute for Occupational Safety and Health (NIOSH) also lists hazardous materials that may pose a health hazard to emergency responders. They publish the *NIOSH Pocket Guide to Chemical Hazards*, which can be obtained free from the agency. During the statistical period from 1996 to 2000, over 6000 transportation accidents involving toxic chemicals were reported in the United States. Those accidents resulted in more than 50 deaths, hundreds of injuries, and thousands of persons evacuated from their homes or sheltered in place. Some of these chemicals are listed in detail in the appropriate chapters of this book.

According to the Association of American Railroads (AAR), over 70% of hazardous materials are transported in tank cars. In 2000, there were 994 nonaccident releases of hazardous materials from railcars. Tank cars were responsible for 88% of those releases. Figure 1.3 shows the top 23 hazardous materials shipped by rail. In 2000, there were 72 train accidents involving hazardous materials cars. During these accidents, there were no fatalities and one injury. The tank car shipments by DOT hazard class, with number of leaks by hazard class, are listed in Figure 1.4.

TANK CAR LEAKS BY HAZARD CLASS

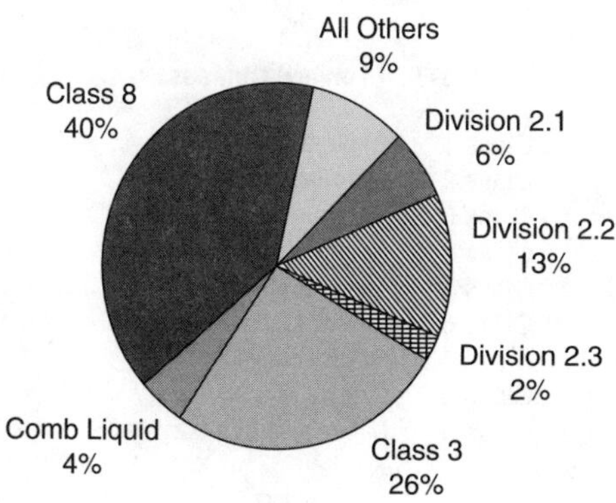

Figure 1.4

Top 10 DOT Transportation Incidents

1. Corrosive Liquids
2. Flammable Liquids
3. Poisonous Materials
4. Miscellaneous Hazardous Materials
5. Non-Flammable Gases
6. Oxidizers
7. Combustible Liquids
8. Flammable Gases
9. Organic Peroxides
10. Infectious Substances

Figure 1.5

The DOT reports that there were 15 deaths, 247 injuries, and damage costs of over $74 million from 17,552 incidents involving hazardous materials in the year 2000. This is a decrease of over 140 incidents from 1999; the number of deaths increased and injuries decreased during the same period. The incidents are broken down by mode of transportation: highway: 15,047 and railway: 1066. Deaths, injuries, and damage occurred most often in incidents involving corrosive and flammable liquids. The top 10 commodities released during the incidents were all flammable or corrosive liquids. The top 10 chemicals released during 2000 are shown in Figure 1.5.

DOT/UN HAZARD CLASSES OF HAZARDOUS MATERIALS

Chemicals regulated by the DOT are listed in the hazardous materials tables in 49 CFR 100–199. This book will use the nine DOT/UN hazard classes to group hazardous materials. The hazard classes are shown in Figure 1.6.

The hazard classes recognized by the DOT are: Class 1, Explosives; Class 2, Compressed Gases; Class 3, Flammable Liquids; Class 4, Flammable Solids; Class 5, Oxidizers; Class 6, Poisons; Class 7, Radioactives; Class 8, Corrosives; and Class 9, Miscellaneous Hazardous Materials. Class 1 Explosives are subdivided into six subclasses: 1.1 through 1.6. Class 2 Compressed Gases have three

DOT/UN Hazard Classes

Class 1	Explosives
Class 2	Compressed Gases
Class 3	Flammable Liquids
Class 4	Flammable Solids
Class 5	Oxidizers
Class 6	Poisons
Class 7	Radioactives
Class 8	Corrosives
Class 9	Miscellaneous HazMat

Figure 1.6

DOT Classification of Materials with More Than one Hazard

1. Radioactive
2. Division 2.3: Poison Gas
3. Division 2.1: Flammable Gas
4. Division 2.2: Non-Flammable Gas
5. Division 6.1: Poisonous Liquid, Inhalation Hazard

The above chart does not apply to the following: Class 1 explosives, 5.2 organic peroxides, Division 6.2 infectious substances, and wetted explosives such as picric acid.

Figure 1.7

subclasses: 2.1, Flammable; 2.2, Non-flammable, and 2.3, Poison. Class 4 Flammable Solids have three subclasses: 4.1, Flammable solid; 4.2, Spontaneously combustible; and 4.3, Dangerous when wet. Class 5 Oxidizers have two subclasses: 5.1, Oxidizers and 5.2, Organic peroxide. Class 6 Poisons have two subclasses and some special classifications of placards. Subclass 6.1 is poisons that are liquids or solids. Subclass 6.2 is infectious substances. There are also placards in Class 6 for *keep away from foodstuffs,* and *marine pollutant.* Class 7 Radioactives have no subclasses, but the placard is used only when materials bearing the Radioactive III label are shipped. Class 8 Corrosive materials do not have any subclasses, however, there are two distinctive chemicals in Class 8: acids and bases. Class 9 Miscellaneous Hazardous Materials do not have any subclasses.

Each hazard class has associated placards and labels identifying the hazards of the class during transportation. Each hazard class and associated placard has a color, which indicates a particular hazard. Hazardous materials may have more than one hazard. It is important to note that the placard on a transport vehicle depicts only the most severe hazard of a material as determined by DOT hazard class definitions. When a material has more than one hazard, the DOT prioritizes the hazard that will be placarded. These hazards are listed by the DOT in 49 CFR 173.2a (see Figure 1.7) to determine which hazard will be assigned to a particular material when the material has multiple hazards.

Almost every hazardous material has more than one hazard. As an emergency responder, you must be familiar with other potential, and often hidden, hazards that chemicals may present. Figure 1.8 shows the potential hidden hazards of the nine hazard classes. Across the top of the chart are all the potential hazards a chemical could have that would affect emergency responders. Down the left side are all of the colors representing the UN/DOT hazard classes. An X is used to identify the DOT hazard designated for the material and the color of the placard, which will be on the shipment. An asterisk (*) is used to identify all of the other potential hazards of the materials. That does not mean that a particular chemical has all of the hazards, but until you are able to obtain additional information, precautions must be taken for each. For example, some corrosive materials are classified as oxidizers, such as perchloric acid above 50% concentration. Perchloric acid above 50% concentration

BURKE PLACARD HAZARD CHART

X-Primary hazard
*-Secondary hazard

Placard Color \ HAZARD	Explosive	Fire Hazard	Water React	Air Reactive	Toxic	Corrosive	Comp Gas	Thermal	Polymerize	Spont Combust	Asphyxiant	Oxidizer	Heat-Shock Sen	Radioactive
Orange	X	*			*	*							*	
Red		X			*	*	*	* Cold	*		*			
Green		*				*	X	* Cold			*	*		
Red & White Stripes	*	X		*	*	*								
White over Red		*			*	*				X		*		
Blue		*	X	*	*	*								
Yellow	*		*	*	*	*	*	* Cold				X		
White		X	X		X	*	X	* Cold	X			X		
Yellow over White						*				*		*		X
White over Black		*	*		*	X				*		*		
Black & White Stripes		*	*	*	*	*		* Hot	*	*		*		*
Red White Red	*	*	*	*	*	*	*	* Cold	*	*	*	*	*	*

Figure 1.8

is placarded as an oxidizer. Perchloric acid is also a strong corrosive material, but will not be placarded corrosive.

Do not focus only on the hazard depicted by the placard. Thoroughly examine all hazardous materials encountered to determine all physical and chemical characteristics and hazards associated with the chemical. An incident occurred in Kansas City, Missouri, in which six firefighters were killed fighting a fire involving commercial-grade ammonium nitrate, a 1.5 Blasting Agent under the DOT hazard class

NFPA 704 diamonds are used to mark hazardous material locations in fixed facilities.

system. The storage containers on a construction site were not marked as to the type of hazardous materials inside. As a result of this tragic loss of firefighters' lives, OSHA adopted the DOT placarding and labeling system to be continued in use for fixed storage until the chemicals are used up. OSHA lists and regulates chemicals that are considered potentially dangerous in the workplace as part of worker right-to-know regulations.

NFPA 704 MARKING SYSTEM

The National Fire Protection Association (NFPA) has developed a fixed-facility marking system to designate general hazards of chemicals (see Figure 1.9), referred

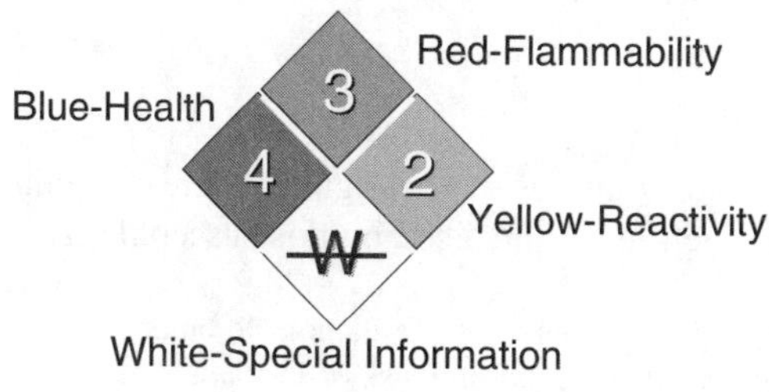

Figure 1.9

to as the NFPA 704 marking system. It is designed to warn emergency responders of the presence of hazardous materials in fixed facilities and to give them some general information about the hazards of the materials. NFPA 704 placards do not identify any specific chemicals or DOT hazard classes other than flammable.

The system utilizes a diamond-shaped placard with four colored sections. Each section indicates a particular hazard: the blue section indicates a health hazard; the red section flammability; and the yellow section reactivity. The white section contains special information, such as "oxy" for oxidizer, "cor" for corrosive, a "W" with a slash through it for water reactivity, and a radioactive propeller indicating radioactivity. Numbers 0 to 4 are placed in the blue, red, and yellow sections, indicating the degree of hazard. The numbers indicate the hazard from 0, which means no hazard, to 4, which indicates the most severe hazard.

Health

In general, the health hazard in firefighting is that of a single exposure, which can vary from a few seconds up to an hour. The physical exertion demanded in firefighting, or other emergency conditions, intensifies the effects of any exposure. Only hazards arising out of an inherent property of the material are considered. The following explanations of the four hazards are based upon protective equipment normally used by firefighters.

4. Materials too dangerous to health to expose firefighters. A few whiffs of the vapor could cause death, or the vapor or liquid could be fatal upon penetrating the firefighter's normal full-protective clothing. The normal full-protective clothing and breathing apparatus available to the average fire department will not provide adequate protection against inhalation or skin contact with these materials.
3. Materials extremely hazardous to health, but areas may be entered with extreme care. Full protective clothing, including self-contained breathing apparatus, coat, pants, gloves, boots, and bands around legs, arms, and waist should be provided. No skin surface should be exposed.
2. Materials hazardous to health, but areas may be entered freely with full-faced mask and self-contained breathing apparatus that provides eye protection.
1. Materials only slightly hazardous to health. It may be desirable to wear self-contained breathing apparatus.
0. Materials that on exposure under fire conditions would offer no hazard beyond that of ordinary combustible material.

Flammability

Susceptibility to burning is the basis for assigning degrees within this category. The method of attacking the fire is influenced by this susceptibility factor.

4. Very flammable gases or very volatile flammable liquids. Shut off flow and keep cooling water streams on exposed tanks or containers.
3. Materials that can be ignited under almost all normal temperature conditions. Water may be ineffective because of the low flash point.

2. Materials that must be moderately heated before ignition will occur. Water spray may be used to extinguish the fire because the material can be cooled below its flashpoint.
1. Materials that must be preheated before ignition can occur. Water may cause frothing if it gets below the surface of the liquid and turns to steam. However, water fog gently applied to the surface will cause a frothing that will extinguish the fire.
0. Materials that will not burn.

Reactivity (Stability)

The assignment of degrees in the reactivity category is based upon the susceptibility of materials to release energy either by themselves or in combination with water. Fire exposure is one of the factors considered along with conditions of shock and pressure.

4. Materials that (in themselves) are readily capable of detonation or of explosive decomposition or explosive reaction at normal temperatures and pressures. Includes materials that are sensitive to mechanical or localized thermal shock. If a chemical with this hazard rating is in an advanced or massive fire, the area should be evacuated.
3. Materials that (in themselves) are capable of detonation or of explosive decomposition or of explosive reaction, but which require a strong initiating source or which must be heated under confinement before initiation. Includes materials that are sensitive to thermal or mechanical shock at elevated temperatures and pressures or that react explosively with water without requiring heat or confinement. Firefighting should be done from a location protected from the effects of an explosion.
2. Materials that (in themselves) are normally unstable and readily undergo violent chemical change, but do not detonate. Includes materials that can undergo chemical change with rapid release of energy at normal temperatures and pressures or that can undergo violent chemical change at elevated temperatures and pressures. Also includes those materials that may react violently with water or that may form potentially explosive mixtures with water. In advanced or massive fires, firefighting should be done from a safe distance or from a protected location.
1. Materials that (in themselves) are normally stable, but which may become unstable at elevated temperatures and pressures or which may react with water with some release of energy, but not violently. Caution must be used in approaching the fire and applying water.
0. Materials that (in themselves) are normally stable, even under fire-exposure conditions, and that are not reactive with water. Normal firefighting procedures may be used.*

NFPA 704 diamond placards are placed on the outside of buildings, storage tanks, storage sheds, and doors leading to areas where hazardous materials are

 This warning system is intended to be interpreted and applied only by properly trained individuals to identify fire, health, and reactivity hazards of chemicals. The user is referred to a certain limited number of chemicals with recommended classifications in NFPA 49 and NFPA 325, which would be used as a guideline only. Whether the chemicals are classified by NFPA or not, anyone using the 704 systems to classify chemicals does so at their own risk.

Almost 50% of all trucks on the highways are transporting hazardous materials.

present. This information is provided as a type of "stop sign" for response personnel. It says hazardous materials are present, but responders still have to obtain specific information and the identity of the chemicals. Chemicals must still be identified for specific hazards before responders can enter. NFPA 704 designations are not available for all hazardous materials. Those chemicals in this book that have been assigned NFPA 704 designations will have the numerical designation listed with the chemical characteristics. These designations are found in the *NFPA Fire Protection Guide to Hazardous Materials*.

CHEMICAL CHARACTERISTIC LISTINGS AND INCIDENTS

Chapters 3 through 11 will list the nine DOT hazard classes and the types of chemicals found in each class. Specific chemicals are listed, along with their physical and chemical characteristics. Examples are provided to illustrate the similarities of some chemicals in the same families and the differences of chemicals in other families. These examples are also provided because they are among the most common hazardous materials encountered by emergency responders. When reading through specific chemicals, note the multiple hazards, many of which are not placarded for the hazard class, and others that are not assigned to a hazard class at all. Along with the listings of chemicals, there are excerpts from incidents that have occurred involving many of the chemicals listed. These are provided to show the dangers of the hazardous materials and how the physical and chemical characteristics contributed to the incident outcome.

CHEMICAL AND PHYSICAL CHARACTERISTICS TRAINING COMPETENCIES

The following discussion of the competencies required both by OSHA 1910.120 and NFPA 472 compares the level of training and the level of knowledge of chemistry that is necessary for each level. OSHA requires emergency responders to demonstrate competency at each level for which they are required to perform. OHSA identifies five levels of training for emergency responders to hazardous materials incidents. They are: awareness, operations, technician, specialist, and the incident commander. OSHA regulations are not nearly as specific with competency requirements as is NFPA 472. Excerpts from the OSHA regulation are included in this section for review.

OSHA 1910.120

"Competent" means possessing the skills, knowledge, experience, and judgment to perform assigned tasks or activities satisfactorily as determined by the employer.

"Demonstration" means the showing by actual use of equipment or procedures.

"Hazardous substance" means any substance designated or listed under [A] through [D] of this definition, exposure to which results, or may result, in adverse effects on the health or safety of employees:

[A] Any substance defined under section 101(14) of CERCLA.
[B] Any biologic agent and other disease-causing agent, which, after release into the environment and upon exposure, ingestion, inhalation, or assimilation into any person, either directly from the environment or indirectly by ingestion through food chains, will or may reasonably be anticipated to cause, death, disease, behavioral abnormalities, cancer, genetic mutation, physiological malfunctions (including malfunctions in reproduction), or physical deformations in such persons or their offspring.
[C] Any substance listed by the U.S. Department of Transportation as hazardous materials under 49 CFR 172.101 and appendices.
[D] Hazardous waste as herein defined.

"Hazardous waste" means:

[A] A waste or combination of wastes as defined in 40 CFR 261.3.
[B] Those substances defined as hazardous wastes in 49 CFR 171.8.

Preliminary evaluation. A preliminary evaluation of a site's characteristics shall be performed prior to site entry by a qualified person in order to aid in the selection of appropriate employee protection methods prior to site entry. Immediately after initial site entry, a more detailed evaluation of the site's specific characteristics shall be performed by a qualified person in order to further identify existing site hazards and to further aid in the selection of the appropriate engineering controls and personal protective equipment for the tasks to be performed.

Hazard identification. All suspected conditions that may pose inhalation or skin absorption hazards that are immediately dangerous to life or health (IDLH) or other conditions that may cause death or serious harm shall be identified during the preliminary survey and evaluated during the detailed survey. Examples of such hazards include, but are not limited to, confined space entry, potentially explosive or flammable situations, visible vapor clouds, or areas where biological indicators, such as dead animals or vegetation, are located.

Based upon the results of the preliminary site evaluation, an ensemble of personal protective equipment (PPE) shall be selected and used during initial site entry, which will provide protection to a level of exposure below permissible exposure limits and published exposure levels for known or suspected hazardous substances and health hazards, and which will provide protection against other known and suspected hazards identified during the preliminary site evaluation. If there is no permissible exposure limit or published exposure level, the employer may use other published studies and information as a guide to appropriate personal protective equipment.

Risk identification. Once the presence and concentrations of specific hazardous substances and health hazards have been established, the risks associated with these substances shall be identified. Employees who will be working on the site shall be informed of any risks that have been identified. In situations covered by the Hazard Communication Standard, 29 CFR 1910.1200, training required by that standard need not be duplicated.

NOTE TO (C)(7). Risks to consider include, but are not limited to:

[a] Exposures exceeding the permissible exposure limits and published exposure levels.
[b] IDLH concentrations.
[c] Potential skin absorption and irritation sources.
[d] Potential eye irritation sources.
[e] Explosion sensitivity and flammability ranges.
[f] Oxygen deficiency.

Operations Level

(q)(6)(ii)(C)

Understand basic hazardous materials terms.

Technician

(q)(6)(iii)(I)

Understand basic chemical and toxicological terminology and behavior.

Specialist

(q)(6)(iv)(I)

Understand chemical, radiological, and toxicological terminology and behavior.

Recommendations from Appendix 1910.120, not mandatory.

b. Technical knowledge

(1) Type of potential exposures to chemical, biological, and radiological hazards; types of human responses to these hazards and recognition of those responses; principles of toxicology and information about acute and chronic hazards; health and safety considerations of new technology.
(2) Fundamentals of chemical hazards, including but not limited to, vapor pressure, boiling points, flashpoints, pH and other physical and chemical properties.
(3) Fire and explosion hazards of chemicals.
(4) General safety hazards, such as but not limited to, electrical hazards, powered equipment hazards, motor vehicle hazards, walking- and working-surface hazards, excavation hazards, and hazards associated with working in hot and cold temperature extremes.

Awareness

First responders at the awareness level are not required to have a great deal of knowledge about the chemistry of hazardous materials. However, there are many shortcuts in street chemistry that could assist first responders in recognizing hazards through family characteristics. After all, first responders have the least amount of resources at their disposal when arriving on the scene of a hazardous materials release. The knowledge they bring with them through previous training and experience may be all they have to assist in the recognition of the presence of hazardous materials. Recognition of the existence of hazardous materials is the most important thing any emergency responder can do upon arrival at an incident scene. First responders at the awareness level have four basic responsibilities: recognition, notification, isolation, and protection. A basic knowledge of street chemistry can assist first responders in carrying out their responsibilities safely.

Operations

Operations-level personnel are required to have awareness training as well as an additional 8 hours of operations-level training. According to OSHA and NFPA, they must be able to predict the behavior of hazardous materials and their containers. Knowledge of the physical and chemical characteristics of hazardous materials is of great benefit in making those predictions. Responders at this level are expected to obtain information from Material Safety Data Sheets (MSDS). Much of the information on the MSDS is in the form of physical and chemical terminology. Once again, knowledge of street chemistry will benefit operations-level personnel in researching information needed to deal safely with hazardous materials. Operations-level personnel should understand the following chemical terms, which are recommended in NFPA 472.

1. Boiling point
2. Chemical reactivity
3. Corrosivity (pH)
4. Flammable range
5. Flashpoint
6. Vapor pressure

7. Ignition temperature
8. Physical state
9. Specific gravity
10. Toxic products of combustion
11. Vapor density
12. Water solubility

Also required of operations-level personnel is the knowledge of certain toxicological terms. Those include the following, which are recommended in NFPA 472.

1. Asphyxiant
2. Chronic health hazard
3. Convulsant
4. Irritant/Corrosive
5. Sensitizer/Allergen

Technician

Technician-level responders are members of hazardous materials response teams. They will enter the "hot zone" and work in close proximity to hazardous materials. Technician-level personnel should have an extensive knowledge of hazardous materials chemistry. While OSHA only requires 24 hours of training above the operations level, it is difficult to adequately prepare personnel with just 24 hours of training. Most hazardous materials teams have developed their own training standards for team membership, and many require hundreds of hours of training. Part of that training usually includes a chemistry of hazardous materials course ranging from 3 days to 2 weeks in length. Technicians should have an in-depth knowledge of explosives; cryogenic liquids; compressed gases; liquefied gases; flammable solids, liquids, and gases; poisons; oxidizers; and corrosives. Their understanding of chemistry is critical in determining the proper type of monitoring instruments, protective clothing, decontamination, and tactical options. Terminology required as a minimum level of knowledge of the chemistry of hazardous materials is in the following list. These competencies are recommended in NFPA 472.

1. Acid, caustic
2. Air reactivity
3. Boiling point
4. Catalyst
5. Chemical interactions
6. Chemical reactivity
7. Compound, mixture
8. Concentration
9. Corrosivity (pH)
10. Critical temperature and pressure
11. Expansion ratio
12. Flammable range
13. Fire point
14. Flashpoint

15. Halogenated hydrocarbon
16. Ignition temperature
17. Inhibitor
18. Instability
19. Ionic and covalent compounds
20. Maximum safe storage temp (MSST)
21. Melting point
22. Miscibility
23. Organic and inorganic
24. Oxidation potential
25. pH
26. Physical state
27. Polymerization
28. Radioactivity
29. Saturated, unsaturated, and aromatic hydrocarbons
30. Self-accelerating decomposition
31. Specific gravity
32. Strength
33. Sublimation
34. Temperature of product
35. Toxic products of combustion
36. Vapor density
37. Vapor pressure
38. Viscosity
39. Volatility
40. Water reactivity
41. Water solubility temperature (SADT)

Also required of technician personnel is the knowledge of certain toxicological terms. Those include the following, which are recommended in NFPA 472.

1. Parts per million
2. Parts per billion
3. Lethal Dose (LD50)
4. Lethal Concentration (LC50)
5. Threshold Limit Value (TLV)
6. Threshold Limit Value (STEL)
7. Threshold Limit Value (TLV-C)
8. Immediately Dangerous to Life and Health (IDLH)

In addition, technician personnel should be familiar with radiological terminology and physical characteristics of radiological materials. The following are the competencies recommended in NFPA 472.

1. Alpha
2. Beta
3. Gamma
4. Activity
5. Quantity gamma

6. Absorbed dose
7. Half-life
8. Inverse square law
9. Time, distance, and shielding
10. Curie
11. Roentgen*

Incident Commander

The incident commander requires training to a minimum of the operations level and competencies outlined in NFPA 472 for the incident commander. Toxicology and radiological terms the incident commander should be able to describe are the same as those required for the technician-level personnel listed previously. In addition, the incident commander should be aware of the routes of exposure for hazardous materials, acute and delayed toxicity, local and systemic effects, dose response, and synergistic effects. Detailed information about the training competencies for responders to hazardous materials incidents can be found in NFPA Standard 472. Additional information concerning hazardous materials response can be found in NFPA 471. Competencies for emergency medical personnel are outlined in NFPA 473. Many of the chemical and physical characteristics required of operations- and technician-level personnel are also required of EMS responders. However, the EMS responder is more concerned about the effect the hazardous material will have on the victims and how they can protect themselves while dealing with the treatment of the patients than mitigating the incident.

This street chemistry book covers all of the chemistry requirements of NFPA 472, OSHA 1910.120, and much more to prepare emergency responders at each level to effectively deal with hazardous materials released into the atmosphere.

REVIEW QUESTIONS

1. Which of the following people developed the first Periodic Table of Elements?
   A. Davy
   B. Mendeleev
   C. Plato
   D. Bush
2. Of the following federal agencies, which developed the term "hazardous material?"
   A. EPA
   B. OSHA
   C. DOT
   D. FEMA
3. Following the Bhopal, India, incident Congress passed which law?
   A. CERCLA
   B. Clean Water Act

C. Occupational Safety Act
D. EPCRA

4. Which emergency responders would benefit from the study of chemistry?
A. Firefighters
B. Police officers
C. EMS
D. All of the above
5. How many hazard classes does the DOT recognize?
A. 7
B. 10
C. 9
D. 8
6. DOT commonly placards which of the following?
A. The most severe hazard
B. The most common hazard
C. Hazards most dangerous to responders
D. None of the above
7. Hazard Class 2 chemicals present which of the following hazards?
A. Explosive, flammable, and corrosive
B. Toxic, oxidizer, and radioactive
C. Flammable, poison, and oxidizer
D. Poison, non-flammable, and flammable
8. Class 1 contains how many hazard subclasses?
A. 4
B. 2
C. 6
D. 1
9. Which of the following is not one of the three hazards of Hazard Class 4?
A. Flammable oxidizer
B. Spontaneously combustible
C. Flammable solid
D. Dangerous when wet
10. Which of the following is not one of the four colors of the NFPA 704 marking system?
A. Yellow
B. Orange
C. Blue
D. Red
11. Of the following federal regulations, which one covers hazmat training?
A. 29 CFR
B. 49 CFR
C. 37 CFR
D. 10 CFR
12. Hazardous materials incident commanders must be trained to which level?
A. Awareness
B. Operations
C. Technician
D. Specialist

CHAPTER 2

Basics of Chemistry

Chemistry is the study of matter. It can be divided into two sections: inorganic and organic. Inorganic chemistry involves acids, bases, salts, elements, and the physical state of matter in which they are found. Organic chemistry involves compounds that contain carbon. The definition of matter is anything that occupies space and has mass. Matter can exist as a solid, liquid, or gas. Temperature and pressure can affect the physical state of a chemical, but not its properties. The hazards presented by a chemical may not be the same, depending on the physical state of the material. For example, only gases burn; solids and liquids do not burn, even though they may be listed as flammable. A solid or liquid must be heated until it produces enough vapors to burn. It is important to understand the states of matter in order to better understand the physical and chemical characteristics of hazardous materials. There can be some intermediate steps in the process of classifying solids, liquids, and gases. Some solids may have varying particle sizes, from large blocks to filings, chips, and dusts. Particle sizes of vapors may vary, from vapors that are small enough to be invisible to mists that are readily visible.

A molecule is the smallest particle of a compound that can normally exist by itself. Molecules of compounds contain different types of atoms bonded together in fixed proportions. The molecules of solids are packed together closely in an organized pattern. Because the molecules are packed tightly together, they can only vibrate gently in a small space. This is why solids have a definite size and shape. When particles are this close together, they attract each other, and it takes a lot of energy to pull them apart. When a solid is heated, the molecules start to vibrate faster and eventually pull apart. The particles in liquids are farther apart, but are still able to attract each other. They are not arranged in a regular pattern. Liquids do not have a shape of their own, so they conform to the shape of the container in which they are placed. Particles of a gas are moving rapidly and are not attracted to each other. Gases have no shape of their own and conform to the space in which they exist.

Hazardous materials may undergo both chemical and physical changes. A chemical change involves a reaction that alters the composition of the substance and thereby alters its chemical identity. Chemical properties include reactivity, stability, corrositivity, toxicity, and oxidation potential. A new compound may be formed that may have different characteristics than the compounds or elements that make it up. Chlorine, for example, is a poison gas; sodium is a reactive metal. When they are combined, they form sodium chloride, which is neither a poison nor a reactive chemical.

Physical changes involve changes in the physical state of the chemical, but do not produce a new substance, such as the physical transformation from a liquid to a gas or a liquid to a solid. Physical properties include specific gravity, vapor pressure, boiling point, vapor density, melting point, solubility, flash point, fire point, auto-ignition temperature, flammable range, heat content, pH, threshold limit value (TLV), and permissible exposure level (PEL).

THE PERIODIC TABLE

The basics of chemistry cannot be effectively discussed without studying the Periodic Table of Elements (Figure 2.1). Dmitri Mendeleev, a Russian chemist, is considered the father of the Periodic Table based upon his work in cataloging facts about the 63 known elements in the mid-1800s. Mendeleev was convinced that groups of elements had similar, "periodic" properties. He arranged the known elements according to their increasing atomic mass (weight). He left blank spaces on the Periodic Table where he thought other unknown elements would fit.

British scientist Henry Moseley continued Mendeleev's work on the Periodic Table when he discovered that the number of protons in the nucleus of a particular

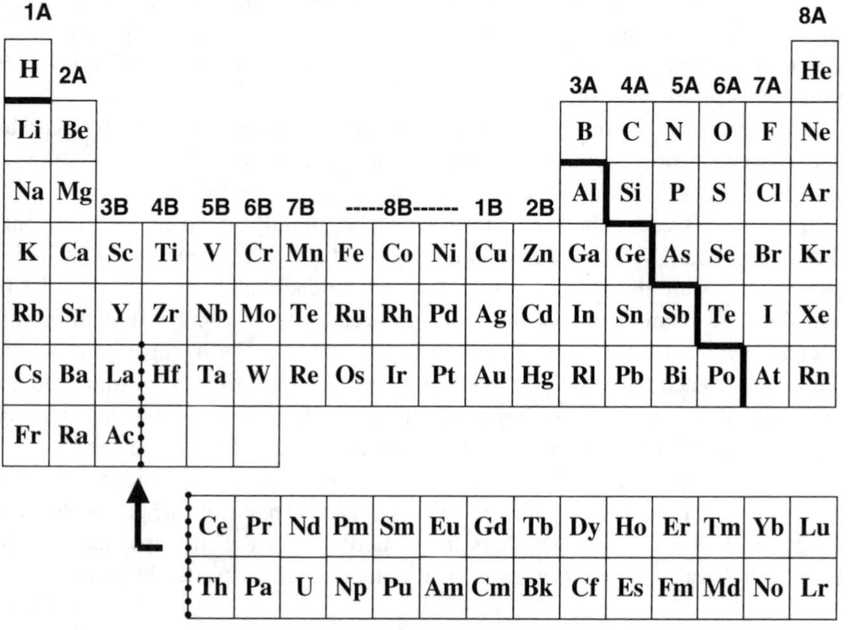

Figure 2.1

type of atom was always the same. It is a result of Moseley's work that the modern Periodic Table is based upon the atomic numbers of the elements. Elements on Moseley's Periodic Table are arranged in order of increasing atomic number, which is also the number of protons in the atom of that element. The properties of the elements repeat in a regular way when the elements are arranged by increasing atomic number.

The Periodic Table is a method of organizing everything that is known about chemistry on one piece of paper. It shows the relationship between the elements by revealing the tendency of their properties to repeat at regular intervals. All chemicals are derived from the elements or from combinations of elements from the Periodic Table. Symbols are used to represent the elements. The table is composed of a series of blocks representing each element. Within each block is a symbol that corresponds to the name of the element, a type of shorthand for the name of the element. For example, the element gold is represented by the symbol Au; chlorine is represented by the symbol Cl; and potassium is represented by the symbol K. Each symbol represents one atom of that element. The symbols may be made up of a single letter or two letters together. A single letter is always capitalized; when there are two letters, the first is capitalized and the second is lowercased. When the second letter is also capitalized, this indicates that the material described is not an element, but a compound. For example:

CO is the molecular formula for the compound carbon monoxide.
Co is the symbol for the element cobalt.

The symbols for the elements are derived from a number of sources. They may have been named for the person who discovered the element. For example, W, the symbol for tungsten, was named for Wolfram, the discoverer. Other elements are named for famous scientists, universities, cities, and states: Es is the symbol for einsteinium, named for Albert Einstein; Cm is the symbol for curium, named for Madame Curie; Bk is the symbol for berkelium, named for the city of Berkeley, CA; and Cf is the symbol for the element californium, named after the state of California. Other element names come from Latin, German, Greek, and English: in the case of sodium, Na comes from the Latin for *natrium*; Au, the symbol for gold, comes from *aurum*, meaning "shining down" in Latin; Cu (copper) is from the Latin *cuprum*, or *cyprium*, because the Roman source for copper was the island of Cyprus. Fe (iron) is from the Latin *ferrum.* Bromine means "stench" in Greek; rubidium means "red" (in color); mercury is sometimes referred to as quicksilver; sulfur is referred to as brimstone in the Bible.

THE ELEMENTS

There are 90 naturally occurring and 20 manmade elements. Not all of the elements on the Periodic Table are common or particularly hazardous to responders. There are, however, some 39 elements that we will call the HazMat elements. These elements are important to the study of the chemistry of hazardous materials. Most

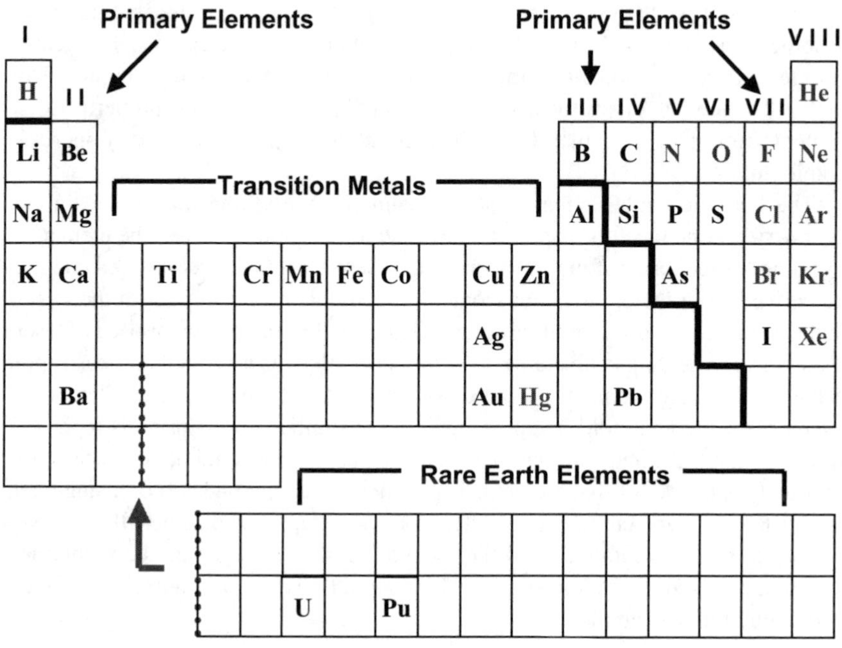

Figure 2.2

of the hazardous materials that response personnel will encounter include or are produced from these 39 elements (see Figure 2.5). Hazardous materials personnel and students of HazMat chemistry should be familiar with these 39 elements by symbol and name. Elements with 83 or more protons are radioactive; many are rare and probably will not be encountered. The manmade elements are the result of nuclear reactions and research. These elements may have existed on earth at one time, but because they are radioactive and many half-lives have passed, they no longer exist naturally.

The elements on the Periodic Table can be divided into three groups: the primary elements, the transition elements, and the rare-earth elements (Figure 2.2). The primary elements have a definite number of electrons in the outer shell. This number is the number at the top of each column in the towers at each end of the Periodic Table. The transition metals may have different numbers of electrons in the outer shell. They are located in the valleys between the towers. The rare-earth metals are relatively uncommon and all are radioactive. The horizontal rows are called periods and are numbered from 1 to 7. Atomic numbers increase by one as you go across the periods from left to right.

ATOMIC NUMBER

The other number on the Periodic Table is a whole number located either above or below the symbol, and is known as the atomic number of the element. The atomic

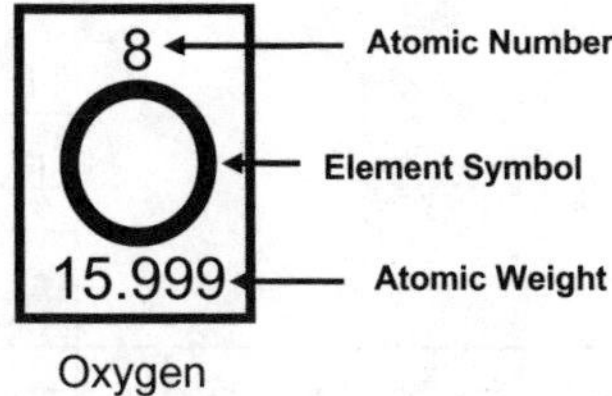

Figure 2.3

number is equal to the number of protons in the nucleus. The atomic number also equals the total number of electrons in the orbits outside the nucleus of the atom. Protons have a positive (+) charge, and electrons have a negative (–) charge. There must be an equal number of protons and electrons in the atom of the element to maintain an electrical balance. It is the number of protons that identifies a specific element. If you change the number of protons, you change the element. Protons act as a kind of "Social Security number" to identify a specific element.

ATOMIC WEIGHT

The atomic weight of an element is also listed on the Periodic Table. The atomic weight is the sum of the weight of the protons and neutrons in the nucleus of the atom. All of the weight of the element occurs in the nucleus of the atom. For the purposes of "street chemistry," electrons do not have weight.

The atomic weight is located on the Periodic Table above or below the symbol of the element. It is the number that is *not* a whole number; the location varies among Periodic Tables, so be sure you look for the number with the decimal point. When using atomic weight to determine molecular weight of a compound, always round it off to the nearest whole number. For example, oxygen has an atomic weight of 15.999; round the number to 16, so the atomic weight of oxygen becomes 16. Nitrogen has an atomic weight of 14.007, so you round the number to 14. The atomic weight is referred to in terms of atomic mass units (AMU).

The Periodic Table is divided into two sections by a stair-stepped line (Figure 2.4). The line starts under hydrogen, goes over to boron, and then stair-steps down one element at a time to astatine or radon, depending on which Periodic Table is used. The 81 elements to the left and below the stair-stepped line are metals. Metals make up about 75% of all the elements. Metals lose their outer-shell electrons easily to nonmetals when forming compounds. Metals are malleable (they can be flattened), ductile (they can be drawn into a wire), and conduct heat and electricity quite well. The farther to the left of the line you go, the more metallic the properties of the element; the closer to the line, the less metallic the properties of the element. Metallic properties increase as you go down a column on the Periodic Table. All metals are solids, except gallium, mercury, francium, and cesium, which are liquids under normal conditions.

I	II											III	IV	V	VI	VII	VIII
H																	He
Li	Be											B	C	N	O	F	Ne
Na	Mg											Al	Si	P	S	Cl	Ar
K	Ca	Sc	Ti	V	Cr	Mn	Fe	Co	Ni	Cu	Zn	Ga	Ge	As	Se	Br	Kr
Rb	Sr	Y	Zr	Nb	Mo	Te	Ru	Rh	Pd	Ag	Cd	In	Sn	Sb	Te	I	Xe

Group 1: All have 1 electron in the outer shell. Group II: Have 2 electrons in the outer shell. Group III: All have 3 electrons in the outer shell. Group IV: All have 4 electrons in the outer shell. Group V: Have 5 electrons in the outer shell. Group VI: All have 6 electrons in the outer shell. Group VII: All have 7 electrons in the outer shell. Group VIII: Has 8 electrons in the outer shell.

Figure 2.4

The 17 elements to the right and above the line are nonmetals. Nonmetals have a strong tendency to gain electrons when forming a chemical bond. Nonmetals may be solids, liquids, or gases; all are poor conductors of heat and electricity. Solid nonmetals are either hard and brittle or soft and crumbly.

The numbers at the top of the vertical columns on the Periodic Table indicate the number of electrons in the outer shell of the elements in that column. The exception to this is the transition elements between the two towers on either end of the table. Unlike the primary elements, the numbers above the transition metal columns do not indicate the number of electrons in the outer shell of those elements. The transition elements can have differing numbers of electrons in their outer shells.

The Periodic Table is organized into vertical columns, called groups or families, and horizontal rows, called periods. Groups are numbered from left to right on the table from 1–18. Periods are numbered from top to bottom on the table from 1–7. The vertical columns of the Periodic Table contain elements that have similar chemical characteristics in their pure elemental form. These elements have the same number of electrons in the outer shell, which is why they have similar chemical behaviors. These similar elements are sometimes referred to as "families." Some of the more important families include the alkali metals in column 1, the alkaline-earth metals in column 2, the halogens in column 7, and the noble or inert gases in column 8. The transition elements in the center of the Periodic Table are also similar in that most of them have the possibility of differing numbers of electrons in their outer shells.

The alkali metals in column 1 begin with lithium and continue through sodium, potassium, rubidium, cesium, and francium. Hydrogen is also located in column 1, but it is above the stair-stepped line and thus is a nonmetal. It is placed in column

1 because it has one electron in the outer shell. Alkali metals are all solids, except for cesium and francium, which are liquids at normal temperatures. The alkali metals are water-reactive: they react violently with water, producing flammable hydrogen gas and enough heat to ignite the hydrogen gas. These elements are so reactive that they do not exist in nature in pure form, but are found as compounds of the metal, such as potassium oxide and sodium chloride. Some isotopes of cesium and all isotopes of francium are radioactive. These elements are somewhat rare, so you are not likely to see them on the street.

The alkaline-earth metals in column 2 are less reactive than the alkali metals in column 1. Beryllium does not react with water at all. The others have varying reactions with water. The alkaline-earth metals are all solids. They have to be burning or in a smaller physical form before they become water-reactive. Magnesium, for example, is violently water-reactive when it is involved in fire. The application of water to a magnesium fire will cause violent explosions that can endanger responders. If it is necessary to fight fires involving magnesium, water should be applied from a safe distance with the use of unmanned appliances. If magnesium is in a particulate form, it can react with water even if it is not on fire.

The other elements in column 2 are also water-reactive to varying degrees. These include calcium, strontium, barium, and radium, which is radioactive. The halogens in column 7 are nonmetals. They may be solids, liquids, or gases. Fluorine and chlorine are gases at normal temperatures and pressures. Bromine is a liquid to 58°C and produces vapor rapidly when above that temperature. Iodine is a solid. Astatine is also in column 7. It is radioactive; however, such a small amount has ever been found that you are not likely to encounter it.

The halogens are all toxic and are also strong oxidizers. Fluorine is a much stronger oxidizer than oxygen; in fact, fluorine is the strongest oxidizer known. In their pure elemental form, the halogens do not burn; however, they will accelerate combustion much like oxygen because they are oxidizers. Some halogen compounds are components of fire-extinguishing agents that are called halons, which are being phased out because of the damage they cause to the ozone layer of the atmosphere.

The elements in column 8 are all gases. They are nonflammable, nontoxic, and nonreactive. Family 8 elements are referred to as the inert, or noble gases. Normally, they do not react chemically with themselves or any other chemicals. Helium is used for weather balloons and airships. Neon is used for lighting and beacons. Argon is used in lightbulbs. Krypton and xenon are used in special lightbulbs for miners and in lighthouses. Radon is radioactive and is used in tracing gas leaks and treating some forms of cancer.

The noble gases have a complete outer shell of electrons: two in helium and eight in the rest: an "octet" is eight electrons in the outer shell; a "duet" is two electrons in the outer shell. Both the octet and duet are stable configurations of electrons in an atom. It is because of the complete outer shell of electrons that the noble gases do not react chemically. All of the other elements on the Periodic Table try to reach the same stable electron arrangement as the nearest noble gas, which is usually in the same period. That is why there are chemical reactions: the elements are trying to reach stability.

The HazMat Elements

H - Hydrogen	Cu - Copper	O - Oxygen
Li - Lithium	Ag - Silver	S - Sulfur
Na - Sodium	Au - Gold	F - Fluorine
K - Potassium	Zn - Zinc	Cl - Chlorine
Be - Beryllium	Hg - Mercury	Br - Bromine
Mg - Magnesium	B - Boron	I - Iodine
Ca - Calcium	Al - Aluminum	U - Uranium
Ba - Barium	C - Carbon	He - Helium
Ti - Titanium	Si - Silicon	Ne - Neon
Cr - Chromium	N - Nitrogen	Ar - Argon
Mn - Manganese	P - Phosphorus	Kr - Krypton
Fe - Iron	As - Arsenic	Xe - Xenon
Co - Cobalt	Pu - Plutonium	Pb - Lead

Figure 2.5

The column 8 elements are all nonhazardous gases; however, they can displace the oxygen in the air inside of buildings or in confined spaces and cause asphyxiation. Inert gases are commonly shipped and stored as compressed gases or cryogenic liquids. These materials are quite cold and can cause thermal burns. They also have large expansion ratios. DOT does not place them in a hazard class unless they are under pressure.

The following paragraphs list the 39 HazMat elements and their characteristics. Information provided includes the symbol, physical state, family, atomic number, atomic weight, location on the Periodic Table, outer shell electrons, DOT hazard class, NFPA 704 designation, and other important characteristics. Also included is some brief information on the history, properties, sources, important compounds, uses, and isotopes of the elements.

THE HAZMAT ELEMENTS

Hydrogen

Atomic Number:	1	Column Number/Family:	IA
Symbol:	H	Metal/Non-Metal	Non-Metal
Atomic Weight:	1.0079	Outer Shell Electrons	1
Family:	None	Physical State:	Gas
DOT Class:	2.1	NFPA 704:	0-4-0
4-Digit Number:	1049	Placard:	Flammable Gas

Figure 2.6

History

From the Greek, *hydro*, meaning water, and *genes*, meaning forming. Named by Lavoisier, hydrogen is the most abundant of all elements in the universe. Hydrogen is also the lightest gas known and was used for airships during the early 1900s.

Sources

Hydrogen is estimated to make up more than 90% of all the atoms, or three-quarters of the mass of the universe. Liquid hydrogen is important in cryogenics and in the study of superconductivity, as its melting point is only 20 degrees above absolute zero. Tritium, an isotope of hydrogen, is readily produced in nuclear reactors and is used in the production of the hydrogen bomb.

Important Compounds

Hydrogen is one of the key elements in hydrocarbon compounds. It is also present in the hydrocarbon derivatives and inorganic acids.

Uses

Hydrogen is used for the hydrogenation of fats and oils and in the formation of hydrocarbon compounds. It is also used in large quantities in methanol production. Other uses include rocket fuel, welding, producing hydrochloric acid, reducing metallic ores, and filling balloons. The hydrogen fuel cell is a developing technology that will allow great amounts of electrical power to be obtained using a source of hydrogen gas. Located in remote regions, power plants would electrolyze seawater; the hydrogen produced would travel to distant cities by pipelines. Pollution-free hydrogen could replace natural gas, gasoline, etc., and could serve as a reducing agent in metallurgy, chemical processing, refining, etc. It could also be used to convert trash into methane and ethylene.

Isotopes

The ordinary isotope of hydrogen, H, is known as protium, the other two isotopes are deuterium (a proton and a neutron) and tritium (a protron and two neutrons). Hydrogen is the only element whose isotopes have been given different names. Deuterium and tritium are both used as fuel in nuclear fusion reactors. One atom of deuterium is found in about 6000 ordinary hydrogen atoms.

Deuterium is used as a moderator to slow down neutrons. Tritium atoms are also present, but in much smaller proportions. Tritium is readily produced in nuclear reactors and is used in the production of the hydrogen (fusion) bomb. It is also used as a radioactive agent in making luminous paints, and as a tracer.

Lithium

Atomic Number:	3	Column Number:	IA
Symbol:	**Li**	Metal/Non-Metal	Metal
Atomic Weight:	6.9412	Outer Shell Electrons	1
Family:	Alkali Metal	Physical State:	Solid
DOT Class	4.3	NFPA 704:	Not Listed
4-Digit Number:	1415	Placard:	Dangerous When Wet

Figure 2.7

History

Derived from the Greek, meaning *lithos*, or stone. Discovered by Arfvedson in 1817. Lithium is the lightest of all metals, with a density only about half that of water.

Sources

It does not occur free in nature; combined, it is found in small units in nearly all igneous rocks and in the waters of many mineral springs. Lithium is silvery in appearance, much like Na and K, other members of the alkali metal series. It reacts with water, but not as vigorously as sodium. Lithium imparts a beautiful crimson color to a flame, but when the metal burns strongly, the flame is a dazzling white.

Uses

Because the metal has the highest specific heat of any solid element, it has found use in heat-transfer applications; however, it is corrosive and requires special handling. The metal has been used as an alloying agent, is of interest in synthesis of organic compounds, and has nuclear applications. It ranks as a leading contender as a battery anode material, as it has a high electrochemical potential. Lithium is used in special glasses and ceramics. Lithium chloride is one of the most lyproscopic materials known, and it, as well as lithium bromide, is used in air conditioning and industrial drying systems. Lithium stearate is used as an all-purpose and high-temperature lubricant. Other lithium compounds are used in dry cells and storage batteries.

Sodium

Atomic Number:	11	Column Number:	IA
Symbol:	**Na**	Metal/Non-Metal	Metal
Atomic Weight:	22.98977	Outer Shell Electrons	1
Family:	Alkali Metal	Physical State:	Solid
DOT Class:	4.3	NFPA 704:	Not Listed
4-Digit Number:	1428	Placard:	Dangerous When Wet

Figure 2.8

History

Comes from English for soda and from medieval Latin, *sodanum*, a headache remedy. Long recognized in compounds, sodium was first isolated by Davy in 1807 by electrolysis of caustic soda.

Sources

Sodium is the fourth most-abundant element on earth, comprising about 2.6% of the earth's crust; it is the most abundant of the alkali group of metals.

Compounds

The most common compound is sodium chloride, but it occurs in many other minerals. It makes up many salt compounds with nonmetal materials. Among the many compounds that are of the greatest industrial importance are common salt (NaCl), soda ash (Na_2CO_3), baking soda ($NaHCO_3$), caustic soda (NaOH), Chile saltpeter ($NaNO_3$), di- and tri-sodium phosphates, sodium thiosulfate (hypo, $Na_2S_2O_3$.), and borax ($Na_2B_4O_7$.).

Properties

Sodium, like every reactive element, is never found free in nature. Sodium is a soft, bright, silvery metal, which floats on water, decomposing it with the evolution of hydrogen and the formation of hydroxide. It may or may not ignite spontaneously on water, depending on the amount of oxide and metal exposed to the water. It normally does not ignite in air at temperatures below 115°C. Sodium metal should be handled with great care. It cannot be maintained in an inert atmosphere, and contact with water and other substances with which sodium reacts should be avoided.

Uses

Metallic sodium is vital in the manufacture of esters and in the preparation of organic compounds. The metal may be used to improve the structure of certain alloys, to descale metal, and to purify molten metals.

Potassium

Atomic Number:	19	Column Number:	IA
Symbol:	**K**	Metal/Non-Metal	Metal
Atomic Weight:	39.098	Outer Shell Electrons	1
Family:	Alkali Metal	Physical State:	Solid
DOT Class:	4.3	NFPA 704:	Not Listed
4-Digit Number:	2257	Placard:	Dangerous When Wet

Figure 2.9

History

Taken from the English, meaning potash — pot ashes; Latin, *kalium*; and Arab, *qali*, or alkali. Discovered in 1807 by Davy, who obtained it from caustic potash (KOH); this was the first metal isolated by electrolysis.

Sources

The metal is the seventh most abundant and makes up about 2.4%, by weight, of the earth's crust. Most potassium minerals are insoluble, and the metal is obtained from them only with great difficulty.

Production

Potassium is never found free in nature, but is obtained by electrolysis of the hydroxide, much in the same manner as was prepared by Davy. Thermal methods also are commonly used to produce potassium.

Uses

The greatest demand for potash has been in its use for fertilizers. Potassium is an essential constituent for plant growth and is found in most soils. An alloy of sodium and potassium (NaK) is used as a heat-transfer medium. Many potassium salts are of utmost importance, including hydroxide, nitrate, carbonate, chloride, chlorate, bromide, iodide, cyanide, sulfate, chromate, and dichromate.

Properties

It is one of the most reactive and electropositive of the metals. Except for lithium, it is the lightest known metal. It is soft, easily cut with a knife, and is silvery in appearance immediately after a fresh surface is exposed. It rapidly oxidizes in air and must be preserved in a mineral oil, such as kerosene. As with other metals of the alkali group, it decomposes in water with the evolution of hydrogen. It catches fire spontaneously on water. Potassium and its salts impart a violet color to flames.

Isotopes

Seventeen isotopes of potassium are known. Ordinary potassium is composed of three isotopes, one of which is 40°K (0.0118%), a radioactive isotope with a half-life of 1.28×10^9 years.

Beryllium

Atomic Number:	4	Column Number:	IIA
Symbol:	**Be**	Metal/Non-Metal	Metal
Atomic Weight:	9.01218	Outer Shell Electrons	2
Family:	Alkaline Earth Metal	Physical State:	Solid
DOT Class:	Not Listed	NFPA 704:	Not Listed
4-Digit Number:	Not Listed	Placard:	Not Listed

Figure 2.10

History

Comes from the Greek, meaning *beryllos*, or beryl; also called glucinium or glucinum; this term is taken from the Greek word *glykys*, sweet. Discovered by Vauquelin as the oxide in beryl and in emeralds in 1798. Wohler and Bussy independently isolated the metal in 1828 by the action of potassium on beryllium chloride.

Sources

Aquamarine and emerald are precious forms of beryl. Beryl and bertrandite are the most important commercial sources of the element and its compounds. Most of the metal is now prepared by reducing beryllium fluoride with magnesium metal. Beryllium metal did not become readily available to industry until 1957.

Properties

The metal, steel gray in color, has many desirable properties. As one of the lightest of all metals, it has one of the highest melting points. It resists attack by concentrated nitric acid, has excellent thermal conductivity, and is nonmagnetic. At ordinary temperatures, beryllium resists oxidation in air, although its ability to scratch glass is probably due to the formation of a thin layer of the oxide. Beryllium and its salts are toxic and should be handled with the greatest of care. Beryllium and its compounds should not be tasted to verify the sweetish nature of beryllium.

Uses

Beryllium is used as an alloying agent in producing beryllium copper, which is extensively used for springs, electrical contacts, spot-welding electrodes, and non-sparking tools. It is applied as a structural material for high-speed aircraft, missiles, spacecraft, and communication satellites. Other uses include windshield frames, brake discs, support beams, and other structural components of the space shuttle. It is used in gyroscopes, computer parts, and instruments where lightness, stiffness, and dimensional stability are required. The oxide has a high melting point and is also used in nuclear work and ceramic applications.

Magnesium

Atomic Number:	12	Column Number:	IIA
Symbol:	**Mg**	Metal/Non-Metal	Metal
Atomic Weight:	24.305	Outer Shell Electrons	2
Family:	Alkaline Earth Metal	Physical State:	Solid
DOT Class:	Not Listed	NFPA 704:	Not Listed
4-Digit Number:	Not Listed	Placard:	Not Listed

Figure 2.11

History

Named after Magnesia, a district in Thessaly, Black first recognized magnesium as an element in 1755. Davy isolated it in 1808, and Bussy prepared it in coherent form in 1831. Magnesium is the eighth most-abundant element in the earth's crust. It does not occur uncombined, but is found in large deposits in the form of magnesite, dolomite, and other minerals.

Sources

The metal is now principally obtained in the U.S. by electrolysis of fused magnesium chloride derived from brines, wells, and seawater.

Properties

Magnesium is a light, silvery-white, and fairly tough metal. It tarnishes slightly in air, and finely divided magnesium readily ignites upon heating in air and burns with a dazzling white flame. Because serious fires can occur, great care should be taken in handling magnesium metal, especially in the finely divided state. Water should not be used on burning magnesium or on magnesium fires.

Uses

Uses include flashlight photography, flares, and pyrotechnics, including incendiary bombs. It is one-third lighter than aluminum, and in alloys is essential for airplane and missile construction. The hydroxide (milk of magnesia), chloride, sulfate (Epsom salts), and citrate are used in medicine.

Compounds

Compounds of magnesium are generally salts.

Calcium

Atomic Number:	20	Column Number:	IIA
Symbol:	**Ca**	Metal/Non-Metal	Metal
Atomic Weight:	40.08	Outer Shell Electrons	2
Family:	Alkaline Earth Metal	Physical State:	Solid
DOT Class:	4.3	NFPA 704:	Not Listed
4-Digit Number:	1401	Placard:	Dangerous When Wet

Figure 2.12

History

Latin for *calx*, meaning lime. Although the Romans prepared lime in the first century under the name *calx*, the metal was not discovered until 1808.

Sources

Calcium is a metallic element, fifth in abundance in the earth's crust, of which it forms more than 3%. Calcium is an essential constituent of leaves, bones, teeth, and shells. Never found in nature uncombined, it occurs abundantly as limestone, gypsum, and fluorite.

Properties

The metal has a silvery color, and is rather hard. Chemically, it is one of the alkaline-earth elements; it readily forms a white coating of nitride in air, reacts with water, and burns with a yellow-red flame.

Uses

It is used as an alloying agent for aluminum, beryllium, copper, lead, and magnesium alloys. It is also used as a fertilizer ingredient.

Compounds

Its natural and prepared compounds are widely used. Quicklime (CaO), which is made by heating limestone that is changed into slaked lime by carefully adding water, is a great base of chemical refinery with countless uses. Mixed with sand, it hardens as mortar and plaster by taking up carbon dioxide from the air. Calcium from limestone is an important element in Portland cement.

Barium

Atomic Number:	56	Column Number:	IIA
Symbol:	**Ba**	Metal/Non-Metal	Metal
Atomic Weight:	137.34	Outer Shell Electrons	2
Family:	Alkaline Earth Metal	Physical State:	Solid
DOT Class:	4.3	NFPA 704:	Not Listed
4-Digit Number:	1400	Placard:	Dangerous When Wet

Figure 2.13

History

Comes from the Greek word *barys*, meaning heavy. Baryta was distinguished from lime by Scheele in 1774; the element was discovered by Sir Humphrey Davy in 1808.

Sources

Barium is found only in combination with other elements, chiefly with sulfate and carbonate, and is prepared by electrolysis of the chloride.

Properties

Barium is a metallic element, soft, and when pure is silvery white, like lead; it belongs to the alkaline-earth metal group, resembling calcium chemically. The metal oxidizes easily and should be kept under petroleum or other suitable oxygen-free liquids to exclude air. It is decomposed by water or alcohol.

Compounds

The most important compounds are peroxide, chloride, sulfate, carbonate, nitrate, and chlorate.

Uses

The sulfate, as permanent white, is also used in paint, in x-ray diagnostic work, and in glassmaking. Barite is extensively used as a weighing agent in oil-well drilling fluids, and is used in making rubber. The carbonate has been used as a rat poison, while the nitrate and chlorate give colors in pyrotechny. All barium compounds that are water- or acid-soluble are poisonous.

Titanium

Atomic Number:	22	Column Number:	IVB
Symbol:	**Ti**	Metal/Non-Metal	Metal
Atomic Weight:	47.90	Outer Shell Electrons	4, 3
Family:	Transitional Metal	Physical State:	Solid
DOT Class:	Not Listed	NFPA 704:	Not Listed
4-Digit Number:	Not Listed	Placard:	Not Listed

Figure 2.14

History

Latin for *titans*, the first sons of the Earth according to Greek mythology. Discovered by Gregor in 1791, and named by Klaproth in 1795. Impure titanium was prepared by Nilson and Pettersson in 1887; however, the pure metal (99.9%) was not made until 1910 by Hunter by heating $TiCl_4$ with *sodium* in a steel bomb.

Sources

Titanium is present in meteorites and in the sun. Rocks obtained during the Apollo 17 lunar mission showed the presence of 12.1% TiO_2, and rocks obtained during earlier Apollo missions show lower percentages. The element is the ninth most abundant in the crust of the earth. Titanium is almost always present in igneous rocks and in the sediments derived from them. Titanium is present in the ash of coal, in plants, and in the human body. The metal was a laboratory curiosity until Kroll, in 1946, showed that titanium could be produced commercially by reducing titanium tetrachloride with *magnesium*. This method is largely used for producing the metal today.

Properties

Titanium, when pure, is a lustrous, white metal. It has a low density, good strength, is easily fabricated, and has excellent corrosion resistance. It is ductile only when it is free of *oxygen*. The metal, which burns in air, is the only element that burns in *nitrogen*. Titanium is resistant to dilute sulfuric and hydrochloric acid, most organic acids, most chlorine gas, and chloride solutions. Titanium metal is considered to be physiologically inert.

Uses

Titanium is important as an alloying agent with *aluminum*, *molybdenum*, *manganese*, *iron*, and other metals. Alloys of titanium are principally used for aircraft and missiles, where lightweight strength and ability to withstand extremes of temperature are important. Titanium is as strong as steel, but 45% lighter. It is 60% heavier than aluminum, but twice as strong. Titanium has potential use in desalination plants for converting seawater into fresh water. The metal has excellent resistance to seawater and is used for propeller shafts, rigging, and other parts of ships exposed to salt water. A titanium anode coated with platinum has been used to provide cathodic protection from corrosion by salt water. Titanium dioxide is extensively used for both house paint and artist's paint, because it is permanent and has good covering power. Titanium oxide pigment accounts for the largest use of the element. Titanium paint is an excellent reflector of infrared and is extensively used in solar observatories, where heat causes poor visibility conditions.

Chromium

Atomic Number:	24	Column Number:	VIB
Symbol:	**Cr**	Metal/Non-Metal	Metal
Atomic Weight:	51.996	Outer Shell Electrons	6,3,2
Family:	Transitional Metal	Physical State:	Solid
DOT Class:	Not Listed	NFPA 704:	Not Listed
4- Digit Number:	Not Listed	Placard:	Not Listed

Figure 2.15

History

From the Greek word, *chroma*, meaning color. Discovered in 1797 by Vauquelin, who prepared the metal the next year, chromium is a steel-gray, lustrous, hard metal that takes a high polish.

Sources

The principal ore is chromite, which is found in Zimbabwe, Russia, Transvaal, Turkey, Iran, Albania, Finland, Democratic Republic of Madagascar, and the Philippines.

Uses

Chromium is used to harden steel, to manufacture stainless steel, and to form many useful alloys. It is used in plating to produce a hard, beautiful surface and to prevent corrosion. Chromium gives glass an emerald green color and is widely used as a catalyst. The refractory industry has found chromite useful for forming bricks and shapes, as it has a high melting point, moderate thermal expansion, and stability of crystalline structure.

Compounds

All compounds of chromium are colored; the most important are the chromates of *sodium* and *potassium* and the dichromates and the *potassium* and ammonium chrome sulfates. The dichromates are used as oxidizing agents in quantitative analysis, and also in tanning leather. Other compounds are of industrial value; lead chromate is chrome-yellow, a valued pigment. Chromium compounds are used in the textile industry as mordants (substances capable of binding dyes to textile fibers), and by the aircraft and other industries for anodizing aluminum. Chromium compounds are toxic and should be handled with proper safeguards.

Manganese

Atomic Number:	25	Column Number:	VIIB
Symbol:	**Mn**	Metal/Non-Metal	Metal
Atomic Weight:	54.9380	Outer Shell Electrons	7,6,4,2,3
Family:	Transitional Metal	Physical State:	Solid
DOT Class:	Not Listed	NFPA 704:	Not Listed
4-Digit Number:	Not Listed	Placard:	Not Listed

Figure 2.16

History

From the Latin *magnes*, meaning magnet, due to the magnetic properties of pyrolusite; also from Italian meaning manganese, a corrupt form of magnesia. Recognized by Scheele, Bergman, and others as an element, and isolated by Gahn in 1774 by reduction of the dioxide with *carbon*.

Sources

Manganese minerals are widely distributed; oxides, silicates, and carbonates are the most common. The discovery of large quantities of manganese nodules on the floor of the oceans may become a source of manganese. These nodules contain about 24% manganese, together with many other elements in lesser abundance. Most manganese today is obtained from ores found in Russia, Brazil, Australia, Republic of South Africa, Gabon, and India.

Properties

It is gray-white, resembling *iron*, but is harder and quite brittle. The metal is reactive chemically, and decomposes cold water slowly. Manganese is used to form many important alloys. In steel, manganese improves the rolling and forging qualities, strength, toughness, stiffness, wear resistance, hardness, and hardenability. With *aluminum* and *antimony*, and especially with small amounts of *copper*, it forms highly ferromagnetic alloys. Manganese metal is ferromagnetic only after special treatment. The pure metal exists in four allotropic forms. The alpha form is stable at ordinary temperature; gamma manganese, which changes to alpha at ordinary temperatures, is flexible, soft, easily cut, and capable of being bent. Exposure to manganese dusts, fume, and compounds should not exceed the ceiling value of 5 mg/m^3 for even short periods because of the element's toxicity level.

Uses

The dioxide (pyrolusite) is used as a depolarizer in dry cells, and is used to "decolorize" glass that is colored green by impurities of iron. Manganese by itself colors glass an amethyst color, and is responsible for the color of true amethyst. The dioxide is also used in the preparation of oxygen and chlorine, and in drying black paints. The permanganate is a powerful oxidizing agent and is used in quantitative analysis and in medicine. Manganese is widely distributed throughout the animal kingdom. It is an important trace element and may be essential for utilization of vitamin B1.

Iron

Atomic Number:	26	Column Number:	VIII
Symbol:	**Fe**	Metal/Non-Metal	Metal
Atomic Weight:	55.847	Outer Shell Electrons	2,3
Family:	Transitional Metal	Physical State:	Solid
DOT Class:	Not Listed	NFPA 704:	Not Listed
4-Digit Number:	Not Listed	Placard:	Not Listed

Figure 2.17

History

From the Anglo-Saxon, iron; also *L. ferrum*. Iron was used prehistorically: Genesis mentions that Tubal-Cain, seven generations from Adam, was "an instructor of every artificer in brass and iron." A remarkable iron pillar, dating to about A.D. 400, remains standing today in Delhi, India. Corrosion to the pillar has been minimal, although it has been exposed to the weather since its erection.

Sources

Iron is a relatively abundant element in the universe. It is found in the sun and many types of stars in considerable quantity. Its nuclei are quite stable. The core of the earth, 2150 miles in radius, is thought to be largely composed of iron, with about 10% occluded hydrogen. The metal is the fourth most-abundant element, by weight, that makes up the crust of the earth. The most common ore is hematite, which is frequently seen as black sands along beaches and banks of streams.

Uses

Iron is a vital constituent of plant and animal life, and appears in hemoglobin. The pure metal is not often encountered in commerce, but is usually alloyed with carbon or other metals.

Properties

The pure metal is highly reactive chemically and rapidly corrodes, especially in moist air or at elevated temperatures. The relations of these forms are peculiar. Pig iron is an alloy containing about 3% carbon, with varying amounts of *sulfur*, *silicon*, *manganese*, and *phosphorus*. Iron is hard, brittle, fairly fusible, and is used to produce other alloys, including steel. Wrought iron contains only a few tenths of a percent of *carbon*, is tough, malleable, less fusible, and has usually a "fibrous" structure. Carbon steel is an alloy of iron with small amounts of Mn, S, P, and Si. Alloy steels are carbon steels with other additives, such as *nickel*, *chromium*, *vanadium*, etc. Iron is a cheap, abundant, useful, and important metal.

Cobalt

Atomic Number:	27	Column Number:	VIII
Symbol:	**Co**	Metal/Non-Metal	Metal
Atomic Weight:	58.9332	Outer Shell Electrons	2,3
Family:	Transitional Metal	Physical State:	Solid
DOT Class:	Not Listed	NFPA 704:	Not Listed
4-Digit Number:	Not Listed	Placard:	Not Listed

Figure 2.18

History

From the German, *Kobald*, meaning goblin or evil spirit; also from *cobalos*, the Greek word for mine. Brandt discovered cobalt in about 1735.

Sources

Cobalt occurs in the minerals cobaltite, smaltite, and erythrite, and is often associated with *nickel*, *silver*, *lead*, *copper*, and *iron* ores, from which it is most frequently obtained as a by-product. It is also present in meteorites. Important ore deposits are found in Zaire, Morocco, and Canada. The U.S. Geological Survey has announced that the bottom of the north central Pacific Ocean may have cobalt-rich deposits at relatively shallow depths in water close to the Hawaiian Islands and other U.S. Pacific territories.

Properties

Cobalt is a brittle, hard metal, resembling iron and nickel in appearance. It has a metallic permeability of about two-thirds that of iron. The transformation is sluggish and accounts in part for the wide variation in reported data on physical properties of cobalt. Exposure to cobalt (metal fumes and dust) should be limited to 0.05 mg/m^3 (8-hour time-weighted average in a 40-hour week).

Uses

It is alloyed with iron, nickel, and other metals to make alnico, an alloy of unusual magnetic strength with many important uses. Stellite alloys, containing cobalt, *chromium*, and *tungsten*, are used for high-speed, heavy-duty, high temperature cutting tools and for dies. Cobalt is also used in other magnetic steels and stainless steels, and in alloys used in jet turbines and gas turbine generators. The metal is used in electroplating because of its appearance, hardness, and resistance to oxidation. The salts have been used for centuries to produce brilliant and permanent blue colors in porcelain, glass, pottery, tiles, and enamels. A solution of the chloride is used as a sympathetic ink. Cobalt carefully used in the form of the chloride, sulfate, acetate, or nitrate has been found effective in correcting a certain mineral-deficiency disease in animals.

Copper

Atomic Number:	29	Column Number:	IB
Symbol:	**Cu**	Metal/Non-Metal	Metal
Atomic Weight:	63.546	Outer Shell Electrons	2,1
Family:	Transitional Metal	Physical State:	Solid
DOT Class:	Not Listed	NFPA 704:	Not Listed
4-Digit Number:	Not Listed	Placard:	Not Listed

Figure 2.19

History

Latin for *cuprum,* from the island of Cyprus. It is believed that copper has been mined for 5000 years.

Properties

Copper is reddish and takes on a bright metallic luster. It is malleable, ductile, and a good conductor of heat and electricity (second only to *silver* in electrical conductivity).

Sources

Large copper ore deposits are found in the U.S., Chile, Zambia, Zaire, Peru, and Canada. The most important copper ores are the sulfides, the oxides, and carbonates. From these, copper is obtained by smelting, leaching, and by electrolysis.

Uses

The electrical industry is one of the greatest users of copper. Copper's alloys, brass and bronze, are important: all American coins are copper alloys, and gun metals also contain copper. Copper has wide use as an agricultural poison and as an algicide in water purification. Copper compounds, such as Fehling's solution, are widely used in analytical chemistry tests for sugar.

Silver

Atomic Number:	47	Column Number:	IB
Symbol:	**Ag**	Metal/Non-Metal	Metal
Atomic Weight:	107.868	Outer Shell Electrons	1, 2
Family:	Transitional Metal	Physical State:	Solid
DOT Class:	Not Listed	NFPA 704:	Not Listed
4-Digit Number:	Not Listed	Placard:	Not Listed

Figure 2.20

History

This word has Anglo-Saxon origins, *Seolfor siolfur*, as well as Latin, *argentums*. Silver has been known since ancient times. It is mentioned in Genesis. Slag dumps in Asia Minor and on islands in the Aegean Sea indicate that man learned to separate silver from lead as early as 3000 B.C.

Sources

Silver occurs both in a native form and in ores, such as argentite (Ag_2S) and horn silver ($AgCl$); lead, lead-zinc, copper, gold, and copper-nickel ores are principal sources. Mexico, Canada, Peru, and the U.S. are the primary silver producers in the Western Hemisphere. Silver is also recovered during electrolytic refining of copper. Commercial fine silver contains at least 99.9% silver. Purities of 99.999+% are available commercially.

Properties

Pure silver has a brilliant, white, metallic luster. It is a little harder than gold and is quite ductile and malleable, being exceeded only by gold and perhaps palladium. Pure silver has the highest electrical and thermal conductivity of all metals, and possesses the lowest contact resistance. It is stable in pure air and water, but tarnishes when exposed to ozone, hydrogen sulfide, or air containing sulfur. The alloys of silver are important. While silver itself is not considered to be toxic, most of its salts are poisonous. Exposure to silver (metal and soluble compounds, as Ag) in air should not exceed 0.01 mg/m^3 (8-hour time-weighted average in a 40-hour week). Silver compounds can be absorbed in the circulatory system, and reduced silver deposited in the various tissues of the body. A condition, known as argyria, results with a grayish pigmentation of the skin and mucous membranes. Silver has germicidal effects and kills many lower organisms effectively without harm to higher animals.

Uses

Sterling silver is used for jewelry, silverware, etc. where appearance is paramount. This alloy contains 92.5% silver, the remainder being copper or some other metal. Silver is of the utmost importance in photography, about 30% of the U.S. industrial consumption going into this application. It is also used for dental alloys. Silver is used in making solder and brazing alloys, electrical contacts, and high-capacity silver-zinc and silver-cadmium batteries. Silver paints are used for making printed circuits. It is used in mirror production and may be deposited on glass or metals by chemical deposition, electrode position, or by evaporation. When freshly deposited, it is the best reflector of visible light known, but tarnishes rapidly and loses much of its reflectance. Silver fulminate, a powerful explosive, is sometimes formed during the silvering process. Silver iodide is used in seeding clouds to produce rain. Silver chloride has interesting optical properties, as it can be made transparent; it also is a cement for glass. Silver nitrate, or lunar caustic, the most important silver compound, is used extensively in photography. For centuries, silver has been used traditionally for coinage by many countries of the world. In recent times, however, consumption of silver has greatly exceeded the output.

Gold

Atomic Number:	79	Column Number:	IB
Symbol:	**Au**	Metal/Non-Metal	Metal
Atomic Weight:	196.9665	Outer Shell Electrons	3,1
Family:	Transitional Metal	Physical State:	Solid
DOT Class:	Not Listed	NFPA 704:	Not Listed
4-Digit Number:	Not Listed	Placard:	Not Listed

Figure 2.21

History

From the Sanskrit, *jval*; Anglo-Saxon, gold; and Latin, *aurum.* Known and highly valued from earliest times, gold is found in nature as the free metal and in tellurides; it is widely distributed and is almost always associated with quartz or pyrite.

Sources

It occurs in veins and alluvial deposits, and is often separated from rocks and other minerals by mining and panning operations. About two-thirds of the world's gold output comes from South Africa, and about two-thirds of the total U.S. production comes from South Dakota and Nevada. The metal is recovered from its ores by cyaniding, amalgamating, and smelting processes. Refining is also frequently done by electrolysis. Gold occurs in seawater to the extent of 0.1 to 2 mg/ton, depending on the location where the sample is taken. As yet, no method has been found for recovering gold from seawater profitably.

Properties

It is estimated that all the gold in the world, so far refined, could be placed in a single cube 60 ft on a side. Of all the elements, gold in its pure state is undoubtedly the most beautiful. It is metallic, having a yellow color when in a mass, but when finely divided, it may be black, ruby, or purple. The Purple of Cassius is a delicate test for auric gold. It is the most malleable and ductile metal; 1 oz of gold can be beaten out to 300 $ft.^2$ It is a soft metal and is usually alloyed to give it more strength. It is a good conductor of heat and electricity, and is unaffected by air and most reagents.

Uses

It is used in coinage and is a standard for monetary systems in many countries. It is also extensively used for jewelry, decoration, dental work, and for plating. It is used for coating certain space satellites, as it is a good reflector of infrared and is inert.

Zinc

Atomic Number:	30	Column Number:	IIB
Symbol:	**Zn**	Metal/Non-Metal	Metal
Atomic Weight:	65.38	Outer Shell Electrons	2
Family:	Transitional Metal	Physical State:	Solid
DOT Class:	Not Listed	NFPA 704:	Not Listed
4-Digit Number:	Not Listed	Placard:	Not Listed

Figure 2.22

History

From the German, *zink*, a word of obscure origin. Centuries before zinc was recognized as a distinct element, zinc ores were used for making brass. Tubal-Cain, seven generations from Adam, is mentioned as being an "instructor in every artificer in brass and iron." An alloy containing 87% zinc has been found in prehistoric ruins in Transylvania. Metallic zinc was produced in the 13th century A.D. in India by reducing calamine with organic substances, such as wool. The metal was rediscovered in Europe by Marggraf in 1746, who showed that it could be obtained by reducing calamine with charcoal.

Sources

The principal ores of zinc are sphalerite (sulfide), smithsonite (carbonate), calamine (silicate), and franklinite (zine, manganese, iron oxide). One method of zinc extraction involves roasting its ores to form the oxide and reducing the oxide with coal or carbon, with subsequent distillation of the metal.

Properties

Zinc is a bluish-white, lustrous metal. It is brittle at ordinary temperatures, but malleable at 100 to 150°C. It is a fair conductor of electricity, and burns in air at high red heat, with evolution of white clouds of the oxide. It exhibits superplasticity. Neither zinc nor *zirconium* is ferromagnetic; but $ZrZn_2$ exhibits ferromagnetism at temperatures below 35°K. It has unusual electrical, thermal, optical, and solid-state properties that have not been fully investigated. Zinc is not considered toxic, but when freshly formed ZnO is inhaled, a disorder known as the oxide shakes or zinc chills sometimes occurs. Where zinc oxide is encountered, recommendations include providing good ventilation to avoid concentration exceeding 5 mg/m^3 (time-weighted over an 8-hour exposure, 40-hour workweek).

Uses

The metal is employed to form numerous alloys with other metals. Brass, nickel silver, typewriter metal, commercial bronze, spring bronze, German silver, soft solder, and aluminum solder are some of the more important alloys. Large quantities of zinc are used to produce die-castings, which are used extensively by the automotive, electrical, and hardware industries. An alloy called Prestal®, consisting of 78% zinc and 22% *aluminum*, is reported to be almost as strong as steel and as easy to mold as plastic. In fact, it is so moldable that it can be molded into form using inexpensive ceramics or cement die-casts. Zinc is also used extensively to galvanize other metals, such as *iron,* to prevent corrosion. Zinc oxide is a unique and useful material for modern civilization. It is widely used in the manufacture of paints, rubber products, cosmetics, pharmaceuticals, floor coverings, plastics, printing inks, soap, storage batteries, textiles, electrical equipment, and other products. Lithopone, a mixture of zinc sulfide and barium sulfate, is an important pigment. Zinc sulfide is used in making luminous dials, x-ray and TV screens, and fluorescent lights. The chloride and chromate are also important compounds. Zinc is an essential element in the growth of human beings and animals. Tests show that zinc-deficient animals require 50% more food to gain the same weight as an animal supplied with sufficient zinc.

Mercury

Atomic Number:	80	Column Number:	IIB
Symbol:	**Hg**	Metal/Non-Metal	Metal
Atomic Weight:	200.59	Outer Shell Electrons	2,1
Family:	Transitional Metal	Physical State:	Solid
DOT Class:	8	NFPA 704:	Not Listed
4-Digit Number:	2809	Placard:	Corrosive

Figure 2.23

History

Named after the planet Mercury; known to ancient Chinese and Hindus; found in Egyptian tombs of 1500 B.C. Mercury is the only common metal forming a liquid at ordinary temperatures. It only rarely occurs free in nature. The chief ore is cinnabar. Spain and Italy produce about 50% of the world's supply of the metal. The commercial unit for handling mercury is the "flask," which weighs 76 lbs. The metal is obtained by heating cinnabar in a current of air and condensing the vapor. It is a heavy, silvery-white metal; a rather poor conductor of heat, as compared with other metals; and a fair conductor of electricity. It easily forms alloys with many metals, such as gold, silver, and tin, which are called amalgams. Its ease in amalgamating with gold is made use of in the recovery of gold from its ores. The most important salts are mercury chloride (corrosive sublimate — a violent poison), mercurous chloride (calomel, occasionally still used in medicine), mercury fulminate (a detonator widely used in explosives), and mercuric sulfide (vermilion, a high-grade paint pigment). Organic mercury compounds are important. Mercury is a virulent poison and is readily absorbed through the respiratory tract, the gastrointestinal tract, or through unbroken skin. It acts as a cumulative poison, and dangerous levels are readily attained in air. Air saturated with mercury vapor at 20°C contains a concentration that exceeds the toxic limit many times. The danger increases at higher temperatures. It is therefore important that mercury be handled with care. Containers of mercury should be securely covered, and spillage should be avoided.

If it is necessary to heat mercury or mercury compounds, it should be done in a well-ventilated hood. Methyl mercury is a dangerous pollutant and is now widely found in ground water and streams.

Uses

The metal is widely used in laboratory work for making thermometers, barometers, diffusion pumps, and many other instruments. It is used in making mercury-vapor lamps and advertising signs, and is used in mercury switches and other electronic apparatus. Other uses are in making pesticides, mercury cells for caustic soda and chlorine production, dental preparations, antifouling paint, batteries, and catalysts.

Boron

Atomic Number:	5	Column Number:	III
Symbol:	**B**	Metal/Non-Metal	Non-Metal
Atomic Weight:	10.81	Outer Shell Electrons	3
Family:	Non-Metal	Physical State:	Solid
DOT Class:	Not Listed	NFPA 704:	Not Listed
4-Digit Number:	Not Listed	Placard:	Not Listed

Figure 2.24

History

From the Argentine word, *buraq*, and the Persian, *burah*. Boron compounds have been known for thousands of years, but the element was not discovered until 1808 by Sir Humphry Davy and by Gay-Lussac and Thenard.

Sources

The element is not found free in nature, but occurs as orthoboric acid, usually found in certain volcanic springwaters, and as borates in boron and colemantie. Ulexite, another boron mineral, is interesting, as it is nature's own version of "fiber optics." Important sources of boron are ore rasorite (kernite) and tincal (borax ore). Both of these ores are found in the Mojave Desert. Tincal is the most important source of boron from the Mojave. Extensive borax deposits are also found in Turkey. High-purity crystalline boron may be prepared by the vapor-phase reduction of boron trichloride or tribromide with hydrogen on electrically heated filaments. The impure, or amorphous, boron, a brownish-black powder, can be obtained by heating the trioxide with magnesium powder. Boron of 99.9999% purity has been produced and is available commercially.

Properties

Optical characteristics include transmitting portions of the infrared. Boron is a poor conductor of electricity at room temperature, but a good conductor at high temperature. Elemental boron and the borates are not considered toxic, and they do not require special care in handling. However, some of the more exotic boron hydrogen compounds are definitely toxic and do require care.

Uses

Amorphous boron is used in pyrotechnic flares to provide a distinctive green color, and in rockets as an igniter. By far, the most commercially important boron compound in terms of dollar sales is $Na_2B_4O_7{\cdot}5H_2O$. This pentahydrate is used in large quantities in the manufacture of insulation fiberglass and sodium perborate bleach. Boric acid is also an important boron compound, with major markets in textile products. Use of borax as a mild antiseptic is minor in terms of dollars and tons. Boron compounds are also extensively used in the manufacture of borosilicate glasses. Other boron compounds show promise in treating arthritis. The isotope boron-10 is used as a control for nuclear reactors, as a shield for nuclear radiation, and in instruments used for detecting neutrons. Boron nitride has remarkable properties and can be used to make a material as hard as diamond. The nitride also behaves like an electrical insulator, but conducts heat like a metal. It also has lubricating properties similar to graphite. The hydrides are easily oxidized, with considerable energy liberation, and have been studied for use as rocket fuels. Demand is increasing for boron filaments, a high-strength, lightweight material chiefly employed for advanced aerospace structures. Boron is similar to carbon in that it has a capacity to form stable, covalently bonded molecular networks. Carbonates, metalloboranes, phosphacarboranes, and other families comprise thousands of compounds.

Aluminum

Atomic Number:	13	Column Number:	III
Symbol:	**Al**	Metal/Non-Metal	Metal
Atomic Weight:	26.98	Outer Shell Electrons	3
Family:	Metal	Physical State:	Solid
DOT Class:	Not Listed	NFPA 704:	Not Listed
4-Digit Number:	Not Listed	Placard:	Not Listed

Figure 2.25

History

From the Latin, *alumen*, for alum. The ancient Greeks and Romans used alum as an astringent and as a mordant in dyeing. In 1761, de Morveau proposed the name alumine for the base in alum, and Lavoisier, in 1787, thought this to be the oxide of a still-undiscovered metal. Wohler is generally credited with having isolated the metal in 1827, although an impure form was prepared by Oersted two years earlier. In 1807, Davy proposed the name aluminium for the metal, undiscovered at that time, and later agreed to change it to aluminum. Shortly thereafter, the name aluminium was adopted to conform to the "ium" ending of most elements, and this spelling is now in use elsewhere in the world. Aluminium was also the accepted spelling in the U.S. until 1925, at which time the American Chemical Society officially decided to use the name aluminum thereafter in their publications.

Sources

The method of obtaining aluminum metal by the electrolysis of alumina dissolved in cryolite was discovered in 1886 by Hall in the U.S., and at about the same time by Heroult in France. Cryolite, a natural ore found in Greenland, is no longer widely used in commercial production, but has been replaced by an artificial mixture of sodium, aluminum, and calcium fluorides. Aluminum can now be produced from clay, but the process is not economically feasible at present. Aluminum is the most abundant metal in the earth's crust (8.1%), but is never found free in nature. In addition to the minerals mentioned previously, it is found in granite and in many other common minerals.

Properties

Pure aluminum, a silvery-white metal, possesses many desirable characteristics. It is light, nonmagnetic, and nonsparking; stands second among metals in the scale of malleability, and sixth in ductility.

Uses

It is extensively used for kitchen utensils, outside building decoration, and in thousands of industrial applications where a strong, light, easily constructed material is needed. Although its electrical conductivity is only about 60% that of copper, it is used in electrical transmission lines because of its light weight. Pure aluminum is soft and lacks strength, but it can be alloyed with small amounts of copper, magnesium, silicon, manganese, and other elements to impart a variety of useful properties. These alloys are of vital importance in the construction of modern aircraft and rockets. Aluminum, evaporated in a vacuum, forms a highly reflective coating for both visible light and radiant heat. These coatings soon form a thin layer of the protective oxide and do not deteriorate as do silver coatings. They are used to coat telescope mirrors and to make decorative paper, packages, and toys.

Compounds

The compounds of greatest importance are aluminum oxide, the sulfate, and the soluble sulfate with potassium (alum). The oxide, alumina, occurs naturally as ruby, sapphire, corundum, and emery, and is used in glassmaking and refractories. Synthetic ruby and sapphire are used in lasers for producing coherent light.

Carbon

Atomic Number:	6	Column Number:	IV
Symbol:	**C**	Metal/Non-Metal	Non-Metal
Atomic Weight:	12.01	Outer Shell Electrons	4
Family:	Non-Metal	Physical State:	Solid
DOT Class:	4.2	NFPA 704:	None Listed
4-Digit Number:	1362	Placard:	Spontaneously Combustible

Figure 2.26

History

From the Latin, *carbo*, meaning charcoal. Carbon, an element of prehistoric discovery, is widely distributed in nature. It is found in abundance in the sun, stars, comets, and atmospheres of most planets. Carbon in the form of microscopic diamonds is found in some meteorites. Natural diamonds are found in kimberlite of ancient volcanic "pipes" found in South Africa, Arkansas, and elsewhere. Diamonds are now also being recovered from the ocean floor off the Cape of Good Hope. About 30% of all industrial diamonds used in the U.S. are now made synthetically.

Forms

Carbon is found free in nature in three allotropic forms: amorphous, graphite, and diamond. A fourth form, known as "white" carbon, is now thought to exist. Ceraphite is one of the softest known materials, while diamond is one of the hardest. "White" carbon is a transparent birefringent material. Little information is presently available about this allotrope.

Compounds

In combination, carbon is found as carbon dioxide in the atmosphere of the earth and dissolved in all natural waters. It is a component of great rock masses in the form of carbonates of calcium (limestone), magnesium, and iron. Coal, petroleum, and natural gas are chiefly hydrocarbons. Carbon is unique among the elements in the vast number and variety of compounds it can form. With hydrogen, oxygen, nitrogen, and other elements, it forms a large number of compounds, carbon atom often linking to carbon atom. There are close to 10 million known carbon compounds, many thousands of which are vital to organic and life processes. Without carbon, the basis for life would be impossible. While it has been thought that silicon might take the place of carbon in forming a host of similar compounds, it is now not possible to form stable compounds with long chains of silicon atoms. The atmosphere of Mars contains 96.2% CO_2. Some of the most important compounds of carbon are carbon dioxide (CO_2), carbon monoxide (CO), carbon disulfide (CS_2), chloroform ($CHCl_3$), carbon tetrachloride (CCl_4), methane (CH_4), ethylene (C_2H_4), acetylene (C_2H_2), benzene (C_6H_6), acetic acid (CH_3COOH), and their derivatives.

Isotopes

Carbon has seven isotopes. In 1961, the International Union of Pure and Applied Chemistry adopted the isotope carbon-12 as the basis for atomic weights. Carbon-14, an isotope with a half-life of 5715 years, has been widely used to date such materials as wood, archaeological specimens, etc.

Silicon

Atomic Number:	14	Column Number:	IV
Symbol:	**Si**	Metal/Non-Metal	Non-Metal
Atomic Weight:	28.09	Outer Shell Electrons	4
Family:	Non-Metal	Physical State:	Solid
DOT Class:	None Listed	NFPA 704:	None Listed
4-Digit Number:	None Listed	Placard:	None Listed

Figure 2.27

History

From the Latin words *silex* or *silicis*, meaning flint. Davy, in 1800, thought silica to be a compound and not an element; later, in 1811, Gay Lussac and Thenard probably prepared impure amorphous silicon by heating potassium with silicon tetrafluoride. In 1824, Berzelius, generally credited with the discovery, prepared amorphous silicon by the same general method and purified the product by removing the fluosilicates by repeated washings. Deville, in 1854, first prepared crystalline silicon, the second allotropic form of the element.

Sources

Silicon is present in the sun and stars, and is a principal component of a class of meteorites known as aerolites. It is also a component of tektites, a natural glass of uncertain origin. Silicon makes up 25.7% of the earth's crust, by weight, and is the second most-abundant element, exceeded only by *oxygen*. Silicon is not found free in nature, but occurs chiefly as the oxide and as silicates. Sand, quartz, rock crystal, amethyst, agate, flint, jasper, and opal are some of the forms in which the oxide appears. Granite, hornblende, asbestos, feldspar, clay, mica, etc. are but a few of the numerous silicate minerals. Silicon is prepared commercially by heating silica and carbon in an electric furnace using carbon electrodes. Several other methods can be used for preparing the element.

Uses

Silicon is one of man's most useful elements. In the form of sand and clay, it is used to make concrete and brick; it is a useful refractory material for high-temperature work, and in the form of silicates, it is used in making enamels and pottery. Silica, as sand, is a principal ingredient of glass, one of the most inexpensive of materials with excellent mechanical, optical, thermal, and electrical properties. Glass can be made in a great variety of shapes, and is used as containers, window glass, insulators, and thousands of other uses. Hyperpure silicon can be doped with boron, gallium, phosphorus, or arsenic to produce silicon for use in transistors, solar cells, rectifiers, and other solid-state devices, which are used extensively in the electronics and space-age industries. Hydrogenated amorphous silicon has shown promise in producing economical cells for converting solar energy into electricity. Silicon is important to plant and animal life. Diatoms, in both fresh and salt water, extract silica from the water to build their cell walls. Silica is present in the ashes of plants and in the human skeleton. Silicon is an important ingredient in steel; silicon carbide is one of the most important abrasives and has been used in lasers to produce coherent light of 4560 A. Silcones are important products of silicon.

Properties

Crystalline silicon has a metallic luster and grayish color. Silicon is a relatively inert element, but it is attacked by halogens and dilute alkali. Most acids, except hydrofluoric, do not affect it. Elemental silicon transmits more than 95% of all wavelengths of infrared, from 1.3 to 6.y micro-m.

Nitrogen

Atomic Number:	7	Column Number:	V
Symbol:	**N**	Metal/Non-Metal	Non-Metal
Atomic Weight:	14.01	Outer Shell Electrons	5
Family:	Non-Metal	Physical State:	Gas
DOT Class:	2.2	NFPA 704:	Not Listed
4-Digit Number:	1066	Placard:	Non-Flammable Gas

Figure 2.28

History

From the Latin *nitrum* and Greek *nitron*, meaning native soda, and *genes*, meaning forming. Chemist and physician Daniel Rutherford discovered nitrogen in 1772. He removed oxygen and carbon dioxide from air and showed that the residual gas would not support combustion or living organisms. At the same time, there were other noted scientists working on the problem of identifying and explaining the behavior of nitrogen. They called it "burnt [or] dephlogisticated air," which means air without oxygen.

Sources

Nitrogen gas (N_2) makes up 78.1% of the Earth's air, by volume. The atmosphere of Mars, by comparison, is only 2.6% nitrogen. From an exhaustible source in our atmosphere, nitrogen gas can be obtained by liquefaction and fractional distillation. Nitrogen is found in all living systems as part of the makeup of biological compounds.

Properties

French chemist Antoine Laurent Lavoisier named nitrogen *azote*, meaning without life. However, nitrogen compounds are found in foods, fertilizers, poisons, and explosives. Nitrogen as a gas is colorless, odorless, and generally considered an inert element. As a liquid (boiling point = minus 195.8°C), it is also colorless and odorless, and is similar in appearance to water. Nitrogen gas can be prepared by heating a water solution of ammonium nitrite (NH_4NO_3). The nitrogen cycle is one of the most important processes in nature for living organisms. Although nitrogen gas is relatively inert, bacteria in the soil are capable of "fixing" the nitrogen into a usable form (as a fertilizer) for plants. In other words, nature has provided a method to produce nitrogen for plants to grow. Animals eat the plant material, and the nitrogen is incorporated into their system, primarily as protein. The cycle is completed when other bacteria convert the waste nitrogen compounds back to nitrogen gas. Nitrogen has become crucial to life, being a component of all proteins.

Nitrogen Compounds

Sodium nitrate ($NaNO_3$) and potassium nitrate (KNO_3) are formed by the decomposition of organic matter, with compounds of these metals present. In certain dry areas of the world, these saltpeters are found in quantity and are used as fertilizers. Other inorganic nitrogen compounds are nitric acid (HNO_3), ammonia (NH_3), the oxides (NO, NO_2, N_2O_4, N_2O), cyanides (CN^-), etc.

Ammonia

Ammonia (NH_3) is the most important commercial compound of nitrogen. It is produced by the Haber Process. This process is the synthesis of ammonia by the water-gas reaction from hot coke, air, and steam. Natural gas (methane, CH_4) is reacted with steam to produce carbon dioxide and hydrogen gas (H_2) in a two-step process. Hydrogen gas and nitrogen gas are then reacted in the Haber Process to produce ammonia. This colorless gas with a pungent odor is easily liquefied. In fact, the liquid is used as a nitrogen fertilizer. Ammonia is also used in the production of urea, NH_2CONH_2, which is used as a fertilizer, in the plastic industry, and in the livestock industry as a feed supplement. Ammonia is often the starting compound for many other nitrogen compounds.

Phosphorus

Atomic Number:	15	Column Number:	V
Symbol:	**P**	Metal/Non-Metal	Non-Metal
Atomic Weight:	30.97	Outer Shell Electrons	5
Family:	Non-Metal	Physical State:	Solid
DOT Class:	4.1	NFPA 704:	Not Listed
4-Digit Number:	1338	Placard:	Flammable Solid

Figure 2.29

History

From the Greek, *phosphoros*, meaning light-bearing; also an ancient name for the planet Venus when appearing before sunrise. Brand discovered phosphorus in 1669 by preparing it from urine.

Properties

Phosphorus exists in three or more allotropic forms: white (or yellow), red, and black (or violet). Ordinary phosphorus is a waxy, white solid; when pure, it is colorless and transparent. It is insoluble in water, but soluble in carbon disulfide. It takes fire spontaneously in air, burning to the pentoxide. It is highly poisonous, 50 mg constituting an approximate fatal dose. Exposure to white phosphorus should not exceed 0.1 mg/m^3 (8-hour time-weighted average in a 40-hour workweek). White phosphorus should be kept under water, as it is dangerously reactive in air, and it should be handled with forceps, as contact with the skin may cause severe burns. When exposed to sunlight or when heated in its own vapor to 250°C, it is converted to the red variety, which does not phosphoresce in air, as does the white variety. This form does not ignite spontaneously and is not as dangerous as white phosphorus. It should, however, be handled with care, as it does convert to the white form at some temperatures and it emits highly toxic fumes of the oxides of phosphorus when heated. The red modification is fairly stable, sublimes with a vapor pressure of 1 atm at 17°C, and is used in the manufacture of safety matches, pyrotechnics, pesticides, incendiary shells, smoke bombs, tracer bullets, etc.

Sources

Never found free in nature, it is widely distributed in combination with minerals. Phosphate rock, which contains the mineral apatite, an impure tricalcium phosphate, is an important source of the element. Large deposits are found in Russia, Morocco, Florida, Tennessee, Utah, Idaho, and elsewhere.

Uses

In recent years, concentrated phosphoric acids, which may contain as much as 70% to 75% P_2O_5 content, have become of great importance to agriculture and farm production. Worldwide demand for fertilizers has caused record phosphate production. Phosphates are used in the production of special glasses, such as those used for sodium lamps. Bone ash, calcium phosphate, is used to create fine chinaware and to produce monocalcium phosphate, used in baking powder. Phosphorus is also important in the production of steels, phosphor bronze, and many other products. Trisodium phosphate is important as a cleaning agent, as a water softener, and for preventing boiler scale and corrosion of pipes and boiler tubes. Phosphorus is also an essential ingredient of all cell protoplasm, nervous tissue, and bones.

Arsenic

Atomic Number:	33	Column Number:	V
Symbol:	**As**	Metal/Non-Metal	Non-Metal
Atomic Weight:	74.92	Outer Shell Electrons	5
Family:	Non-Metal	Physical State:	Solid
DOT Class:	6.1	NFPA 704:	Not Listed
4-Digit Number:	1558	Placard:	Poison

Figure 2.30

History

From the Latin *arsenicum* and the Greek *arsenikon*, referring to yellow orpiment (arsenic trisulfide), identified with *arenikos* (Greek for male), from the belief that metals were different sexes; also from the Arabic, *az-zernikh*, referring to the orpiment from Persian, *zerni-zar*, or gold. Elemental arsenic occurs in two solid modifications: yellow and gray, or metallic, with specific gravities of 1.97 and 5.73, respectively. It is believed that Albertus Magnus obtained the element in A.D. 1250. In 1649, Schroeder published two methods of preparing the element. Mispickel, arsenopyrite, (FeSAs) is the most common mineral, from which, on heating, the arsenic sublimes, leaving ferrous sulfide.

Properties

The element is a steel-gray, brittle, crystalline, semimetallic solid; it tarnishes in air and, when heated, is rapidly oxidized to arsenious oxide with the odor of garlic. Arsenic and its compounds are poisonous.

Uses

Arsenic is used in bronzing, pyrotechny, and for hardening and improving the sphericity of shot. The most important compounds are white arsenic, the sulfide, Paris green, calcium arsenate, and lead arsenate; the last three have been used as agricultural insecticides and poisons.

Plutonium

Atomic Number:	94	Column Number:	V
Symbol:	**Pu**	Metal/Non-Metal	Metal
Atomic Weight:	244	Outer Shell Electrons	6,5,4,3
Family:	Metal	Physical State:	Solid
DOT Class:	Not Listed	NFPA 704:	Not Listed
4-Digit Number:	Not Listed	Placard:	Not Listed

Figure 2.31

History

Named for the planet Pluto, plutonium was the second transuranium element of the actinide series to be discovered. The isotope ^{238}Pu was produced in 1940 by Seaborg, McMillan, Kennedy, and Wahl by deuteron bombardment of uranium in the 60-inch cyclotron at Berkeley, California. Plutonium also exists in trace quantities in naturally occurring uranium ores. It is formed in much the same manner as neptunium, by irradiation of natural uranium with the neutrons that are present.

Isotopes

By far of greatest importance is the isotope ^{239}Pu, with a half-life of 24,100 years, produced in extensive quantities in nuclear reactors from natural uranium: ^{238}U(n, gamma) — > ^{239}U — (beta) — > ^{239}Np — (beta) — > ^{239}Pu. Fifteen isotopes of plutonium are known. Plutonium forms binary compounds with oxygen: PuO, PuO_2, and intermediate oxides of variable composition; with the halides: PuF_3, PuF_4, $PuCl_3$, $PuBr_3$, and PuI_3; and with carbon, nitrogen, and silicon: PuC, PuN, and $PuSi_2$. Oxyhalides are also well known: PuOCl, PuOBr, and PuOI.

Uses

Plutonium has assumed a position of dominant importance among the transuranium elements because of its successful use as an explosive ingredient in nuclear weapons and the place it holds as a key material in the development of industrial use of nuclear power. One kilogram is equivalent to about 22 million-kilowatt hours of heat energy. The complete detonation of a kilogram of plutonium produces an explosion equal to about 20,000 tons of chemical explosive. Its importance depends on the nuclear property of being readily fissionable with neutrons and its availability in quantity. The world's nuclear-power reactors are now producing about 20,000 kg of plutonium a year. By 1982, it was estimated that about 300,000 kg had accumulated. The various nuclear applications of plutonium are well known. ^{238}Pu has been used in the Apollo lunar missions to power seismic and other equipment on the lunar surface. As with neptunium and uranium, plutonium metal can be prepared by reduction of the trifluoride with alkaline-earth metals.

Properties

The metal has a silvery appearance and takes on a yellow tarnish when slightly oxidized. It is chemically reactive. A relatively large piece of plutonium is warm to the touch because of the energy given off in alpha decay. Larger pieces will produce enough heat to boil water. The metal readily dissolves in concentrated hydrochloric acid, hydroiodic acid, or perchloric acid. Because of the high rate of emission of alpha particles and the element being specifically absorbed on bone surface and collected in the liver, plutonium, as well as all of the other transuranium elements, except neptunium, are radiological poisons and must be handled with special equipment and precautions. Plutonium is a highly dangerous radiological hazard. Precautions must also be taken to prevent the unintentional formulation of a critical mass. Plutonium in liquid solution is more likely to become critical than solid plutonium. The shape of the mass must also be considered where criticality is concerned.

Oxygen

Atomic Number:	8	Column Number:	VI
Symbol:	**O**	Metal/Non-Metal	Non-Metal
Atomic Weight:	16	Outer Shell Electrons	6
Family:	Non-Metal	Physical State:	Gas
DOT Class:	2.2	NFPA 704:	Not Listed
4-Digit Number:	1073	Placard:	Non-Flammable Gas, Oxidizer

Figure 2.32

History

From the Greek, *oxys*, meaning sharp or acid, and *genes*, meaning forming: acid former. For many centuries, workers occasionally realized air was composed of more than one component. The behavior of oxygen and nitrogen as components of air led to the advancement of the phlogiston theory of combustion, which captured the minds of chemists for a century. Oxygen was prepared by several workers, including Bayen and Borch, but they did not know how to collect it, did not study its properties, and did not recognize it as an elementary substance. Priestley is generally credited with its discovery, although Scheele also discovered it independently. Its atomic weight was used as a standard of comparison for each of the other elements until 1961, when the International Union of Pure and Applied Chemistry adopted carbon-12 as the new basis.

Sources

Oxygen is the third most-abundant element found in the sun, and it plays a part in the carbon-nitrogen cycle, the process once thought to give the sun and stars their energy. Oxygen under excited conditions is responsible for the bright red and yellow-green colors of the Aurora Borealis. A gaseous element, oxygen forms 21% of the atmosphere by volume and is obtained by liquefaction and fractional distillation. The atmosphere of Mars contains about 0.15% oxygen. The element and its compounds make up 49.2%, by weight, of the earth's crust. About two-thirds of the human body and nine-tenths of water is oxygen.

Properties

The gas is colorless, odorless, and tasteless. The liquid and solid forms are a pale blue color and are strongly paramagnetic.

Forms

Ozone (O_3), a highly active compound, is formed by the action of an electrical discharge or ultraviolet light on oxygen. Ozone's presence in the atmosphere (amounting to the equivalent of a layer 3 mm thick under ordinary pressures and temperatures) helps prevent harmful ultraviolet rays of the sun from reaching the earth's surface. Pollutants in the atmosphere may have a detrimental effect on this ozone layer. Ozone is toxic, and exposure should not exceed 0.2 mg/m^3 (8-hour time-weighted average in a 40-hour work week). Undiluted ozone has a bluish color. Liquid ozone is bluish-black, and solid ozone is violet-black.

Compounds

Oxygen, which is highly reactive, is a component of hundreds of thousands of organic compounds and combines with most elements.

Uses

Plants and animals rely on oxygen for respiration. Hospitals frequently prescribe oxygen for patients with respiratory ailments.

Isotopes

Oxygen has nine isotopes. Natural oxygen is a mixture of three isotopes. Naturally occurring oxygen-18 is stable and available commercially, as is water (H_2O with 15% ^{18}O). Commercial oxygen consumption in the U.S. is estimated at 20 million short tons per year, and the demand is expected to increase substantially. Oxygen enrichment of steel blast-furnaces accounts for the greatest use of the gas. Large quantities are also used in making synthesis gas for ammonia and methanol, ethylene oxide, and for oxyacetylene welding. Air separation plants produce about 99% of the gas, while electrolysis plants produce about 1%.

Sulfur

Atomic Number:	16	Column Number:	VI
Symbol:	S	Metal/Non-Metal	Non-Metal
Atomic Weight:	32.07	Outer Shell Electrons	6
Family:	Non-Metal	Physical State:	Solid
DOT Class:	4.1	NFPA 704:	2-1-0
4-Digit Number:	1350	Placard:	Flammable Solid

Figure 2.33

History

From the Sanskrit, *sulvere*; L. sulpur. Known to the ancients; referred to in Genesis as brimstone.

Sources

Sulfur is found in meteorites. R.W. Wood suggests that the dark area near the crater Aristarchus is a sulfur deposit. Sulfur occurs naturally in the vicinity of volcanos and hot springs. It is widely distributed in nature as iron pyrites, galena, sphalerite, cinnabar, stibnite, gypsum, epsom salts, celestite, barite, etc.

Production

Sulfur is commercially recovered from wells sunk into the salt domes along the Gulf Coast of the U.S. Using the Frasch process, heated water is forced into the wells to melt the sulfur, which is then brought to the surface. Sulfur also occurs in natural gas and petroleum crudes and must be removed from these products. Formerly, this was done chemically, which wasted the sulfur; new processes now permit recovery. Large amounts of sulfur are also recovered from Alberta gas fields.

Properties

Sulfur is a pale yellow, odorless, brittle solid, which is insoluble in water, but soluble in carbon disulfide. In every state, whether gas, liquid, or solid, elemental sulfur occurs in more than one allotropic form or modification; these present a confusing multitude of forms whose relations are not yet fully understood. In 1975, University of Pennsylvania scientists reported synthesis of polymeric sulfur nitride, which has the properties of a metal, although it contains no metal atoms. The material has unusual optical and electrical properties. High-purity sulfur is commercially available in purities of 99.999+%. Amorphous, or "plastic," sulfur is obtained by fast cooling of the crystalline form. Crystalline sulfur seems to be made of rings, each containing eight sulfur atoms, which fit together to give a normal x-ray pattern. Carbon disulfide, hydrogen sulfide, and sulfur dioxide should be handled carefully. Hydrogen sulfide in small concentrations can be metabolized, but in higher concentrations, it quickly can cause death by respiratory paralysis. It quickly deadens the sense of smell. Sulfur dioxide is a dangerous component in atmospheric air pollution.

Isotopes

Eleven isotopes of sulfur exist. None of the four isotopes that exist in nature are radioactive. A finely divided form of sulfur, known as flowers of sulfur, is obtained by sublimation.

Compounds

Organic compounds containing sulfur are important. Calcium sulfur, ammonium sulfate, carbon disulfide, sulfur dioxide, and hydrogen sulfide are but a few of the many important compounds of sulfur.

Uses

Sulfur is a component of black gunpowder, and is used in the vulcanization of natural rubber and as a fungicide. It is also used extensively in making phosphatic fertilizers.

A tremendous tonnage is used to produce sulfuric acid, the most important manufactured chemical. It is used to make sulfite paper and other papers, to fumigants, and to bleach dried fruits. The element is a good insulator. Sulfur is essential to life. It is a minor constituent of fats, body fluids, and skeletal minerals.

Fluorine

Atomic Number:	9	Column Number:	VII
Symbol:	F	Metal/Non-Metal	Non-Metal
Atomic Weight:	19	Outer Shell Electrons	7
Family:	Halogen	Physical State:	Gas
DOT Class:	2.3	NFPA 704:	Not Listed
4-Digit Number:	1045	Placard:	Poison Gas, Oxidizer

Figure 2.34

History

From Latin and French for *fluere*, meaning flow or flux. In 1529, Georigius Agricola described the use of fluorspar as a flux, and as early as 1670, Schwandhard found that glass was etched when exposed to fluorspar treated with acid. Scheele and many later investigators, including Davy, Gay-Lussac, Lavoisier, and Thenard, experimented with hydrofluoric acid, some experiments ending in tragedy. The element was finally isolated in 1866 by Moissan after nearly 74 years of continuous effort.

Properties

Fluorine is the most electronegative and reactive of all elements. It is a pale yellow, corrosive gas, which reacts with most organic and inorganic substances. Finely divided metals, glass, ceramics, carbon, and even water burn in fluorine with a bright flame. Until World War II, there was no commercial production of elemental fluorine. The nuclear bomb project and nuclear energy applications, however, made it necessary to produce large quantities.

Uses

Fluorine and its compounds are used in producing uranium (from the hexafluoride) and more than 100 commercial fluorochemicals, including many well-known high-temperature plastics. Hydrofluoric acid etches the glass of lightbulbs, etc. Fluorochlorohydrocarbons are extensively used in air conditioning and refrigeration. The presence of fluorine as a soluble fluoride in drinking water to the extent of 2 ppm may cause mottled enamel in teeth when used by children acquiring permanent teeth; in smaller amounts, however, fluorides are added to water supplies to prevent dental cavities. Elemental fluorine has been studied as a rocket propellant, as it has an exceptionally high specific-impulse value.

Compounds

One hypothesis says that fluorine can be substituted for hydrogen wherever it occurs in organic compounds, which could lead to an astronomical number of new fluorine compounds. Compounds of fluorine with rare gases have now been confirmed in fluorides of xenon, radon, and krypton.

Handling

Elemental fluorine and the fluoride ion are highly toxic. The free element has a characteristic pungent odor, detectable in concentrations as low as 20 ppb, which is below the safe working level. The recommended maximum allowable concentration for a daily 8-hour, time-weighted exposure in a 40-hour workweek is 1 ppm. Safe handling techniques enable the transport of liquid fluorine by the ton.

Chlorine

Atomic Number:	17	Column Number:	VII
Symbol:	**Cl**	Metal/Non-Metal	Non-Metal
Atomic Weight:	35.45	Outer Shell Electrons	7
Family:	Halogen	Physical State:	Gas
DOT Class:	2.3	NFPA 704:	Not Listed
4-Digit Number:	1017	Placard:	Poison Gas

Figure 2.35

History

From the Greek, *chloros*, meaning greenish-yellow. Discovered in 1774 by Scheele, who thought it contained oxygen. Chlorine was named in 1810 by Davy, who insisted it was an element.

Sources

In nature, it is found in the combined state only, chiefly with sodium as common salt (NaCl), carnallite, and sylvite.

Properties

It is a member of the halogen (salt-forming) group of elements, and is obtained from chlorides by the action of oxidizing agents and, more often, by electrolysis; it is a greenish-yellow gas, combining directly with nearly all elements.

Uses

Chlorine is widely used in making many everyday products. It is used for producing safe drinking water the world over. Even the smallest water supplies are now usually chlorinated. It is also extensively used in the production of paper products, dyestuffs, textiles, petroleum products, medicines, antiseptics, insecticides, food, solvents, paints, plastics, and many other consumer products. Most of the chlorine produced is used in the manufacture of chlorinated compounds for sanitation, pulp bleaching, disinfectants, and textile processing. Further use is in the manufacture of chlorates, chloroform, carbon tetrachloride, and in the extraction of bromine. Organic chemistry demands much from chlorine, both as an oxidizing agent and in substitution, since it often brings many desired properties in an organic compound when substituted for hydrogen, as in one form of synthetic rubber.

Bromine

Atomic Number:	35	Column Number:	VII
Symbol:	**Br**	Metal/Non-Metal	Non-Metal
Atomic Weight:	75.90	Outer Shell Electrons	7
Family:	Halogen	Physical State:	Liquid
DOT Class:	8	NFPA 704:	Not Listed
4-Digit Number:	1744	Placard:	Corrosive, Poison

Figure 2.36

History

From the Greek, *bromos*, meaning stench. Discovered by Balard in 1826, but not prepared in quantity until 1860.

Sources

A member of the halogen group of elements, it is obtained from natural brines from wells in Michigan and Arkansas. Little bromine is extracted today from seawater, which contains only about 85 ppm.

Properties

Bromine is the only liquid nonmetallic element. It is a heavy, mobile, reddish-brown liquid, volatilizing readily at room temperature to a red vapor with a strong disagreeable odor, resembling chlorine, and having an irritating effect on the eyes and throat; it is readily soluble in water or carbon disulfide, forming a red solution, and is less active than chlorine, but more so than iodine; it unites readily with many elements and has a bleaching action; when spilled on the skin, it produces painful sores. It presents a serious health hazard, and maximum safety precautions should be taken when handling it.

Production

Much of the bromine output in the U.S. was used in the production of ethylene dibromide, a lead scavenger used in making gasoline antiknock compounds. Lead in gasoline, however, has been drastically reduced, due to environmental considerations. This will greatly affect future production of bromine.

Uses

Bromine is used in making fumigants, flameproofing agents, water purification compounds, dyes, medicinals, sanitizers, inorganic bromides for photography, etc. Organic bromides are also important.

Iodine

Atomic Number:	53	Column Number:	VII
Symbol:	I	Metal/Non-Metal	Non-Metal
Atomic Weight:	126.9	Outer Shell Electrons	7
Family:	Halogen	Physical State:	Solid
DOT Class:	Not Listed	NFPA 704:	Not Listed
4-Digit Number:	Not Listed	Placard:	Not Listed

Figure 2.37

History

From the Greek, *iodes*, meaning violet. Discovered by Courtois in 1811, iodine, a halogen, occurs sparingly in the form of iodides in seawater, from which it is assimilated by seaweeds, in Chilean saltpeter, and nitrate-bearing earth (known as caliche), in brines from old sea deposits, and in brackish waters from oil and salt wells.

Sources

Ultrapure iodine can be obtained from the reaction of potassium iodide with copper sulfate. Several other methods of isolating the element are known.

Properties

Iodine is a bluish-black, lustrous solid, volatizing at ordinary temperatures into a blue-violet gas with an irritating odor; it forms compounds with many elements, but is less active than the other halogens, which displace it from iodides. Iodine exhibits some metallic-like properties. It dissolves readily in chloroform, carbon tetrachloride, or carbon disulfide to form beautiful purple solutions. It is only slightly soluble in water. Care should be taken in handling and using iodine, as contact with the skin can cause lesions; iodine vapor is intensely irritating to the eyes and mucus membranes. The maximum allowable concentration of iodine in air should not exceed 1 mg/m^3 (8-hour time-weighted average in a 40-hour workweek).

Isotopes

Thirty isotopes are recognized. Only one stable isotope, ^{127}I, is found in nature. The artificial radioisotope, ^{131}I, with a half-life of eight days, has been used in treating the thyroid gland. The most common compounds are the iodides of sodium and potassium (KI) and the iodates (KIO_3). Lack of iodine is the cause of goiter.

Uses

Iodine compounds are important in organic chemistry and are quite useful in medicine. Iodides and thyroxine, which contains iodine, are used internally in medicine and as a solution of KI, and iodine in alcohol is used for external wounds. Potassium iodide finds use in photography. The deep blue color in a starch solution is characteristic of the free element.

Uranium

Atomic Number:	92	Column Number:	VII
Symbol:	**U**	Metal/Non-Metal	Metal
Atomic Weight:	238	Outer Shell Electrons	6,5,4,3
Family:	Metal	Physical State:	Solid
DOT Class:	7	NFPA 704:	Not Listed
4-Digit Numbers:	2979	Placard:	Radioactive

Figure 2.38

History

Named after the planet Uranus. Yellow-colored glass, containing more than 1% uranium oxide and dating back to A.D. 79 has been found near Naples, Italy. Klaproth recognized an unknown element in pitchblende (uraninite or uranium oxide) and attempted to isolate the metal in 1789. The metal was first isolated in 1841 by Peligot, who reduced the anhydrous chloride using potassium.

Sources

Uranium, not as rare as once thought, is now considered to be more plentiful than mercury, antimony, silver, or cadmium, and is about as abundant as molybdenum or arsenic. It occurs in numerous minerals, such as pitchblende, uraninite, carnotite, autunite, uranophane, and tobernite. It is also found in phosphate rock, lignite, and monazite sands, and can be recovered commercially from these sources. The United States Department of Energy purchases uranium in the form of acceptable U_3O_8 concentrates. This incentive program has greatly increased the known uranium reserves. Uranium can be prepared by reducing uranium halides with alkali or alkaline-earth metals or by reducing uranium oxides by calcium, aluminum, or carbon at high temperatures. The metal can also be produced by electrolysis of KUF_5 or UF_4, dissolved in a molten mixture of $CaCl_2$ and NaCl. High-purity uranium can be prepared by the thermal decomposition of uranium halides on a hot filament.

Properties

Uranium is a heavy, silvery-white metal, which is pyrophoric when finely divided. It is a little softer than steel, and is attacked by cold water in a finely divided state. It is malleable, ductile, and slightly paramagnetic. In air, the metal becomes coated with a layer of oxide. Acids dissolve the metal, but it is unaffected by alkalis. Finely divided uranium metal, being pyrophoric, presents a fire hazard. Working with uranium requires the knowledge of the maximum allowable concentrations that may be inhaled or ingested. Recently, the natural presence of uranium in many soils has become of concern to homeowners because of the generation of radon and its daughters.

Isotopes

Uranium has sixteen isotopes, all of which are radioactive. Naturally occurring uranium nominally contains 99.28305, by weight, ^{238}U, 0.7110% ^{235}U, and 0.0054% ^{234}U. Natural uranium is sufficiently radioactive to expose a photographic plate in an hour or so. Much of the internal heat of the earth is thought to be attributable to the presence of uranium and thorium. Uranuim-238, with a half-life of 4.51×10^9 years, has been used to estimate the age of igneous rocks. The origin of uranium, the highest member of the naturally occurring elements — except perhaps for traces of neptunium or plutonium, is not clearly understood. However, it may be presumed that uranium is a decay product of elements with higher atomic weight, which may have once been present on earth or elsewhere in the universe. These original elements may have been formed as a result of a primordial creation, known as the big bang, in a supernova or in some other stellar processes.

Uses

Uranium is of great importance as a nuclear fuel. Uranium-235, ^{235}U, while occurring in natural uranium to the extent of only 0.71%, is so fissionable with slow neutrons that a self-sustaining fission chain reaction can be made in a reactor constructed from natural uranium and a suitable moderator, such as heavy water or graphite, alone. One pound of completely fissioned uranium has the fuel value of over 1500 tons of coal. The uses of nuclear fuels to generate electrical power, to make isotopes for peaceful purposes, and to make explosives are well known. The estimated worldwide capacity of the 429 nuclear-power reactors in operation in January 1990 amounted to about 311,000 megawatts. Uranium in the U.S. is controlled by the U.S. Department of Energy. New uses are being found for depleted uranium, i.e., uranium with the percentage of ^{235}U lowered to about 0.2%. Uranium is used in inertial guidance devices, in gyrocompasses, as counterweights for aircraft control surfaces, as ballast for missile reentry vehicles, and as a shielding material. Uranium metal is used for x-ray targets for production of high-energy x-rays; the nitrate has been used as a photographic toner, and the acetate is used in analytical chemistry. Uranium salts have also been used for producing yellow "vaseline" glass and glazes. Uranium and its compounds are highly toxic, both from a chemical and radiological standpoint.

Helium

Atomic Number:	2	Column Number:	VIII
Symbol:	**He**	Metal/Non-Metal	Non-Metal
Atomic Weight:	4.004	Outer Shell Electrons	8
Family:	Noble Gases	Physical State:	Gas
DOT Class:	2.2	NFPA 704:	Not Listed
4-Digit Number:	1046	Placard:	Non-Flammable Gas

Figure 2.39

History

From the Greek, *helios*, meaning the sun. Janssen obtained the first evidence of helium during the solar eclipse of 1868, when he detected a new line in the solar spectrum.

Lockyer and Frankland suggested the name helium for the new element. In 1895, Ramsay discovered helium in the uranium mineral clevite, while it was independently discovered in cleveite by the Swedish chemists Cleve and Langlet at about the same time. Rutherford and Royds in 1907 demonstrated that alpha particles are helium nuclei.

Sources

Except for *hydrogen*, helium is the most abundant element found throughout the universe. Helium is extracted from natural gas. In fact, all natural gas contains at least trace quantities of helium. The fusion of hydrogen into helium provides the energy of the hydrogen bomb. The helium content of the atmosphere is about 1 part in 200,000. While it is present in various radioactive minerals as a decay product, the bulk of the Free World's supply is obtained from wells in Texas, Oklahoma, and Kansas. The only known helium extraction plants outside the United States in 1984 were in Eastern Europe (Poland), the USSR, and a few in India.

Properties

Helium has the lowest melting point of any element and is widely used in cryogenic research because its boiling point is close to absolute zero. Also, the element is vital in the study of superconductivity. Helium is the only liquid that cannot be solidified by lowering the temperature. It remains liquid down to absolute zero at ordinary pressures, but it can readily be solidified by increasing the pressure. Solid ^3He and ^4He are unusual in that both can be changed in volume by more than 30% by applying pressure. The specific heat of helium gas is unusually high. The density of helium vapor at the normal boiling point is also high, with the vapor expanding greatly when heated to room temperature. Containers filled with helium gas at 5 to 10 K should be treated as though they contained liquid helium due to the large increase in pressure resulting from warming the gas to room temperature. While helium normally has a 0 valence, it seems to have a weak tendency to combine with certain other elements. Means of preparing helium difluoride have been studied, and species such as HeNe and the molecular ions He+ and He++ have been investigated.

Uses

Helium is used as an inert gas shield in arc welding, as a protective gas in growing silicon and germanium crystals and producing titanium and zirconium, as a cooling medium for nuclear reactors, and as a gas for supersonic wind tunnels. A mixture of helium and oxygen is used as an artificial atmosphere for divers and others working under pressure. Different ratios of He/O^2 are used for different depths at which the diver is operating. Helium is extensively used for filling balloons, as it is a much safer gas than hydrogen. One of the recent largest uses for helium has been for pressuring liquid-fuel rockets. A Saturn booster, like the type used on the Apollo lunar missions, required about 13 million ft^3 of helium for a firing, plus more for checkouts. Liquid helium's use in magnetic resonance imaging (MRI) continues to increase as the medical profession accepts and develops new uses for the equipment. This equipment has eliminated some need for exploratory surgery by accurately diagnosing patients. Helium is also used to advertise on blimps for various companies, including Goodyear. Other lifting-gas applications are developed by the Navy and Air Force to detect low-flying cruise missiles. The Drug Enforcement Agency is using radar-equipped blimps to detect drug smugglers along the United States borders. In addition, NASA is currently using helium-filled balloons to sample the atmosphere in Antarctica to determine what is depleting the ozone layer.

Neon

Atomic Number:	10	Column Number:	VIII
Symbol:	**Ne**	Metal/Non-Metal	Non-Metal
Atomic Weight:	20.18	Outer Shell Electrons	8
Family:	Noble Gases	Physical State:	Gas
DOT Class:	2.2	NFPA 704:	Not Listed
4-Digit Number:	1065	Placard:	Non-Flammable Gas

Figure 2.40

History

From the Greek word, *neos*, meaning new. Discovered by Ramsay and Travers in 1898, neon is a rare gaseous element present in the atmosphere to the extent of 1 part in 65,000 of air. It is obtained by liquefaction of air and separated from the other gases by fractional distillation.

Compounds

Neon, an inert element, is, however, said to form a compound with fluorine. It is still questionable if true compounds of neon exist, but evidence is mounting in favor of their existence.

Properties

In a vacuum discharge tube, neon glows reddish-orange. It has over 40 times more refrigerating capacity per unit volume than liquid helium and more than three times that of liquid hydrogen. It is compact, inert, and is less expensive than helium when it meets refrigeration requirements.

Uses

Although neon advertising signs account for the bulk of its use, neon also functions in high-voltage indicators, lightning arrestors, wave meter tubes, and TV tubes. Neon and helium are used in making gas lasers. Liquid neon is now commercially available and is finding important application as an economical cryogenic refrigerant.

Argon

Atomic Number:	18	Column Number:	VIII
Symbol:	**Ar**	Metal/Non-Metal	Non-Metal
Atomic Weight:	39.95	Outer Shell Electrons	8
Family:	Noble Gases	Physical State:	Gas
DOT Class:	2.2	NFPA 704:	Not Listed
4-Digit Number:	1006	Placard:	Non-Flammable Gas

Figure 2.41

History

From the Greek word, *argos*, meaning inactive. Its presence in air was suspected by Cavendish in 1785 and discovered by Lord Rayleigh and Sir William Ramsay in 1894.

Sources

The gas is prepared by fractionation of liquid air, because the atmosphere contains 0.94% argon. The atmosphere of Mars contains 1.6% of ^{40}Ar and 5 ppm of ^{36}Ar.

Properties

Argon is two- and one-half times as soluble in water as nitrogen, having about the same solubility as oxygen. Argon is colorless and odorless, both as a gas and a liquid. Argon is considered to be an inert gas and is not known to form true chemical compounds, as do krypton, xenon, and radon.

Uses

It is used in electric light bulbs and in fluorescent tubes at a pressure of about 400 Pa and in filling photo tubes, glow tubes, etc. Argon is also used as an inert gas shield for arc welding and cutting, as a blanket for the production of titanium and other reactive elements, and as a protective atmosphere for growing silicon and germanium crystals.

Krypton

Atomic Number:	36	Column Number:	VIII
Symbol:	**Kr**	Metal/Non-Metal	Non-Metal
Atomic Weight:	83.80	Outer Shell Electrons	8
Family:	Noble Gases	Physical State:	Gas
DOT Class:	2.2	NFPA 704:	Not Listed
4-Digit Number:	1056	Placard:	Non-Flammable Gas

Figure 2.42

History

From the Greek word, *kryptos*, meaning hidden. It was discovered in 1898 by Ramsay and Travers in the residue left after liquid air had nearly boiled away. In 1960, it was internationally agreed that the fundamental unit of length, the meter, should be defined in terms of the orange-red spectral line of 86Kr. This replaced the standard meter of Paris, which was defined in terms of a bar made of a platinum-iridium alloy. In October 1983, the meter, which originally was defined as one ten-millionth of a quadrant of the earth's polar circumference, was again redefined by the International Bureau of Weights and Measures as the length of a path traveled by light in a vacuum during a time interval of 1/299,792,458 of a second.

Sources

Krypton is present in the air to the extent of about 1 ppm. The atmosphere of Mars has been found to contain 0.3 ppm of krypton. Solid krypton is a white crystalline substance with a face-centered cubic structure, which is common to all the "rare gases."

Properties

It is one of the noble gases. It is characterized by its brilliant green and orange spectral lines.

Uses

Krypton clathrates (an inclusion complex in which molecules of one substance are completely enclosed within the other) have been prepared with hydroquinone and phenol. 85Kr has found recent application in chemical analysis. By imbedding the isotope in various solids, kryptonates are formed. The activity of these kryptonates is sensitive to chemical reactions at the surface. Estimates of the concentration of reactants are therefore made possible. Krypton is used in certain photographic flash lamps for high-speed photography. Uses thus far have been limited because of its high cost.

Xenon

Atomic Number:	54	Column Number:	VIII
Symbol:	**Xe**	Metal/Non-Metal	Non-Metal
Atomic Weight:	131.3	Outer Shell Electrons	8
Family:	Noble Gases	Physical State:	Gas
DOT Class:	2.2	NFPA 704:	Not Listed
4-Digit Number:	2036	Placard:	Non-Flammable Gas

Figure 2.43

History

From the Greek word, *xenon*, meaning stranger. Discovered by Ramsay and Travers in 1898 in the residue left after evaporating liquid air components, xenon is a member of the noble, or inert, gases. It is present in the atmosphere to the extent of about one part in 20 million. Xenon is present in the Martian atmosphere to the extent of 0.08 ppm. The element is also found in the gases evolved from certain mineral springs, and is commercially obtained by extraction from liquid air.

Uses

The gas is used in making electron tubes, stroboscopic lamps, bactericidal lamps, and lamps used to excite ruby lasers for generating coherent light. Xenon is used in the nuclear energy field in bubble chambers, probes, and other applications where a high molecular weight is of value. The element is also available in sealed glass containers of gas at standard pressure. Xenon is not toxic, but its compounds are highly toxic because of their strong oxidizing characteristics.

Lead

Atomic Number:	82	Column Number:	IV
Symbol:	**Pb**	Metal/Non-Metal	Metal
Atomic Weight:	207.2	Outer Shell Electrons	4
Family:	Metal	Physical State:	Solid
DOT Class:	Not Listed	NFPA 704:	Not Listed
4-Digit Number:	Not Listed	Placard:	Not Listed

Figure 2.44

History

From the Anglo-Saxon, lead; Latin for *plumbum*. Long known, mentioned in Exodus. Alchemists believed lead to be the oldest metal, and it was associated with the planet Saturn. Native lead occurs in nature, but it is rare.

Sources

Lead is obtained chiefly from galena (PbS) by a roasting process. Anglesite, cerussite, and minim are other common lead minerals.

Properties

Lead is a bluish-white metal of bright luster, is soft, highly malleable, ductile, and a poor conductor of electricity. It is resistant to corrosion; lead pipes bearing the insignia of Roman emperors used as drains from the baths are still in service. It is used in containers for corrosive liquids (such as sulfuric acid) and may be toughened by the addition of a small percentage of antimony or other metals. Care must be used in handling lead, as it is a cumulative poison. Environmental concerns with lead poisoning have resulted in a national program to eliminate the lead in gasoline.

Forms

Its alloys include solder, type metal, and various antifriction metals. Great quantities of lead, both as the metal and as the dioxide, are used in storage batteries. Much metal also goes into cable covering, plumbing, ammunition, and in the manufacture of lead tetraethyl.

Uses

The metal is effective as a sound absorber, is used as a radiation shield around x-ray equipment and nuclear reactors, and is used to absorb vibration. White lead, the basic carbonate, sublimed white lead, chrome yellow, and other lead compounds are used extensively in paints, although in recent years, the use of lead in paints has been drastically curtailed to eliminate or reduce health hazards. Lead oxide is used in producing fine "crystal glass" and "flint glass" of a high index of refraction for achromatic lenses. The nitrate and the acetate are soluble salts. Lead salts, such as lead arsenate, have been used as insecticides, but their use in recent years has been practically eliminated in favor of less-harmful organic compounds.

COMPOUNDS AND MIXTURES

Two or more elements that combine chemically (exchange or sharing of outer shell electrons) and form chemical bonds are referred to as compounds. All compounds must be electrically neutral. They must fulfill the octet and duet rules of bonding, with two or eight electrons in the outer shell, and must have the same number of positively charged protons in the nucleus as there are negatively charged electrons orbiting outside the nucleus. Compounds can be identified by names and formulas. There are three types of formulae: empherical, molecular, and structural. Structures generally apply to nonmetal compounds, including hydrocarbons and hydrocarbon-derivative families.

Chemical compounds may also be combined to form mixtures. A mixture is two or more compounds blended together without any chemical bond taking place. Each of the compounds retains its own characteristic properties. There are two types of mixtures: homogeneous and heterogeneous. Homogeneous means "the same kind" in Latin. In a homogeneous mixture, every part is exactly like every other part. For example, water has a molecular formula of H_2O. Pure water is homogeneous; it contains no substances other than hydrogen and oxygen. Loosely translated to include mixtures, homogeneous refers to two or more compounds or elements that are uniformly dispersed in each other. A solution is another example of a homogeneous mixture.

Heterogeneous means "different kinds" in Latin. In a heterogeneous mixture, the different parts of the mixture have different properties. A heterogeneous mixture can be separated mechanically into its component parts. Some examples of heterogeneous mixtures are gasoline, the air we breathe, blood, and mayonnaise.

Solubility

Solubility is a term associated with mixing two or more compounds together. The definition of solubility from the *Condensed Chemical Dictionary* is "the ability or tendency of one substance to blend uniformly with another, e.g., solid in liquid, liquid in liquid, gas in liquid, and gas in gas." Solubility may vary from one substance to another. When researching chemicals in reference sources, relative solubility terms may include: very soluble, slightly soluble, moderately soluble, and insoluble. Generally speaking, nothing is absolutely insoluble. Insoluble actually means "very sparingly soluble," i.e., only trace amounts dissolve.

Compounds of the alkali metals of family 1 on the Periodic Table are all soluble. Salts containing NH_4^+, NO_3, ClO_4, and ClO_3, and organic peroxides containing $C_2H_3O_2$ are also soluble. All chlorides (Cl), bromides (Br), and iodides (I) are soluble, except for those containing the metals Ag^+, Pb^{+2}, and Hg^{+2}. All sulfates are soluble, except for those with the metals Pb^{+2}, Ca^{+2}, Sr^{+2}, Hg^{+2}, and Ba^{+2}. All hydroxides (OH^{-1}) and metal oxides (containing O_2) are insoluble, except those of family 1 on the Periodic Table and Ca^{+2}, Sr^{+2}, and Ba^{+2}. When metal oxides do dissolve, they give off hydroxides (their solutions do not contain O^2 ions). The following illustration is an example of a metal oxide and water reaction. (This reaction is shown as an illustration and has not been balanced.)

$$Na_2O + H_2O = NaOH$$

All compounds that contain PO_4, CO_3, SO_3, and SO_2 are insoluble, except for those of family 1 on the periodic table and NH_4. Most hydrocarbon compounds are not soluble, such as gasoline, diesel fuel, pentane, octane, etc. Compounds that are polar, such as the alcohols, ketones, aldehydes, esters, and organic acids, are soluble in water. Several factors affect the solubility of a material. One is particle size: the smaller the particle, the more surface area that is exposed to the solvent; therefore, more dissolving takes place over a shorter period of time. Higher temperatures *usually* increase the rate of dissolving. The term miscibility is often used synonymously with the term solubility. Solubility is also related to polarity insofar as those materials that are polar are generally soluble in other polar materials. Miscibility, solubility, polarity, and mixtures will be discussed as they pertain to specific chemicals and families of chemicals in other chapters of this book.

Some elements do not exist naturally as single atoms. They chemically bond with another atom of that same element to form "diatomic" molecules. The diatomic elements are hydrogen, oxygen, nitrogen, chlorine, bromine, iodine, and fluorine. One way to remember the diatomic elements is by using the acronym HONClBrIF, pronounced honk-le-brif, which includes the symbols for all of the diatomic elements. Oxygen is commonly referred to as O_2, primarily because oxygen is a diatomic element. Two oxygen atoms have covalently bonded together and act as one unit.

Much can be learned about a compound by looking at its elemental composition. Generally speaking, chemicals that contain chlorine in their formula may be toxic to some degree because chlorine is toxic. There are exceptions, such as sodium chloride, with the formula NaCl. Sodium chloride is table salt. The toxicity is low; but even if you did not know sodium chloride was table salt and treated it as a toxic material because of the chlorine, your error would be on the side of safety. If you are going to make errors when dealing with hazardous materials, always attempt to err on the side of safety. You may have "egg on your face" afterward and take some ribbing, but no one has ever died from embarrassment. On the other hand, if you are not cautious and your error is not on the side of safety, it could be fatal!

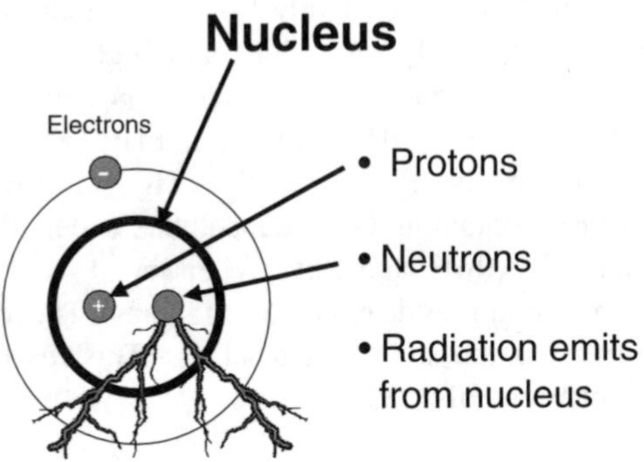

Figure 2.45

THE ATOM

An atom is the smallest particle of an element that retains all of its elemental characteristics. The word *atom* comes from the Greek, meaning "not cut." For example, take a sheet of paper and tear it in half. Keep tearing the paper in half until you cannot tear it any more. You could then take scissors and cut the paper into smaller pieces. Eventually you will not be able to cut the paper into a smaller piece. An atom is like that last piece of paper: you cannot have a smaller piece of an element than an atom.

Symbols of the elements on the Periodic Table represent one atom of each element. A single atom cannot be altered chemically. To create a smaller part of an atom of any element would require that the atom be split in a nuclear reaction. Therefore, a single atom is the smallest particle of an element that would normally be encountered. The atom is comprised of three major parts: electrons, protons, and neutrons (see Figure 2.45). An atom is like a miniature solar system, with the nucleus in the center and the electrons orbiting around it. Parts of the atom are referred to as subatomic particles. The atom is made up of positively (+) charged protons that are located in the nucleus along with neutrons. Neutrons do not have a charge, so they are electrically neutral. Orbiting in shells, or energy levels, around the nucleus are varying numbers of negatively (–) charged electrons. Electrons are important in discussing chemistry for hazardous materials responders. The shells, or orbits, about the nucleus of an atom have important characteristics:

1. Each energy level is capable of containing a specific number of electrons.
2. There is a limit on the number of electrons in the various shells.
3. The inner shell next to the nucleus *never* holds more than two electrons.
4. The last shell *never* holds more than eight electrons.
5. The electrons in the outer shell control chemical reactions.

The only elements that occur naturally with eight electrons in the outer shell are the inert gases in family 8 of the Periodic Table. The actual number of outer-shell electrons that exist in any given element is represented by a Roman numeral at the top of the vertical columns of the Periodic Table. The exceptions to the Roman numeral designation for the number of outer-shell electrons are the transitional element metals. The transitional metals have varying numbers of outer-shell electrons, and the process for identifying them will be discussed with the salt compounds.

Atoms are electrically neutral, so they must have an equal number of protons and electrons. Atoms are held together by a strong attraction between the positive protons and the negative electrons. Inner-shell electrons are held tightly by the nucleus. The farther away the outer shells are from the nucleus, the less control the nucleus will have over those electrons. This makes large atoms tend to lose electrons and small atoms tend to gain electrons. Electrons of metals are generally farther away from the nucleus than are nonmetal electrons; therefore, metals give up electrons and nonmetals take on electrons.

The atomic weight of an element comes from the nucleus. Protons and neutrons in the nucleus each have a weight of 1 AMU. Atoms of elements have varying numbers of protons and neutrons. The total number of neutrons and protons in each atom equals the atomic weight. The atomic number of an element equals the number of protons in that element. The number of neutrons is determined by subtracting the number of protons from the atomic weight; what remains is the number of neutrons. In chemistry, we are concerned more with the electrons orbiting the nucleus than the nucleus itself. We are particularly interested in the electrons in the outer shell of the atom. Chemical activity takes place between the outer-shell electrons of elements; this chemical activity forms compounds. In radioactivity, the concern is with the nucleus, where radiation is emitted. Radioactivity will be discussed further in Chapter 8.

Formulae

Just as an atom is the smallest part of an element, a single element is the smallest portion of a chemical compound that can be encountered. Chemical compounds are made up of two or more elements that have bonded covalently or ionically. Chemical compounds are represented by formulae, much like elements are represented by symbols. According to the *Condensed Chemical Dictionary*, "a formula is a written representation, using symbols, of a chemical entity or relationship." There are three types of chemical formulae: empirical, molecular, and structural.

Empirical formulae indicate composition, not structure, of the relative number and kinds of atoms in a molecule of one or more compounds, e.g., CH is the empirical formula for both acetylene and benzene.

The molecular formula is made up of the symbols of the elements that are in the compound and the number of atoms of each. The number of atoms of each element present are represented by a subscript number behind the symbol. The molecular formula is a kind of recipe for the compound. It lists all the ingredients and the proportions of each so that the compound can be reproduced if the formula is followed. The molecular formula also shows the actual number and kind of atoms in a chemical entity, e.g., the molecular formula for sulfuric acid is H_2SO_4. Sodium chloride, table salt, has the molecular formula NaCl. The compound has one atom of sodium and one atom of chlorine. Aluminum oxide has a molecular formula of Al_2O_3 (two atoms of aluminum and three atoms of oxygen).

The structural formula indicates the location of the atoms in relation to each other in a molecule, as well as the number and location of chemical bonds. The following illustration of the compound butane is an example of a chemical structure:

```
    H   H   H   H
    |   |   |   |
H - C - C - C - C - H
    |   |   |   |
    H   H   H   H
```

Figure 2.46

Electrons are shared or exchanged in the process of a chemical reaction. Once this exchange or sharing of electrons occurs, a chemical compound is formed. Opposite charges of metals and nonmetals make atoms want to come together to form salt compounds. Two basic groups of chemical compounds are formed from elements. The first group is the salts. Salts are made up of a metal and a nonmetal. For example, when combined with the nonmetal chlorine (Cl), the metal sodium (Na) forms the salt compound sodium chloride, with the molecular formula NaCl. Metals generally do not bond together. Metals that are combined are melted and mixed together to form an alloy, e.g., copper and zinc are melted and mixed together to make brass. Brass is not an element. No chemical bond is involved; rather, it is a mixture of zinc and copper.

The second group of compounds is made up totally of nonmetal elements. For example, the nonmetal carbon combines with the nonmetal hydrogen to form a hydrocarbon. A typical hydrocarbon might be methane, with the molecular formula CH_4. Hydrocarbons will be discussed further in Chapters 4 and 5. When the outer-shell electrons of the metal lithium are given up to the nonmetal chlorine, a salt compound is formed through a chemical bond (see Figures 2.47 and 2.48). The outer shell of the lithium atom is now empty, so the next shell becomes the outer shell. This shell will have two or eight electrons, which is a stable configuration; in the case of lithium, there are two electrons. Lithium is now stable and electrically satisfied. Chlorine receives the electron from the lithium and now has eight electrons in its outer shell. Chlorine is now stable and electrically satisfied. The result is that an ionically bonded compound is formed.

IONIC BONDING

The process of gaining or losing electrons is called ionization, or ionic bonding. When the electrons are transferred, ions are formed. Like atoms, compounds must be electrically neutral. There must be an equal number of positive protons in the nucleus and negative electrons outside in each of the atoms. When a metal gives up electrons, the metal has a positive charge because there are now more positive protons in the nucleus than negative electrons outside the nucleus. There is an electrical imbalance. An element cannot exist with an electrical imbalance.

Ionic Bonding

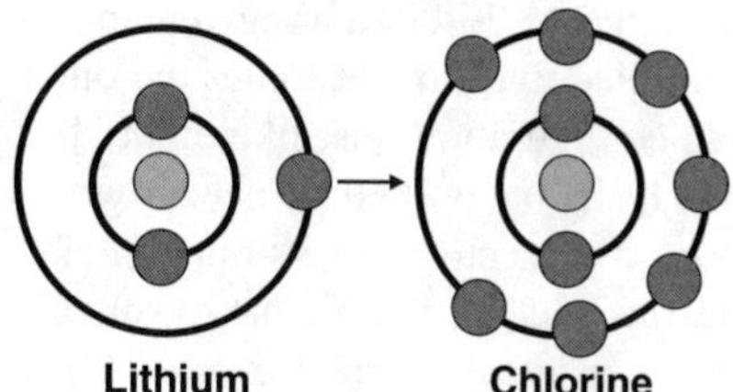

Figure 2.47

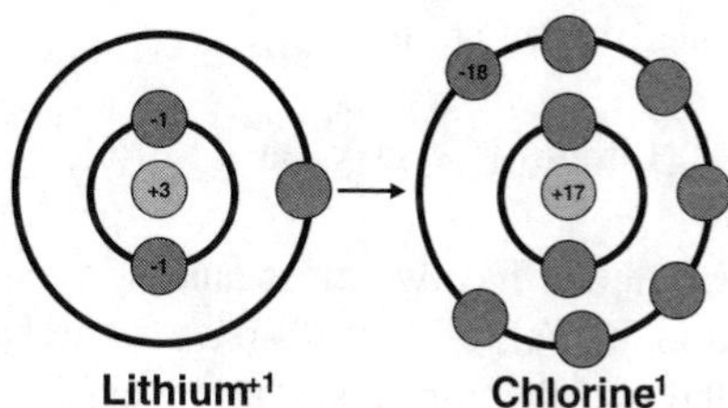

Figure 2.48

The ion formed in the case of the metal is referred to as a positive cation, represented by a plus (+) sign and a superscript number to the right of the element symbol. The number represents the number of electrons that were given up to the nonmetal element. When the nonmetal receives electrons from the metal, there are now more negative electrons outside the nucleus than there are positive protons in the nucleus. Again, there is an electrical imbalance. The ion of the nonmetal is referred to as a negative anion, represented by a minus (–) sign and a superscript number to the right of the symbol. The number represents the electrons received from the metal. Superscript numbers must balance and cancel each other out for the compound to be electrically neutral. If the superscript numbers are not equal, additional atoms of the elements involved are needed to balance the formula. If that happens, the formula will have subscript numbers indicating how many of the atoms of the specific element are present in the compound.

Calcium + Chlorine = Calcium Chloride

$$Ca^{+2} + Cl^{-1} = CaCl_2$$

Figure 2.49

If there are no subscripts in the formula, it is understood that there is just one atom of each of the elements shown. In the following example, each lithium atom has three electrons, two in the inner shell and one in the outer shell. Lithium will give up that one outer-shell electron to oxygen. Lithium now has one less electron than the total number of protons. There is one more positive charge in the nucleus than negative charges around the outside of the nucleus. Because of this imbalance of charges, lithium has a $^{+}1$ charge. Lithium has given up its one outer shell electron, so the inner shell with two electrons now becomes the outer shell. Two electrons in the outer shell is a stable configuration, just like eight electrons in the outer shell. Oxygen has six electrons in its outer shell; it needs two electrons to complete its octet. Oxygen will take on two electrons in its outer shell and will have two more negative charges around the nucleus than it has positive charges in the nucleus. Because of this imbalance, oxygen has a $^{-}2$ charge. Oxygen can get one of the electrons that it needs from lithium. Lithium has only one electron to give; however, oxygen needs another electron, so a second lithium atom is taken, which gives oxygen the two electrons it needs. The balanced formula for lithium oxide, Li_2O, is two lithium atoms and one oxygen atom.

$$\text{Lithium}^{+1} + \text{Oxygen}^{-2} = Li_2O$$

In the next example, calcium has two electrons in its outer shell. Calcium will give up those two electrons to sulfur and will have a $^{+}2$ charge. Sulfur needs two electrons and will take the two electrons given up by calcium and have a $^{-}2$ charge. Since calcium has a $^{+}2$ charge and sulfur has a $^{-}2$ charge, the charges cancel each other out and the balanced formula is CaS, calcium sulfide. Rules for naming will be covered in the Salt, Hydrocarbon, and Hydrocarbon Derivative sections of this chapter.

$$\text{Calcium}^{+2} + \text{Sulfur}^{-2} = CaS$$

It is important to understand how elements combine to form compounds. However, it is not the formula that will be seen most often in the real world of emergency response, but the name of the compound. When researching hazardous materials in reference books, often the formula will be listed. The formula will look more familiar if you have a basic understanding of chemistry. This will help responders to better understand the characteristics of the hazardous material.

COVALENT BONDING

The other family of compounds formed when elements combine is the nonmetal compound. Nonmetals are comprised of two or more nonmetal elements combining to form a compound. Nonmetals may be solids, liquids, or gases. The most frequently encountered groups of hazardous materials are made up of just a few nonmetal materials. They are carbon, hydrogen, oxygen, sulfur, nitrogen, phosphorus, fluorine, chlorine, bromine, and iodine. In elemental form and in compounds, these elements make up the bulk of hazardous materials found by emergency responders in most incidents. In the case of nonmetals, electrons are shared between the nonmetal elements.

Approximately 90% of covalently bonded hazardous materials are made up of carbon, hydrogen, and oxygen; the remaining 10% are composed of chlorine, nitrogen, fluorine, bromine, iodine, sulfur, and phosphorus. It is still necessary that each atom of each element have two or eight electrons in the outer shell. However, there is no exchange of those electrons. When the bonding takes place, each atom of each element brings along its electrons and shares them with the other elements. This process of sharing electrons is called covalent bonding. In the following example, carbon has four electrons in its outer shell; carbon needs to have four more to become stable. Hydrogen has one electron in its outer shell. Hydrogen can share its one electron with carbon and hydrogen will think it has a complete outer shell. Hydrogen gets the second electron it needs by sharing with carbon. This covalent bond fulfills the duet rule of bonding. Carbon still needs three more electrons. One way carbon can get three more electrons is by sharing with three more hydrogen atoms. When this happens, the carbon is complete. Carbon thinks it has eight electrons in the outer shell, and each hydrogen atom thinks it has two. The octet and duet rules of bonding are satisfied. The compound formed between carbon and hydrogen is complete, and the molecular formula is CH_4. In the following structure, the shared pairs of electrons are represented by dots between the hydrogen and the carbon. This is known as the Lewis Dot Structure (Figure 2.50).

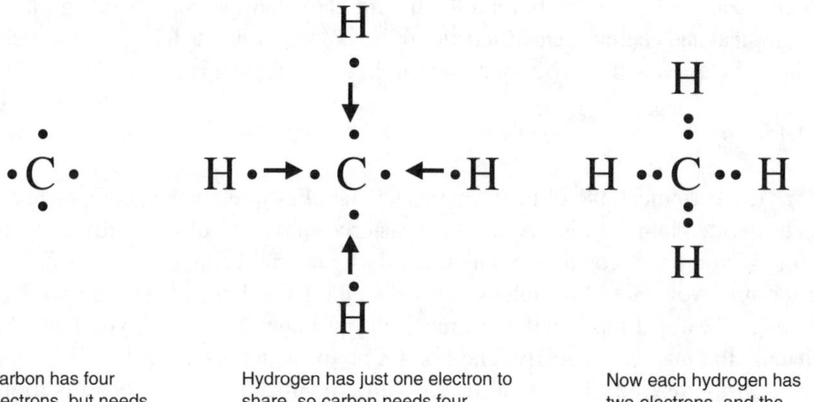

Figure 2.50

Electrons always share in pairs. Another way of representing pairs of shared electrons is by using a dash. The same one-carbon, four-hydrogen compound is shown with dashes representing the pairs of shared electrons (Figure 2.51).

A chemical compound becomes electrically stable through the process of sharing and exchanging electrons. The fact that compounds have become electrically stable does not mean they are no longer hazardous. Quite simply, elements combine and chemical reactions occur so that compounds can become electrically stable. These

Each dash represents a pair of shared electrons

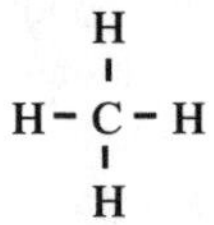

```
     H
     |
 H - C - H
     |
     H
```

Figure 2.51

combinations of elements that form compounds create many new hazardous and nonhazardous chemicals.

Emergency responders may encounter elemental chemicals that are hazardous when released in an accident. However, most of the hazardous chemicals encountered by emergency responders will be in the form of compounds or mixtures. Compounds and mixtures present a broad range of hazards, from explosive to corrosive. It is important for emergency responders at all levels to recognize these dangers. This book will use the Department of Transportation (DOT)/United Nations (UN) system of classifying hazardous materials. The remaining chapters of this book will examine the nine DOT/UN hazard classes and the types of hazardous materials they include. This hazard class system only identifies the most severe hazard presented by a material; almost every hazardous material has more than one hazard. This book will also focus on the hidden hazards of materials beyond the hazard class and the corresponding placard. The physical and chemical characteristics of hazardous materials found within each of the hazard classes will be discussed in detail in the remaining chapters.

SALTS

Just as the Periodic Table of the elements has families of elements, compounds can also be divided into families. A family of materials has particular hazards associated with it. If you can recognize in which family a material belongs from the name or the formula, you should be able to determine the hazard even if you do not know anything else about the specific chemical. If you know the hazard, you know how to handle the material properly. The first family of materials we will look at is the salts (see Figure 2.52). If a metal reacts with a nonmetal, an ionic bond is formed

Salt Families

Binary	**Binary Oxide**	**Peroxide**	**Hydroxide**	**Oxysalts**
M+NM	M+Oxygen	M+$(O2)^{-2}$	M+ $(OH)^{-1}$	M+ Oxy Rad
Not Oxy	Ends in Oxide	Ends in	Ends in	
Ends in	WR = CL, RH	Peroxide	Hydroxide	**Cyanide**
-ide		WR = CL,	WR = CL	M= CN
General		RH, RO2	RH	Ends in
Hazard				Cyanide
				Poison, CL

Figure 2.52

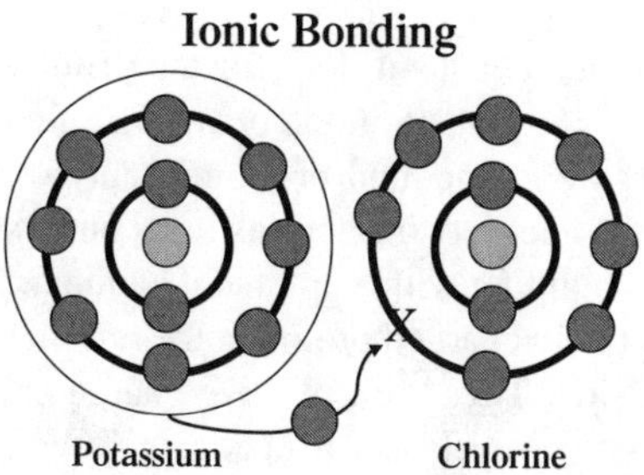

Figure 2.53

and the resulting compound is called a salt. In Figure 2.53, potassium, a metal, reacts with chlorine, a nonmetal. Potassium is located in the alkali metal family in column 1 of the Periodic Table. This means that potassium has one electron in the outer shell; therefore, potassium will give up that one electron and there will be eight electrons in the next shell. That shell will then become the outer shell and will be stable with eight electrons.

Chlorine has seven electrons in its outer shell. Chlorine will take the electron given by potassium to give it eight electrons in its outer shell. Chlorine will become stable. The name of the resulting compound is potassium chloride. Potassium chloride, much like sodium chloride (table salt), does not pose a serious risk to emergency responders.

$$K^{+1} + Cl^{-1} = KCl, \text{ potassium chloride, a binary salt}$$

Transition metals can also be combined with nonmetals to produce salts. Transition metals may have varying numbers of electrons in their outer shells (see Figure 2.54). The Periodic Table does not list the number of electrons in the outer

Transition Metal Charges for HazMat Chemistry

Elements	Possible Charges	Example Compound	Naming of Metal Ion+ Old	Naming of Metal Ion+ New
Copper	+1	CuCl	Copper I	Cuprous
	+2	$CuCl_2$	Copper II	Cupric
Iron	+2	$FeCl_2$	Iron II	Ferrous
	+3	$FeCl_3$	Iron III	Ferric
Mercury	+1	HgCl	Mercury I	Mercurous
	+2	$HgCl_2$	Mercury II	Mercuric
Tin	+2	$SnCl_2$	Tin II	Stannous
	+4	$SnCl_4$	Tin IV	Stannic
Cobalt	+2	$CoCl_2$	Cobalt II	
	+3		Cobalt III	

Figure 2.54

shells of the transitional elements as it does for the primary elements. The number of outer-shell electrons of the transition element in a salt compound can be determined from the name of the compound or from a correct molecular formula. For example, copper may have one or two electrons in its outer shell. When copper loses one electron during ionic bonding with a nonmetal element, it becomes Cu^{+1} because it now has one more positively charged proton in the nucleus than negatively charged electrons around the outside. This extra positive charge creates a $^{+}1$ charge on the element. If copper loses two electrons, it becomes Cu^{+2}, because it now has two more positive charges in the nucleus than negative charges around the outside. If nonmetal chlorine atoms were to pick up these electrons from copper, copper I or II chloride salts are formed. If copper loses only one electron, it is called copper I chloride, with a molecular formula of CuCl. When copper loses two electrons to chlorine, it becomes copper II chloride, with the molecular formula $CuCl_2$. Note that the Roman numeral in the name indicates the charge of the copper metal.

There is an older, alternate naming system for the transitional metals when they ionically bond to form salt compounds. There may be chemicals encountered still using this older naming system, therefore, responders should be familiar with it. In this system, the suffixes "ic" and "ous" are used to indicate the higher and lower valence numbers (outer-shell electrons) of a transitional metal. For example, if copper I combines with chlorine, the name would be cuprous chloride. If copper I combines with oxygen, the name would be cuprous oxide. The lowest number of electrons in the outer shell of copper is one. When the metal with the lowest number of electrons is used, the suffix in the alternate naming system is "ous." When the metal with the highest number of electrons is used, the suffix is "ic." If copper II conbined with phosphorus, it would create cupric phosphide. For example, copper II combined with chlorine would create cupric chloride.

The number of outer-shell electrons can also be determined from the molecular formula. First, look at the nonmetal element in a compound and determine what the charge is on that element.

The charge can be found as the group number on the Periodic Table at the top of the appropriate column. This group number is also the number of outer-shell electrons. The transition metal charge can then be determined by reversing the subscript numbers in the formula to the top of the elements. For example:

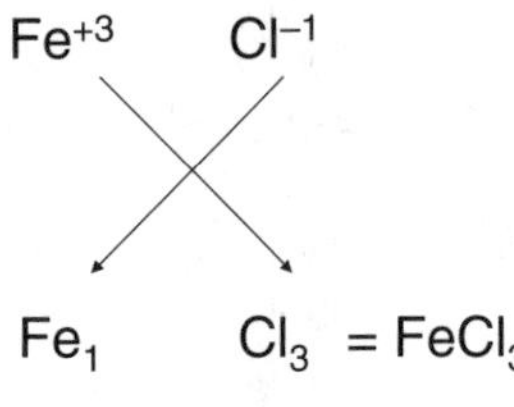

Figure 2.55

The number that ends up at the top of the metal represents the number of electrons in the outer shell of that metal. Another way of viewing the same formula, Fe_1C_3, would be to look at the charge of the chlorine from family 7, which is –1. This –1 is placed above the Cl^{-1} in the formula. There are three atoms of chlorine in the compound. Three times one equals three. There is one atom of iron in the compound. What number times one equals three? The answer is simple, three. So the charge of the iron in the compound is three. This is represented by a +3 above the Fe^{+3} in the formula. The number of electrons in the outer shell of the iron in this compound is three. See the following illustration:

$$\begin{array}{cc} Fe^{+3} & Cl^{-1} \\ X & X \\ Fe_1 & Cl_3 \\ \hline +3 & -3 \end{array}$$

Figure 2.56

There are three positive charges and three negative charges. The compound is in balance by using one atom of iron and three atoms of chlorine. The number above iron indicates the number of electrons in the outer shell of that atom of iron. If a compound containing a transitional metal does not have subscript numbers, this indicates that the charges at the top are balanced. The compound does not require any additional atoms of those elements. In the following example, copper is combined with chlorine. From the Periodic Table, the charge on chlorine is $^{-}1$. Therefore, the charge on the copper must be $^{+}1$ in order to be electrically balanced, which means that the copper in this compound is copper I.

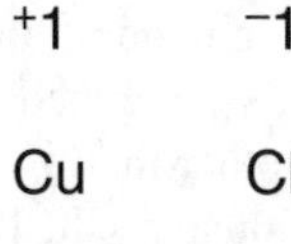

Figure 2.57

As mentioned, in the older naming system, the metal that contains the lower number of outer-shell electrons ends with the suffix "ous." The metal that contains the higher number of outer-shell electrons ends with the suffix "ic." For example, iron has the possibility of two or three electrons in the outer shell. When iron II is used in a compound, the name for the iron is *ferrous*; therefore, a salt compound containing iron II and chlorine would be called *ferrous* chloride. In this case, the "ous" indicates two electrons in the outer shell of the iron. In order to use this system, you must first know the possible numbers of outer-shell electrons to determine which is the higher and which is the lower number of electrons in the outer shell. For example, mercury can have one or two, copper one or two, iron two or three, etc. If iron III were used in a compound, the name would end in "ic"; therefore, the iron III name would be *ferric*. If iron III were combined with chlorine, the name of the salt compound would be *ferric* chloride.

Salts have particular hazards, depending on which salt family they belong to. Salts generally do not burn, but can be oxidizers and support combustion. Some salts are toxic and some may be water-reactive. Salt compounds can also be divided into families. There are six salt families that will be presented in this book: binary, binary oxide, peroxide, hydroxide, oxysalts, and cyanide salts.

Binary Salts

Binary (meaning two) salts are made up of two elements: a metal and a nonmetal, except oxygen. They end in "ide," such as potassium *chloride*. Binary salts, as a family, have varying hazards. They may be water-reactive, toxic, and, in contact with water, may form a corrosive liquid and release heat. Chemical reactions often release heat, which is referred to as an exothermic reaction. The hazard of an individual binary salt cannot be determined by the family. To determine the hazards of the binary salts, they have to be researched in reference materials. This varying hazard applies to all binary salts, except for nitrides, carbides, hydrides, and phosphides.

One helpful way to remember these four binary salts is by using the first letters of the element to form the acronym NCHP, which can be represented by "North Carolina Highway Patrol." These are compounds in which the metal has bonded with one of the nonmetals (nitrogen, carbon, phosphorus, or hydrogen). These compounds have particular hazards associated with them when they are in contact with water: nitrides give off ammonia, carbides produce acetylene, phosphides give off phosphine gas, and hydrides form hydrogen gas. In addition, a corrosive base is formed as a by-product from contact with water. The corrosive base will be the hydroxide of the metal that is attached to the nonmetal. For example, calcium carbide in contact with water will produce acetylene gas and the corrosive liquid calcium hydroxide. You have to look up the remaining binary salts to determine the hazard. In the first example, lithium metal is combined with chlorine. The resulting compound has a metal and a nonmetal other than oxygen, and the name ends in "ide;" therefore, it fits the definition of a binary salt. If lithium chloride is researched in reference sources, it is found to be soluble in water; it is not water-reactive. In fact, lithium chloride does not present any significant hazard in a spill. The DOT does not list lithium chloride on its hazardous materials tables.

$Li^{+1} + Cl^{-1} = LiCl$, lithium chloride, binary salt, varying hazard

The second example combines calcium metal and the nonmetal phosphorus. The compound name ends in "ide;" therefore, this is also a binary salt, which have varying hazards. Calcium phosphide, however, is one of the exceptions. Phosphides are one of the salts that have a known hazard: they give off phosphine gas and form calcium hydroxide liquid, which is a corrosive base. Even so, this is just a preliminary estimate. All materials should still be researched for additional hazards. Phosphine is a dangerous fire risk and highly toxic by inhalation.

$Ca^{+2} + P^{-3} = Ca_3P_2$, calcium phosphide, binary salt

Notice the varying hazards of the following examples: lithium fluoride is a strong irritant to the eyes and skin; potassium bromide is toxic by ingestion and inhalation; sodium chloride is table salt, a medical concern when ingested in excess, but certainly of no significant hazard to emergency responders. However, if sodium chloride is washed into a farmer's field as the result of an incident, the farmer may not be able to grow crops in that field for many years.

Binary Oxides (Metal Oxides)

The next group of salts is known as the binary, or metal, oxides. They are also made up of two elements: a metal and a nonmetal, but in this case, the nonmetal can only be oxygen. They end in "oxide," such as aluminum oxide. As a group, they are water-reactive and, when in contact with water, almost always produce heat and form a corrosive liquid. However, they do not give off oxygen, because there is not an excess of oxygen. There is only one oxygen atom and it is held tightly by the oxide salt and is not released as free-oxygen gas. In the following example, potassium metal is combined with one oxygen atom. The name of the salt compound is potassium oxide. This compound meets the definition of a binary oxide salt.

$K^{+1} + O^{-2} = K_2O$, potassium oxide, binary oxide salt, when in contact with water, releases heat and forms potassium hydroxide, a corrosive liquid

Peroxide Salts

Peroxides are composed of a metal and a nonmetal peroxide radical, O_2^{-2}. The prefix "per" in front of a compound or element name means that the material is "loaded" with atoms of a particular element. In the case of the peroxides, they are loaded with oxygen. A radical in the salt families is two or more nonmetals covalently bonded together, acting as a single unit with a particular negative (–) charge on the radical. The charge will not be found on the Periodic Table. The charges must be found on a listing of radical charges, as shown in Figure 2.54. In the case of the peroxide salt, two oxygen atoms have bonded together and are acting as one unit with a $^-2$ charge. When a peroxide comes in contact with water, heat is produced, a corrosive liquid is formed, and oxygen is released. This makes peroxides particularly dangerous in the presence of fire. Peroxides release oxygen because, unlike the oxide salts, there is an excess of oxygen present. In the following example, sodium metal combines with the peroxide radical to form the compound sodium peroxide. The name ends in "peroxide," so this is a peroxide salt. When in contact with water, sodium peroxide is a dangerous fire and explosion risk, and a strong oxidizing agent.

$Na^{+1} + O_2^{-2} = Na_2O_2$, sodium peroxide, a peroxide salt, gives off heat and a corrosive liquid, sodium hydroxide, and releases oxygen when in contact with water

Hydroxide Salts

Hydroxide salts are made up of a metal and the nonmetal hydroxide radical $-OH^{-1}$. The name always ends with the word "hydroxide." They are water-reactive and, when in contact with water, release heat and form a corrosive liquid. In the following example, calcium metal is combined with the hydroxide radical; the resulting compound is calcium hydroxide, a hydroxide salt. $Ca^{+2} + OH^{-1} = Ca(OH)^2$, calcium hydroxide, a hydroxide salt, releases heat and forms a corrosive liquid, calcium hydroxide, in contact with water.

Oxysalts

The oxysalts are made up of a metal and an oxyradical. The names end in "ate" or "ite," and may have the prefixes "per" or "hypo." Generally, as a group, they do not react with water; they dissolve in water. Some of the "hypo-ites" technically do react with water to release chlorine, but the reaction is mild. Oxysalts are oxidizers as a family; they will release oxygen, which accelerates combustion if fire is present. Another hazard occurs when oxysalts dissolve in water and the water is soaked into another material, such as packaging or firefighter turnouts. Water will evaporate, and the oxysalt will be left in the material. If the material is then exposed to heat or fire, the material will burn rapidly because the oxysalt in the material accelerates the combustion. Nine oxysalt radicals will be presented with this group. There are other oxyradicals, but the ones chosen are considered most important to emergency response personnel.

The first six oxyradicals all have –1 charges: FO_3 (fluorate), ClO_3 (chlorate), BrO_3 (bromate), IO_3 (iodate), NO_3 (nitrate), and MnO_3 (manganate). The next two have –2 charges: CO_3 (carbonate) and SO_4 (sulfate). The last oxyradical is PO_4 (phosphate), which has a –3 charge (see Figure 2.58). All of the radicals listed are

Oxy Radicals

-1	-2	-3
FO_3	CO_3	PO_4
ClO_3	SO_4	
BrO_3		
IO_3		
NO_3		
MnO_3		

Figure 2.58

considered to be in their base state. The base state is the normal number of oxygen atoms present in that oxyradical. When a metal is added to any oxyradical in the base state, the compound suffix is "ate," such as sodium *phosphate*. In the following example, the metal potassium is combined with the oxyradical carbonate; the resulting compound is potassium carbonate. The compound does not have a prefix on the oxyradical and the suffix is "ate;" therefore, it is the base state of the compound.

Naming Oxy Salts

+1 Oxygen prefix	**Per-_______ate**
Base State Ending	**_______ate**
-1 Oxygen Ending	**_______ite**
-2 Oxygen Prefix	**Hypo-_______ite**

Figure 2.59

$K^{+1} + CO_3^{-2} = K_2CO_3$, potassium carbonate

Potassium carbonate is an oxysalt in the base state, an oxidizer. Oxyradicals may be found with varying numbers of oxygen atoms than the base state. Regardless of the number of oxygen atoms in the oxyradical, the charge of the radical does not change. When naming compounds with an additional oxygen atom, the prefix *per* is used to indicate excess oxygen over the base state (see Figure 2.59); the suffix is still "ate." An example is sodium *persulfate*. In the following example, the metal potassium is combined with the oxyradical perchlorate: the resulting compound is potassium perchlorate. The level of oxygen is one above the base state. Notice that the charge on the oxyradical is still –1, even though the number of oxygen atoms has changed. Potassium perchlorate is a fire risk in contact with organic materials, a strong oxidizer, and a strong irritant.

$K^{+1} + ClO_4^{-1} = KClO_4$, potassium perchlorate, an oxidizer

When the number of oxygen atoms is one less than the base state of an oxyradical, the suffix of the oxyradical name will be "ite;" an example is magnesium *sulfite*. In the following example, the metal sodium is combined with the oxyradical phosphite. There is now one less oxygen than the base state. The charge on the oxyradical has not changed. In addition to being an oxidizer, sodium phosphite is also used as an antidote in mercuric chloride poisoning.

$Na^{+1} + PO_3^{-3} = Na_3PO_3$, sodium phosphite, an oxidizer

Finally, an oxyradical can have two less oxygen atoms than the base state. The oxyradical name will now have a *hypo* prefix and the suffix will be "ite." An example would be aluminum *hypo*phosph*ite*. In the following example, calcium is combined with the oxyradical hypochlorite; the resulting compound is calcium hypochlorite, a common swimming pool chlorinator. Calcium hypochlorite is an oxidizer and a fire risk when in contact with organic materials.

$Ca^{+2} + ClO^{-1} = Ca(ClO)_2$, calcium hypoclorite, an oxidizer

Cyanide Salts

The last salt family to be discussed is the cyanide salts. They are made up of a metal and the cyanide radical, CN. The name of the resulting compound ends in the word "cyanide;" an example is potassium cyanide. Cyanide salts are toxic materials that dissolve in water to form a hydrogen cyanide solution. Hydroxide ions are produced that will make the solution basic. Cyanide salts react with acids to produce hydrogen cyanide gas, which has an almond-like odor. Hydrogen cyanide gas is used in gas chambers. Cyanides are deadly poisons and can be found as salts or in solution and may produce a toxic gas when heated. When the cyanide ion enters the body, it forms a complex ion with the copper ion located in the cells. These copper ions are essential to the enzyme that allows the cell to use oxygen from the blood. This enzyme is deactivated by the binding of the cyanide and copper ions. In the following example, sodium metal is combined with the cyanide radical. The resulting compound is sodium cyanide. Sodium cyanide is toxic by ingestion and inhalation.

$$Na^{+1} + CN^{-1} = NaCN\text{, sodium cyanide, a poison}$$

NONMETAL COMPOUNDS

Nonmetal compounds are combinations of nonmetallic elements that combine in a covalent bonding process. When elements bond covalently, the bonding electrons are shared between the elements; the resulting compound is carbon disulfide. Carbon may bond with itself to satisfy its need for electrons. This happens frequently in the hydrocarbon families (to be discussed in detail in Chapters 4 and 5). The naming of the hydrocarbon and hydrocarbon-derivative families in this book will focus primarily on the trivial naming system, which uses prefixes indicating the number of carbons in the compound (see Figure 2.60). The suffix of the hydrocarbon name reflects the family and type of bond between the carbons in the compound (see Figure 2.61). Another naming system, called the IUPAC system, was developed by the International Union of Pure and Applied Chemistry. (An outline of the IUPAC system for naming organic compounds is located in the Appendix of this book.) In the following example, the carbon is sharing four electrons with the two sulfur atoms.

Hydrocarbon Prefixes

- Meth/Form - 1 Carbon
- Eth/Acet - 2 Carbons
- Prop - 3 Carbons
- But - 4 Carbons
- Pent - 5 Carbons
- Hex - 6 Carbons
- Hept - 7 Carbons
- Oct - 8 Carbons
- Non - 9 Carbons
- Dec - 10 Carbons

Figure 2.60

Hydrocarbon Families

Alkanes	**Alkenes**	**Alkynes**	**Aromatics**
Single Bond	Double Bond	Triple Bond	Resonant Bond
Saturated	Unsaturated	Unsaturated	Acts Saturated
Ends in -ane	Ends in -ene	Ends in yne	**BTX**
			Benzene
			Toluene
			Xylene

Figure 2.61

The result is that the carbon thinks it has eight electrons in its outer shell and each sulfur thinks it has eight electrons in its outer shell.

The carbon and the sulfur atoms are satisfied. The compound formed is carbon disulfide, which is a poison by absorption; it is a highly flammable, dangerous fire and explosion risk, has a wide flammable range from 1 to 50%, and can be ignited by friction. Carbon disulfide also has a low ignition temperature and can be ignited by a steam pipe or a light bulb.

S=C=S

Figure 2.62

Nonmetals and their compounds may be solids, liquids, or gases. Some may burn; some are toxic; and they can also be reactive, corrosive, and oxidizers. The largest quantities of hazardous materials encountered are made up of nonmetal (nonsalt) materials. These materials can also be divided into families. Hydrocarbon fuel is the most commonly encountered hazardous material. Products such as gasoline, diesel fuel, and fuel oil are all used to power our vehicles or heat our homes and businesses. Propane, butane, and natural gas are also fuels, but are not mixtures; they are pure compounds. They are transported frequently and stored in large quantities and can present frequent problems to responders. They form a family called the hydrocarbons because they are made up primarily of carbon and hydrogen. Hydrocarbons are flammable, and may be toxic or cause asphyxiation by displacing oxygen in the air.

There are four subfamilies of hydrocarbons, known as alkanes, alkenes, alkynes, and aromatics. (These families will be discussed in detail in Chapters 4 and 5.) The alkane and aromatic families of hydrocarbons occur naturally; the alkenes and alkynes are manmade. Both types of hydrocarbons are used to make other families of chemicals, known as hydrocarbon derivatives. Radicals of the hydrocarbon families are made by removing at least one hydrogen from the hydrocarbon and replacing it with a nonmetal other than carbon or hydrogen. Ten of these hydrocarbon derivatives will be discussed in detail in the appropriate chapters associated with their major hazards: alkyl halides, nitros, amines, ethers, peroxides, alcohols, ketones,

Hydrocarbon Derivative Families

Family	General Formula	Hazard
Alkyl Halide	R–X X=F,Cl,Br,I	Toxic/Flammable
Nitro	$R-NO_2$	Explosive
Amine	$R-NH_2$ R_2-NH R_3-N	Toxic/Flammable
Ether	R–O–R	WFR/Anesthetic
Peroxide	R–O–O–R	Explosive/Oxidizer
Alcohol	R–O–H	WFR/Toxic
Ketone	R–C–O–R	Flammable/Narcotic
Aldehyde	R–C–H–O	WFR/Toxic
Ester	R–C–O–O–R	Polymerize
Organic Acid	R–C–O–O–H	Corrosive/Toxic/Flammable

Figure 2.63

aldehydes, esters, and organic acids (see Figure 2.63). Just a few elements in addition to carbon and hydrogen are combined to make the 10 hydrocarbon derivative families discussed in this book. These materials can be toxic, flammable, corrosive, explosive, or reactive; some may be oxidizers and some may polymerize.

PHYSICAL AND CHEMICAL TERMS

Physical and chemical terms, such as boiling point, flash point, ignition temperature, pH, sublimation, water reactivity, spontaneous combustion, and others will be discussed in the remaining chapters of the book where appropriate. Many of the reference books used by emergency responders contain physical and chemical terms, and it is important that responders understand the significance of each of them. There are numerous hazards that chemicals can present to emergency responders; very few hazardous materials have only one hazard. In addition to the nine DOT/UN hazard classes, some materials have hidden hazards. Not all placards will tell that a material is water-reactive, air-reactive, may be explosive, or may polymerize. These hidden hazards will be discussed throughout the remaining chapters of this book. Many of the hazardous materials and their hazards that will be discussed in this book can be placed in families with similar characteristics. Potential hazardous materials can also be identified by endings and names in compounds. Figure 2.64 identifies some of the hints to hazardous materials. These are just hints to the possible presence of hazardous materials; the chemicals still need to be researched before any tactical operations are undertaken.

Hints to Hazardous Material Names

Names ending in:

-al	-ate	-ane	-azo	-ene	-ine	-ite	-ol
-one	-oyl	-yde	-ane	-yl	-yne		

Names that include:

acet	acid	alkali	amyl	azide	bis	caustic	cis
hepyl	hydride	iso	mono	naptha	oxy	penta	
per	tetra	trans	tris	vinyl	nitrile	cyan	

Figure 2.64

REVIEW QUESTIONS CHAPTER 2

(Answers located in the Appendix)

1. Chemistry is the study of:
   A. Isomers
   B. Matter
   C. Isotopes
   D. Reactions
2. Elements on the Periodic Table are represented by:
   A. The atomic number
   B. The atomic weight
   C. Symbols
   D. Molecules
3. Match the following family names with the column in which they are located on the Periodic Table.

A. Noble gases	____ Family I
B. Bromine family	____ Family II
C. Alkaline-earth metals	____ Family VII
D. Alkali metals	____ Family VIII
E. Halogens	____ Not a Family

4. When metals and nonmetals combine, the bond that is formed is called:
   A. Combination
   B. Ionic
   C. Covalent
   D. Atomic
5. When nonmetals combine, the bond that is formed is called:
   A. Covalent
   B. Non-covalent
   C. Ionic
   D. Isomeric

6. Atoms are made up of three subatomic particles; which of the following is not one of the particles?
   A. Neutrons
   B. Protons
   C. Neurons
   D. Electrons
7. An atom of an element is said to be electrically stable when it has how many electrons in its outer shell?
   A. Six
   B. Eight
   C. Two
   D. Both B and C
8. The numbers in question 7 represent two rules of bonding; which of the following is not one of the names for the bonding rules?
   A. Octet
   B. Isomeric
   C. Duet
   D. None of the above
9. Balance the formulas, if needed; give the name, the salt family, and the hazards for the following salts. (Salts with transitional metals are already balanced.)

NaCl	$CaPO_4$	AlO_2	$CuBr_2$	KOH
LiO	MgClO	HgO_2	NaF	$FeCO_3$

10. Provide a balanced formula, family name, and hazard(s) for the following salts:

Calcium hypochlorite	Aluminum chloride	Lithium hydroxide
Copper II peroxide	Sodium oxide	Potassium iodide
Magnesium phosphide	Mercury I perchlorate	Iron III fluorate

CHAPTER 3

Explosives

The first Department of Transportation/United Nations (DOT/UN) hazard class deals with explosives. The DOT defines an **explosive** in 49 CFR 173.50 as "any substance or article, including a device, which is designed to function by explosion or which, by chemical reaction within itself, is able to function in a similar manner even if not designed to function by explosion." This definition only applies to chemicals designed to create explosions. Another definition, taken from the *National Fire Academy Tactical Considerations Student Manual*, is "a substance or a mixture of substances, which, when subjected to heat, impact, friction, or other suitable initial impulse, undergoes a very rapid chemical transformation, forming other, more stable, products entirely or largely gaseous, whose combined volume is much greater than the original substance."

Other chemicals have explosive potential under certain conditions, but may not be placarded or recognized as explosives. These materials will be found in fixed facilities as well as in transportation, and may not have any markings or warnings that they have explosive properties. Some examples are ethers, potassium metal, formaldehyde, organic peroxides, and perchloric acid. Ethers are Class 3 Flammable Liquids that have a single oxygen atom in their composition; as ether ages, oxygen from the air can combine with this oxygen atom to form a peroxide molecule. The oxygen-to-oxygen single bond that is formed is a highly reactive and unstable bond, as shown in Figure 3.1.

This same type of bond is present in nitro compounds that are explosive and will be presented later in this chapter. Peroxides formed in compounds in this manner are shock- and heat-sensitive; they may explode simply by being moved or shaken. Formaldehyde solutions are Class 3 Flammable Liquids, and potassium metal is a Class 4.3 Flammable Solid, Dangerous When Wet. These materials form explosive peroxides just like ether and are equally dangerous as they age. Organic peroxides are Class 5.2 Oxidizers. Many organic peroxides are temperature-sensitive; they are usually stored under refrigeration. If the temperature is elevated, organic peroxides may decompose explosively. Perchloric acid is a corrosive by hazard class. However, perchloric acid is also a strong oxidizer and will explode when shocked or heated. Chemical oxidizers are one of the necessary components for a chemical explosive to function. These chemical oxidizers should be treated with the same respect as the explosives in Class 1.

Explosions may be followed by fireballs, shock waves, and flying shrapnel.

- O – O -

Figure 3.1

DEFINITION OF EXPLOSION

An **explosion** is defined in the National Fire Protection Association (NFPA) *Fire Protection Handbook* as "a rapid release of high-pressure gas into the environment." This release of high-pressure gas occurs regardless of the type of explosion that has produced it. The high-pressure energy is dissipated by a shock wave that radiates from the blast center. This shock wave creates an overpressure in the surrounding area that can affect personnel, equipment, and structures (see Figure 3.2). An overpressure of just 0.5 to 1 psi can break windows and knock down personnel. At 5 psi, eardrums can rupture and wooden utility poles can be snapped in two. Ninety-nine percent of people exposed to overpressures of 65 psi or more would die.

CATEGORIES OF EXPLOSIONS

According to the NFPA *Fire Protection Handbook*, there are two general categories of explosions, **physical** and **chemical**. In a physical explosion, the high-pressure gas is produced by mechanical means, i.e., even if chemicals are present in the container, they are not affected chemically by the explosion. In a chemical explosion, the high-pressure gas is generated by the chemical reaction that takes place.

Examples of Overpressure Damage to Property	
0.5–1 psi	Window glass breakage
1–2 psi	Buckling of corrugated steel and aluminum Wood siding and framing blown in
2–3 psi	Shattering of concrete or cinder block walls
3–4 psi	Steel panel building collapse Oil storage tank rupture
5 psi	Wooden utility poles snapping failure
7 psi	Overturning loaded rail cars
7–8 psi	Brick walls shearing and flexure failures
Examples of Injuries and Death to Personnel	
1 psi	Knockdown of personnel
5 psi	Eardrum rupture
15 psi	Lung damage
35 psi	Threshold for fatalities
50 psi	50% fatalities
65 psi	99% fatalities

Figure 3.2

PHASES OF EXPLOSIONS

There are two phases of an explosion, the positive and the negative. The positive phase occurs first as the blast wave travels outward, releasing its energy to objects it comes in contact with. This is also known as the blast pressure; generally, the most destructive element of an explosion. If the explosion is a detonation, the waves travel equally in all directions away from the center of the explosion. The negative phase occurs right after the positive phase stops. A partial vacuum is created near the center of the explosion by all of the outward movement of air from the blast pressure. During the negative phase, the debris, smoke, and gases produced by the blast are drawn back toward the center of the explosion origin, then rise in a thermal column vertically into the air, and are eventually carried downwind by the air currents. The negative phase may last up to three times as long as the positive phase. If the explosion is a result of an exothermic (heat-producing) chemical reaction, the shock wave may be preceded by a high-temperature thermal wave that can ignite combustible materials. Some explosives are designed to disperse projectiles when the explosion occurs, e.g., anti-personnel munitions and hand grenades. Other explosives may be in metal or plastic containers to provide the necessary confinement for an explosion to occur. These containers may become projectiles when the explosion occurs. Projectiles from an explosion travel out equally in all directions from the blast center just as the blast pressure does. In order for the waves or projectiles coming off the explosion to travel equally in all directions, the velocity (speed) of the release of the high-pressure gases must be supersonic, or faster than the speed of sound. This occurs only in a detonation, not in a deflagration.

MECHANICAL OVERPRESSURE EXPLOSIONS

The two major explosion types can be divided further into four subtypes that result in the release of high-pressure gas. The first occurs as the result of the physical overpressurization of a container, causing the container to burst, as in the case of a child's balloon bursting when too much air is placed in it. The container fails because it can no longer hold the pressure built up inside. This can occur in containers that may not have pressure-relief valves, or if the pressure-relief valve fails to operate. This overpressure does not have to occur as a result of filling the container. As heat is applied to a container from ambient temperature increases, or from radiant heat, the pressure increases inside the container. If this increase in pressure is not relieved, the container may fail.

MECHANICAL/CHEMICAL EXPLOSIONS

The second type of explosion occurs via physical or chemical means, as in the case of a hot-water heater or boiler explosion. The water inside the heater turns to steam when overheated, which results in a pressure increase inside the container, and the container fails at its weakest point. This is by far the most common type of accidental explosion.

This type of container failure can also occur in containers that hold liquefied compressed gases. As the temperature of the liquid in the container increases, so does the vapor content in the container; as the vapor content increases, so does the vapor pressure. This increase in temperature may be the result of increases in ambient temperature, radiant heat from a fire, or other heat source, such as direct flame impingement on the pressure container. If the container has been damaged in an accident, it may fail at the point of damage, before the relief valve can function to relieve the pressure buildup. The tank can also be weakened by direct flame impingement on the vapor space of the container. This flame impingement weakens the metal; the tank can no longer hold the pressure, and tank failure occurs. When the heat from flame impingement is on the liquid level, it is absorbed by the liquid so the tank is not weakened. However, as the liquid absorbs the heat, more liquid is turned to vapor. This increase in vapor increases the pressure in the container. Pressure-relief valves are designed to relieve increased pressure caused by increases in ambient temperature or from radiant heat sources. The increase in pressure caused by direct flame impingement may overpower the relief valve, and again the container may fail.

CHEMICAL EXPLOSIONS

The third type of explosion involves a chemical reaction: the combustion of a gas mixture. Many of the explosive materials that the DOT regulates and allows for transport can cause this type of explosion. Therefore, most explosions and explosive

Grain elevators are a primary source of dust explosions.

materials in this hazard class involve materials that explode by chemical means, with the high pressure created by a chemical reaction. This type of explosion is really nothing more than a rapidly burning fire. The components that allow this fire to burn rapidly enough to produce an explosion are the presence of a chemical oxidizer and confinement of the material. There are other chemicals offered for transportation and found in fixed facilities that are not classified as explosives, but can explode through chemical reaction. Responders must be aware of the fact that most chemicals have multiple hazards that may not be depicted by placards or information on shipping papers.

DUST EXPLOSIONS

There is a subgroup of potentially explosive materials that involve a chemical-reaction type of explosion (see Figure 3.3), sometimes referred to as combustible dusts (see Figure 3.4). These are, in many cases, ordinary combustible materials or other chemicals that, because of their physical size, have an increased surface area (see Figure 3.5). This increased surface area exposes more of the particles to oxygen when they are suspended in air. When these materials are suspended in air, they can become explosive if an ignition source is present.

Explosive Properties of Common Grain Dusts

Type of Dust	Ign Temp	LEL
Wheat/Corn/Oats	806°F	55
Wheat Flour	380°F	50
Cornstarch	734°F	40
Rice	824°F	50
Soy Flour	1004°F	60

Figure 3.3

Explosive Dusts

Coals	Crude rubber	Peanut hulls
Soy protein	Sugar	Aluminum
Cork	Cornstarch	Flour
Magnesium	Pea flour	Titanium
Zirconium	Walnut shells	Silicon

Figure 3.4

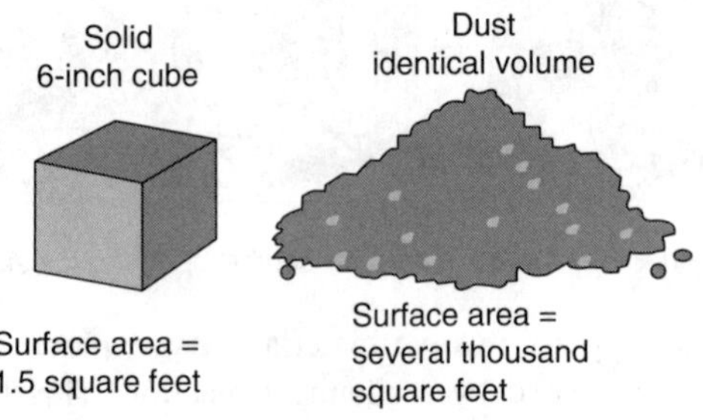

Figure 3.5

One of the major facilities where dust explosions occur is grain elevators; explosions occur when grain dust is suspended in air in the presence of an ignition source. The primary danger area where the explosion is likely to occur within most elevators is the "leg," or the inclined conveyor, the mechanism within the elevator that moves the grain from the entry point to the storage point. For a dust explosion to occur, five factors must be present: an ignition source, a fuel (the dust), oxygen, a mixing of the dust and the oxygen, and confinement. The explosion will not occur unless the dust is suspended in air within an enclosure at a concentration that is above its lower explosive limit (see Figure 3.3).

There are three phases in a dust explosion (see Figure 3.6): initiation, primary explosion, and secondary explosion. Initiation occurs when an ignition source contacts a combustible dust that has been suspended in air. This causes the primary explosion, which shakes more dust loose from the confined area and suspends it in air. The secondary explosion then occurs, which is usually the larger of the two explosions because there is more fuel present.

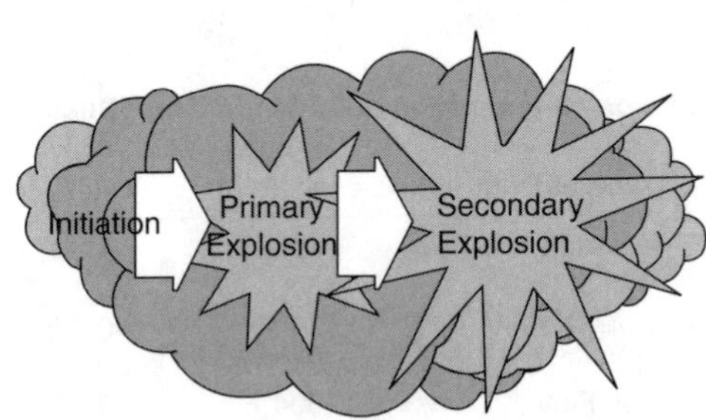

Figure 3.6

Combustible dusts may be present in many different types of facilities. Common places for combustible dusts to be found are in grain elevators, flour mills, woodworking shops, and dry-bulk transport trucks. Dusts in facilities have caused many explosions over the years that have killed and injured employees. An explosion occurred in a facility on the East Coast that had many hazardous materials on site. At first, it was thought that one of the chemicals had exploded. The fire department and the hazmat team were called to the scene. Investigation revealed that the explosion occurred in a dust-collection system; it was a combustible-dust explosion. Dust explosions can be prevented by proper housekeeping and maintenance practices at facilities where these types of dusts are present.

NUCLEAR EXPLOSIONS

The fourth type of explosion is thermonuclear, which is the result of a tactical decision, a weapon that malfunctions, or an act of terrorism. There are two types of nuclear explosions: air burst and ground burst. Both are engaged as a result of a tactical objective. An **air burst** is designed to knock out all electronic equipment, disrupting communications and computer usage. This type of explosion does not create fallout, because it does not reach the ground and does not suck up debris in the negative phase of the explosion. A **ground burst** is designed for mass destruction of everything it contacts. Initially, during the positive phase of the ground blast, there is a thermal wave that is released first, followed by a shock wave. During the negative phase of the explosion, the debris from the explosion is drawn into the cloud, travels downwind, and then falls back to the ground. The debris, while in the cloud, is contaminated with radioactive particles, and radioactive fallout is created.

COMPONENTS OF AN EXPLOSION

In simple terms, an explosive that functions via chemical reaction creates a rapidly burning fire that is made possible by the presence of a chemical oxidizer. Atmospheric oxygen does not provide enough oxygen for a chemical explosion to take place. Four components must be present for a chemical explosion to occur: fuel, heat (initiator, source of ignition), a chemical oxidizer, and confinement of the materials (see Figure 3.7). The materials themselves can provide the confinement. Note the similarity between the requirements for an explosion and the fire triangle. There are other types of explosions that produce high-pressure gases, some of which will be discussed later.

TYPES OF EXPLOSIVES

Explosives can be divided into two primary groups, high explosives and low explosives, based upon the speed with which the chemical transformation takes place, usually expressed in feet per second. Low explosives change physical state from a

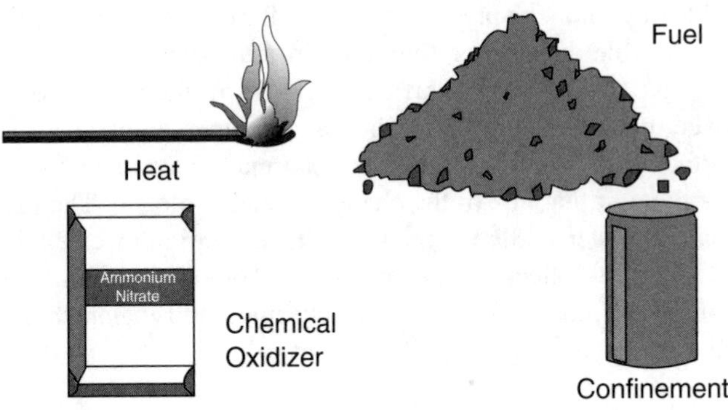

Figure 3.7

Dry-bulk truck transporting Class 1.5 blasting agent on the highway.

solid to a gas rather slowly. The low explosive burns gradually over a somewhat sustained period of time. This action is typically used as a pushing and shoving action on the object against which it is placed. The primary uses of low explosives are as propelling charges and for powder trains, such as in time fuses. Examples of low explosives include black powder and smokeless powder.

High explosives change from a solid to a gas almost immediately, an action referred to as detonation. A high explosive is detonated by heat or shock, which sets up a detonating wave. This wave passes through the entire mass of explosive material instantly. This sudden creation of gases and the extremely rapid extension produces a shattering effect that can overcome great obstructions. Examples of high explosives are trinitrotoluene (TNT) and dynamite.

The DOT explosives hazard class is divided into six subclasses: 1.1 to 1.6 (49 CFR 173.20) (see Figure 3.8). Because of their potential danger, subclasses 1.1 to 1.3 require placarding of the highway transportation vehicle regardless of the quantity of explosives carried (49 CFR 172.504, Table 3.1) (see Figure 3.9). Railroad

Explosive Subclasses

Explosive 1.1	Mass Explosion Hazard
Explosive 1.2	Projection Hazard
Explosive 1.3	Fire, Minor Blast, Minor Projection Hazards
Explosive 1.4	Device with Minor Explosion Hazard
Explosive 1.5	Very Insensitive Explosives
Explosive 1.6	Extremely Insensitive Explosive

Figure 3.8

Table I Materials

Explosives 1.1–1.3
Poison Gas 2.3
Poison 6.1
Dangerous when Wet 4.3
Radioactive Yellow III

Figure 3.9

Table 2 Materials

Explosives 1.4–1.6
Compressed Gases 2.1–2.2
Flammable Liquids 3
Flammable Solids 4.1–4.2
Oxidizers 5.1–5.2
Poison 6.1 (Non-Inhalation Hazard)
Corrosive 8

Figure 3.10

shipments must always be placarded in all hazard classes, regardless of the quantity shipped. Subclasses 1.4 to 1.6 (49 CFR 172.504, Table 3.2) (see Figure 3.10) fall under the 1001-pound rule (49 CFR 172.504(1)), which requires 1001 lbs or more of explosives on the vehicle before a placard is required. This means there could be 1000 pounds or less of a 1.4 to 1.6 explosive and no placard would be required at all!

In the explosives hazard class, next to the hazard subclass number on the placard there will be a letter, known as the compatibility group letter. 49 CFR 173.52, Tables 3.1 and 3.2, identify the procedures for assigning compatibility group numbers to shipments. The compatibility groups are used to prevent the increase in explosive hazard should certain explosives be stored or transported together. This information is designed for the shippers of the explosive materials and has little, if any, emergency response value. All explosives should be treated as if they were Class 1.1, because all explosives will explode under certain conditions. Responders are not likely to recognize when an explosive material is in a condition to cause an explosion. Therefore, responders should treat all explosives as if they were the worst type of explosive they might encounter.

FORBIDDEN EXPLOSIVES FOR TRANSPORTATION

In addition to the explosives that the DOT has approved for transportation, there are forbidden explosives that are too unstable or dangerous to be transported. Some may be transported only if wetted. The following are some examples of explosives that are forbidden to be transported.

Nitrogen triiodide (black unstable crystals) explodes at the slightest touch when dry. When handled, it is kept wet with ether. It is too sensitive to be used as an explosive, because it cannot be stored, handled, or transported. Azides, such as lead azide and hydrazoic azide, are highly unstable. **Lead azide** is a severe explosion risk and should be handled under water; it is also a primary detonating compound. **Hydrazoic acid** or **hydrogen azide** is a dangerous explosion risk when shocked or heated. Metal fulminates, such as **mercury fulminate**, explode readily when dry. They are used in the manufacture of caps and detonators for producing explosions.

Explosives that contain a chlorate along with an ammonium salt or an acidic substance, including a salt of a weak base and a strong acid, are forbidden in transportation, e.g., **ammonium chlorate**, which is shock-sensitive, can detonate when exposed to heat or vibration. It is used in the production of explosives. **Ammonium perchlorate** is also shock-sensitive, and may explode when exposed to heat or by spontaneous chemical reaction. This is the material that was involved in the explosion at the Pepcon plant in Henderson, Nevada. It is also used in the production of explosives, pyrotechnics, etching and engraving, and jet and rocket propellants.

Packages of explosives that are leaking, damaged, unstable, condemned, or contain deteriorated propellants are forbidden in transportation. Nitroglycerine, diethylene glycol dinitrate, or other liquid explosives are not authorized. **Diethylene glycol dinitrate** is a severe explosion hazard when shocked or heated. It is used as a plasticizer in solid rocket propellants. Other forbidden explosives include fireworks that combine an explosive and a detonator or fireworks that contain yellow or white phosphorus. Toy torpedoes exceeding 0.906-inch outside dimension or containing a mixture of potassium chlorate, black antimony (antimony sulfide), and sulfur are prohibited if the weight of the explosive material in the device exceeds 0.01 oz.

The Hazardous Materials Table in CFR 49 Part 172.l0l lists all specific restricted explosives in various modes of transportation and those forbidden from shipment. Even though these materials are not transported, they may be found in fixed facilities of various types, such as research facilities, defense contractors, solid rocket fuel plants, explosives suppliers, and others.

TYPES OF CHEMICAL EXPLOSIONS

Detonation is an instantaneous decomposition of the explosive material in which all of the solid material changes to a gas instantaneously with the release of high heat and pressure shock waves. Detonation is the only type of chemical explosion that will produce a true shock wave. A material that detonates is considered a high-yield explosive (see Figure 3.11). Blast pressures can be as much as 700 tons/sq in.

Characteristics of High and Low Explosives

	High	Low
Initiation Method	Primary by ignition Secondary by detonation	By ignition
Conversion to Gas	Microseconds	Milliseconds
Consumption Velocity	1 to 6 Miles/Second	Few inches to Feet Per Second
Velocity of Flame Front	1 to 6 Miles/Second	1/3 to 1 Mile/Second
Pressure of Explosion	50,000 to 4,000,000 psi	Up to 50,000 psi

Figure 3.11

Pressure and heat waves travel away from the center of the blast equally in all directions. The reaction occurs at supersonic speed or, in other words, faster than the speed of sound, which is 1250 ft/s. Many of the reactions occur between 3300 and 29,900 ft/s, or over 20,300 miles/hr!

Often the terms "explosion" and "detonation" are used interchangeably, which is not accurate. An explosion may be a detonation; however, an explosion can occur that is not a detonation. In either case, each occurs rapidly, and the difference cannot be distinguished easily by the human senses. The only way you can distinguish a detonation from a deflagration is by hearing the sound of the explosion. In a detonation, the explosion will be visualized, and the shock wave sent off before the explosion is actually heard. In a deflagration, the explosion will be heard almost immediately. The two terms only apply to the speeds of the explosions; they do not infer that one is any less dangerous than the other.

Deflagration is a rapid autocombustion that occurs at a subsonic speed, less than 1250 ft/s. The solid material changes to a gas relatively slowly. A material that deflagrates is considered a low-yield explosive. A material that is designed to deflagrate may, however, under the right conditions produce a detonation. The explosion in Kansas City that killed six firefighters involved ammonium nitrate mixed with fuel oil. Ammonium nitrate is listed as a Class 1.5 Insensitive Explosive, designed to produce a deflagration. The ammonium nitrate-fuel oil mixture involved in Kansas City was on fire and, as a result, produced a detonation that may have been caused by the application of water by the firefighters.

EXPLOSIVE EFFECTS

There are primary and secondary effects of explosions. The primary effects are blast pressure, thermal wave, and fragmentation. The blast pressure has two phases, the positive and negative. In the **positive phase**, the high-pressure gas, heat wave, and any projectiles travel outward. During the **negative phase**, a partial vacuum is produced, sucking materials back toward the area of origin.

The next effect is fragmentation. Fragmentation may come from the container that held the explosive material and from materials in close proximity. Objects in

Commercial Explosives

Primary	Secondary
Mercury Fulminate	Nitroglycerine
Lead Azide	Ammonium Nitrate
Diazodinitrophenol	Trinitrotoluene
Lead Styphnate	Dinitrotoluene
Nitromannite	Nitrostarch

Figure 3.12

the path of the explosion are broken into small parts by the force of the blast pressure, creating fragments. These fragments may have jagged or sharp edges. The fragments will travel away from the blast center at high speeds, greater than 2700 ft/s, faster than a speeding bullet!

The final effect of an explosion is thermal, the generation of heat by the explosion. The amount of heat produced will depend on the type of material that is involved. There is a flash and a fireball associated with almost any chemical explosion. The more rapid the explosion, the greater the effects that will be produced by the heat. Although the total heat produced may be similar in each explosion, a detonation will produce the most heat over a larger area, because of the speed of the explosion.

Secondary effects of an explosion are shock-wave modification and fire and shock wave transfer. There are three ways that a shock wave can be modified: it may be reflected, focused, or shielded. Reflection refers to the shock wave striking a solid surface and bouncing off. When a shock wave strikes a concave (curved) surface, the force of the shock wave is focused, or concentrated, on an object or small area once it bounces off the concave surface. This effect is similar to the principle behind satellite dishes. When a signal reaches a satellite dish from the satellite in space, the signal is focused on the electronic sensor protruding out of the front of the satellite dish. Shielding simply means that the shock wave encounters an object too substantial to be damaged by the wave, so the shock wave goes around the object or is absorbed by it. The area immediately behind the object provides a place of shelter from the shock wave. Fire and shock-wave transfer involves the transfer of the shock wave energy and fire to other objects, causing fires and destruction.

YIELD VS. ORDER

The yield of an explosive is associated with the rate or speed in which the explosion occurs. This is an indication of whether an explosive will detonate or deflagrate. A high-yield explosive detonates, and the blast pressure shatters materials that it contacts. Examples of high-yield explosives are dynamite, TNT, nitroglycerine, detchord, C-3, C-4, and explosive bombs (see Figure 3.11). A low-yield explosive deflagrates and is used to push and shove materials. Examples of low-yield explosives are black powder and commercial ammonium nitrate. Deflagrating materials are often used to move rocks in road construction, in quarries, and in mining.

Order has to do with the extent and rate of a detonation. A high-order detonation is one in which all of the explosive material is consumed in the explosion, and the explosion occurs at the proper rate. The proper rate in this case would be supersonic. Thus, a low-order explosion would occur as an incomplete detonation or at less than the desired rate. Yield involves the specific explosive material that is used, and order indicates the way in which the explosive detonated. The hazards to emergency responders are obvious. If an explosion is low order, not all of the explosive material has been consumed, and therefore the remaining material presents a hazard. Whether high or low yield, high or low order, all explosives should be treated as high-yield, high-order Class 1.1 explosives.

EXPLOSIVES SUBCLASSES 1.1 TO 1.3

Division 1.1 explosives present a mass explosion hazard. They are sensitive to heat and shock and they may either detonate or deflagrate when they explode.

Division 1.2 explosives have a projection hazard, but not a mass explosion hazard.

Division 1.3 explosives have a fire hazard and a minor blast hazard or a minor projection hazard or both, but not a mass explosion hazard.

–N(O)(O), with a single bond between the two O atoms

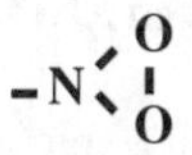

Figure 3.13

NITRO COMPOUNDS

There is one hydrocarbon-derivative family that is classified with an explosive as its primary hazard. However, there are some nitro compounds that have other primary hazards, such as nitrobenzene, which is a poison. This is an exception to the general hazard and, for safety purposes, consider nitros explosive as a group. The nitro group is represented by a nitrogen covalently bonded to two oxygens. Nitrogen must have three connections to complete the octet rule of bonding (see Figure 3.13). The oxygens have a single bond between themselves. This oxygen-to-oxygen single bond is highly unstable and can come apart explosively.

The other bonding spot on the nitrogen is attached to a hydrocarbon or hydrocarbon-derivative backbone of some type. These backbones may include methane, benzene, toluene, or phenol, which is an alcohol, and others. When naming compounds from the nitro group, the word "nitro" is used first and the end is the hydrocarbon to which the nitro is attached. In the following example (see Figure 3.14), one nitro functional group is attached to the hydrocarbon radical methane. The methane has had one hydrogen atom removed to create a place to attach the nitrogen on the nitro functional group. The compound is referred to as nitromethane.

```
      H
      |    O
H  -  C  - N  |
      |    O
      H
```

Nitro-Methane
CH_3NO_2

Figure 3.14

Nitromethane, CH_3NO_2, is a colorless liquid that is soluble in water. The specific gravity is 1.13, which is heavier than water. Nitromethane is a dangerous fire and explosion risk, and is shock- and heat-sensitive. It may detonate from nearby explosions. The boiling point is 213°F, and the flash point is 95ºF. The flammable range only lists a lower explosive limit, which is 7.3% in air; an upper limit has not been established. The ignition temperature is 785ºF. Nitromethane may decompose explosively above 599ºF if confined, and is a dangerous fire and explosion risk, as well as toxic by ingestion and inhalation. The threshold limit value (TLV) is 100 ppm in air. The four-digit UN identification number is 1261; the NFPA 704 designation is health 1, flammability 3, and reactivity 4. Nitromethane is used in drag racing to give the fuel in the engine an extra kick to increase speed. It is also used in polymers and rocket fuel.

Nitroglycerine, $CH_2NO_3CHNO_3CH_2NO_3$, is a pale yellow, viscous liquid. It is slightly soluble in water, with a specific gravity of 1.6, which is heavier than water. It is a severe explosion risk, and will explode spontaneously at 424°F. It is much less sensitive to shock when it is frozen. Nitroglycerine freezes at about 55°F. It is

Dynamite.

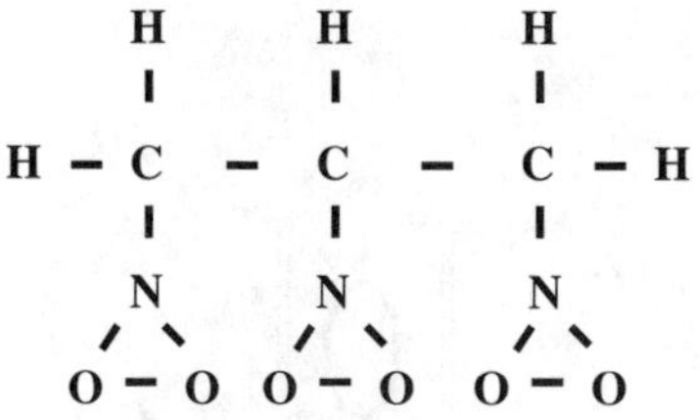

```
        H       H       H
        |       |       |
    H - C   -   C   -   C - H
        |       |       |
        N       N       N
       / \     / \     / \
      O - O   O - O   O - O
```

Nitroglycerine
$CH_2NO_2CHNO_3CH_2NO_3$

Figure 3.15

highly sensitive to shock and heat, and is toxic by ingestion, inhalation, and skin absorption. The TLV is 0.05 ppm in air. Nitroglycerine is forbidden in transportation unless sensitized. When in solution with alcohol at not more than 1% nitroglycerine, the four-digit UN identification number is 1204. When in solution with alcohol and more than 1%, but not more than 5% nitroglycerine, the four-digit UN identification number is 3064. The NFPA 704 designation for nitroglycerine is health 2, flammability 2, and reactivity 4. The primary uses are in explosives and dynamite manufacture, in medicine as a vasodilator, in combating oil-well fires, and as a rocket propellant. The structure for nitroglycerine is shown in Figure 3.15. Notice that there are three nitro functional groups. These nitro groups were attached to the alcohol glycerol after three hydrogens were removed.

Trinitrotoluene (TNT), $CH_3C_6H_2(NO_2)_3$, is flammable, a dangerous fire risk, and a moderate explosion risk. It is light cream to rust in color, and is usually found in 0.5- or 1-lb blocks. It is fairly stable in storage. TNT will detonate only if vigorously shocked or heated to 450°F; it is toxic by inhalation, ingestion, and skin absorption. The TLV is 0.5 mg/m^3 of air. The four-digit UN identification number is 1356, when wetted with not less than 30% water. Other mixtures are listed in the hazardous materials tables with several ID numbers. TNT is one of the common ingredients used in military explosives and is used as a blast-effect measurement for other explosives. Trinitrotoluene is a member of the nitro hydrocarbon-derivative family. In the following structure, toluene is the backbone for TNT. Three hydrogens are removed from the toluene ring and three nitro functional groups are attached.

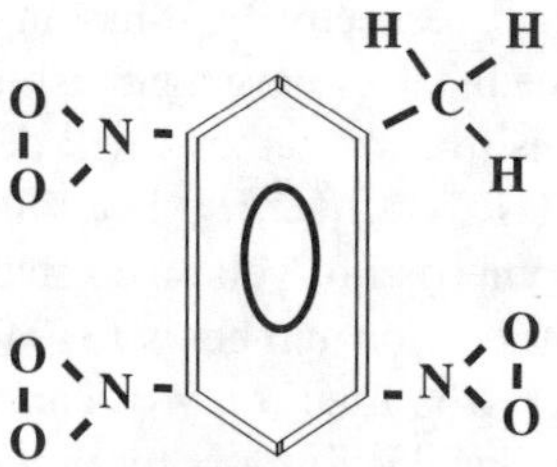

TriNitroToluene (TNT)
$CH_3C_6H_2(NO_2)_3$

Figure 3.16

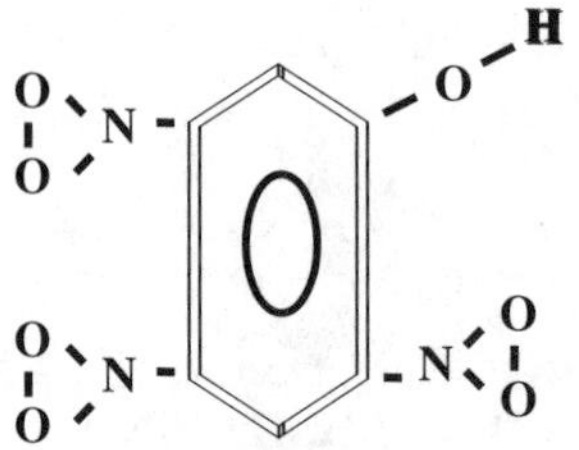

TriNitroPhenol (Picric Acid)
$C_6H_2OH(NO_2)_3$

Figure 3.17

Trinitrophenol (picric acid), $C_6H_2OH(NO_2)_3$ (see Figure 3.17), is composed of yellow crystals that are soluble in water. It is a high explosive, is shock- and heat-sensitive, and will explode spontaneously at 572°F. Trinitrophenol is reactive with metals or metallic salts, and is toxic by skin absorption. The TLV is 0.1 mg/m^3 of air. When shipped in 10 to 30% water, it is stable unless the water content drops below 10% or it dries out completely. The four-digit UN identification number is 1344 when shipped with not less than 10% water. The NFPA 704 designation is health 3, flammability 4, and reactivity 4.

The primary uses are in explosives, matches, electric batteries, etching copper, and textile dyeing. Picric acid is often found in chemical labs in high schools and colleges, and can be a severe explosion hazard if the moisture content of the container is gone. Picric acid was used by the Japanese during World War II as a main charge explosive filler. When in contact with metal, picric acid will form other picrates, which are extremely sensitive to heat, shock, and friction. Great care should be taken when handling World War II souvenirs, because of the possible presence of these picrates.

When the structure of picric acid is compared with the structure of TNT, the only difference is the fuel that the nitro functional groups were placed on; the number of nitro groups is exactly the same. The explosive power of picric acid is similar to that of TNT. There are other Class 1.1 to 1.3 materials that are nitro compounds and some that are made up of other chemicals. **Black powder** is a low-order explosive made up of a mixture of potassium or sodium nitrate, charcoal, and sulfur in 75, 15, and 10% proportions, respectively. It has an appearance of a fine powder to dense pellets, which may be black or have a grayish-black color. It is a dangerous fire and explosion risk, is sensitive to heat, and will deflagrate rapidly.

Ammonium picrate, $C_6H_2(NO_2)_3ONH_4$, is a high explosive when dry and flammable when wet. It is composed of yellow crystals that are slightly soluble in water. The four-digit UN identification number is 1310 for ammonium picrate wetted with not less than 10% water. It is used in pyrotechnics and other explosive compounds. The structure and molecular formula for ammonium picrate are shown in Figure 3.18. Notice the similarity to picric acid and TNT. You may notice that the structure and formula of ammonium picrate do not follow the usual rules of bonding.

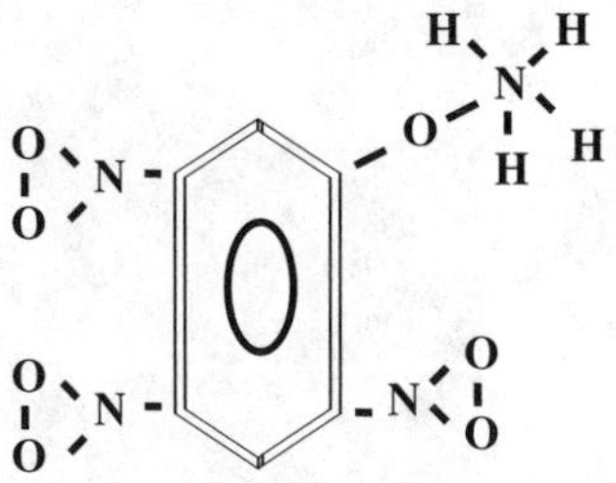

Ammonium Picrate
$C_6H_2(NO_2)_3ONH_4$

Figure 3.18

The ammonium radical has a different bonding configuration, which accounts for the four hydrogen atoms hooked to the nitrogen atom. This is one of those "street chemistry" concepts that it is better to just accept rather than trying to understand why.

INCIDENTS

A number of accidents have occurred over the years involving 1.1 to 1.3 (former Class A and B explosive materials). In Waco, Georgia, an automobile collided with a truck carrying 25,414 lbs of explosives, resulting in a fire and the explosion of the dynamite cargo. The fire started as a result of gasoline and diesel fuels spilling from the vehicles and then igniting. Heat transfer from the fires caused the nitroglycerine-based dynamite to detonate, killing two firefighters, a tow truck driver, and two bystanders; injuring another 33 persons; and causing over $1 million in property damage. The National Transportation Safety Board (NTSB) investigated the incident and, as part of their report, listed the following contributing factors that led to the injuries and deaths:

1. The lack of a workable system to warn everyone within the danger zone of an explosion
2. The failure to notify emergency service personnel promptly and accurately of the hazards
3. The decision of the firefighters to try and contain the hazardous fire
4. Bystanders disregard or lack of understanding of the truck driver's warnings

Dynamite explosions also injured 13 people at Keystone, West Virgina, and 9 at Lancaster, New York. An explosion of up to a ton of smokeless powder, along with some black powder that was stored in the basement of a sporting goods store in Richmond, Indiana killed 41 persons, injured another 100, and caused over $2 million in property damage. Eleven buildings were destroyed by the explosion and four more from the resulting fires.

Blasting caps used to detonate explosive materials.

Dry, shock-sensitive picric acid on the shelf of a high school chemistry classroom.

EXPLOSIVES SUBCLASSES 1.4 TO 1.6

Division 1.4 is made up of explosives that present a minor explosion hazard. The explosive effects are largely confined to the package, and no projection of fragments of appreciable size or range is to be expected. An external fire must not cause virtually instantaneous explosion of almost the entire contents of the package.

Mixer truck used to transport ammonium nitrate and fuel oil and to mix into a blasting agent.

The DOT considers Division 1.5 (Blasting Agents) insensitive explosives. This division comprises substances that have a mass explosion hazard, but are so insensitive that there is little probability of initiation or of transition from burning to detonation under normal conditions of transport. (The probability of transition from burning to detonation is greater when large quantities are transported in a vessel or in storage facilities.) Division 1.6 explosives are considered extremely insensitive and do not have a mass explosion hazard. This division comprises articles that contain only extremely insensitive detonating substances and that demonstrate a negligible probability of accidental initiation or propagation. (The risk from articles of Division 1.6 is limited to the explosion of a single article.)

Most of the listings in 49 CFR 172.101 Hazardous Materials Tables for 1.4 to 1.6 explosives are for small premanufactured explosive devices: small arms ammunition, signal cartridges, tear gas cartridges, detonating cord, detonators for blasting, explosive pest control devices, fireworks, aerial flares, practice grenades, signals, and railway track explosive smoke signals. Recent changes in DOT rules now allow certain Class 1.1 and 1.2 explosives to be shipped in small quantities in special packages as Class 1.4. Normally, Class 1.1 through 1.3 explosives must be placarded regardless of the quantity. The problem was that bomb squads and other explosives experts ordering small amounts of C-4, sheet explosives, DEXS caulk, and slip boosters had to pay high shipping costs, as much as $1000 in some cases, to common carriers. Special packaging was designed and underwent thorough testing by the U.S. Department of Mines. The new DOT rules allow small amounts of explosives to be shipped by UPS, Federal Express, and other small-package shipping companies.

Class 1.4 explosives do not require placarding until 1001 pounds or more are shipped. One of the main chemical explosives listed from groups 1.4 to 1.6 is 1.5 ammonium nitrate. When mixed with fuel oil, it becomes a blasting agent. Subjected to confinement or high heat, it may explode but does not readily detonate. Fertilizer-grade ammonium nitrate, which is a strong oxidizer above 33.5%, may also explode if it becomes contaminated. Fertilizer-grade ammonium nitrate was used in the bombings of the World Trade Center in New York City in 1993 and the Federal Building in Oklahoma City in 1995.

EXPLOSIVE CHEMICALS

Ammonium nitrate is classified as an oxidizer. It is a colorless or white to gray crystal that is soluble in water. It decomposes at 210°C, releasing nitrous oxide gas. The four-digit UN identification number is 1942 with an organic coating, and 2067 as the fertilizer grade. There are a number of other mixtures of ammonium nitrate that have four-digit numbers; they can be found in the Hazardous Materials Tables and in the DOT's *Emergency Response Guide.* There are a number of other chemicals that are not explosives, but have explosive potential under certain conditions. Oxygen is an oxidizer that causes organic materials to burn explosively. Chlorine is also an oxidizer and may be explosive in contact with organic materials. There are three ethers: methyl *tert*-butyl ether, ethylene oxide, and propylene oxide; these can form explosive peroxides in storage as they age. Butadiene may also form explosive peroxides when exposed to air.

INCIDENTS

Several major incidents have occurred with ammonium nitrate: in Texas City, Texas, 2280 tons of prilled ammonium nitrate (a commercial fertilizer) on board a ship in Galveston Harbor detonated after it caught fire. It is estimated that the blast caused 468 deaths, including 27 firefighters (the entire Texas City Fire Department), over 3000 injuries, and more than $50 million in property damage. The ammonium nitrate had been confined below deck on the ship, and steam was injected into the sealed hold in an attempt to extinguish the fire. It is thought that the injection of steam into the confined ammonium nitrate caused the detonation to occur. The shock wave was felt as far away as Colorado. Before the incident was over, two more ships also exploded with results similar to the first explosion, compounding the emergency relief efforts.

In Kansas City, Missouri, commercial-grade ammonium nitrate was inside a highway box trailer used for storage on a construction site. Someone set fire to a pickup truck on the property and the outside of the storage trailer. Firefighters responded to fight the fire. During the firefighting operations, the ammonium nitrate detonated, killing six firefighters and destroying their apparatus. There were no markings on the storage container indicating explosives were present. Because of this incident and others, OSHA now requires the use of the DOT placarding and labeling system for fixed-facility storage of hazardous materials. When they are transported, and until the materials are used up or the containers purged, the placards and labels must remain on the containers.

Fixed-storage bunkers for explosive materials are heavily regulated by state and federal laws.

Fire, thought to be caused by spontaneous heating following the blowout of a tire, quickly spread to the cargo compartment of a 28-ft tractor-trailer truck in Marshall's Creek, Pennsylvania. The driver had disconnected the trailer off the roadway and had driven the tractor several miles to a service station. While the driver was gone, the tires ignited. The truck was carrying 4000 lbs of 60% standard gelatin dynamite in boxes and 26,000 lbs of nitrocarbonitrate blasting agent in 50-lb bags. The compound is a mixture of 850 lbs of ammonium nitrate fertilizer to 7 gal of No. 2 diesel fuel. Another passing tractor-trailer driver reported the fire to the Marshall's Creek Fire Department. The driver reported that there were no markings on the trailer. Three fire engines responded, and an attack line was pulled to fight the blaze. As the firefighters approached the trailer, a detonation occurred. The fire had reached the explosive cargo, and the resulting explosion killed six people, including three firefighters, the truck driver who reported the fire, and two bystanders. Property damage was over $600,000, including all three of the fire engines.

MILITARY EXPLOSIVES

Military explosives are noted for their high shattering power accompanied by rapid detonation velocities (see Figure 3.19). They must be stable, because they are often kept in storage for long periods of time. Because of their intended use, they must detonate dependably after being stored and do so under a variety of conditions. The military explosives used most commonly are TNT, C-3, C-4, and RDX cyclonite. These explosives release large quantities of toxic gases when they explode.

Military Explosives

Primary	Secondary
Mercury Fulminate	Nitroglycerine
Lead Azide	Ammonium Picrate
Diazodinitrophenol	Trinitrotoluene
Lead Styphnate	RDX
Nitromannite	Picric Acid

Figure 3.19

SUMMARY

According to the NFPA *Fire Protection Handbook*, over 90% of explosives in the industrial world are used in mining operations, with the rest used in construction. When responding to fixed or transportation incidents in or around these types of operations, be on the lookout for explosives. When responding to transportation incidents, always consider the possibility of explosives being present. Fire is the principal cause of accidents involving explosive materials. Look for explosive signs, such as placards and labels. Evacuate the area according to the distances listed in the *Emergency Response Guidebook* orange section. If no other evacuation information is available, a 2000-ft minimum distance should be observed, according to the NFPA *Fire Protection Handbook*. There is one rule of thumb in responding to incidents where explosives are involved: **DO NOT FIGHT FIRES IF THE FIRE HAS REACHED THE EXPLOSIVE CARGO.**

REVIEW QUESTIONS CHAPTER 3

1. An explosion has two phases, which of the following is not one of those phases?
   A. Negative
   B. Positive
   C. Thermal
   D. All of the above
2. List the five types of explosions.
3. Name the two types of chemical explosions.
4. Name and draw structures and provide molecular formulas as appropriate for the following nitro compounds.

   CH_3NO_2 Nitro propane $C_6H_2(NO_2)_3CH_3$ Tri-nitro phenol

5. Name the four components necessary for a chemical explosion to occur.
6. A chemical explosion is really nothing more than a rapidly burning fire. Which of the following makes the rapid burning possible?
   A. High-yield explosive material
   B. A chemical oxidizer
   C. The additive "rapid burn"
   D. High-octane fuel

7. A detonation occurs at which of the following speeds?
   A. Supersonic
   B. Faster than light
   C. Subsonic
   D. Slower than light
8. A deflagration occurs at which of the following speeds?
   A. Supersonic
   B. Faster than light
   C. Subsonic
   D. Slower than light
9. Which of the following terms refers to the efficiency of an explosion?
   A. Order
   B. Confinement
   C. Yield
   D. None of the above
10. Which of the following terms refers to the speed of an explosion?
    A. Chemical oxidizer
    B. Order
    C. Yield
    D. All of the above
11. List the six DOT subclasses of explosive materials permitted in transportation.
12. Which of the following is not a Class I Explosive Material?
    A. Dynamite
    B. Picric acid (20% Water)
    C. Ammonium nitrate
    D. Blasting caps
13. Of the following blast waves, which is released first during an explosion?
    A. Shock
    B. Fragmentation
    C. Thermal
    D. None of the above
14. Blast waves travel away from the center of an explosion in which direction?
    A. Downwind
    B. Only in opposite directions
    C. Equally in all directions
    D. Only upwind
15. Which of the following nuclear explosions creates radioactive fallout?
    A. Ground burst
    B. Neutron explosion
    C. Air burst
    D. Electron ionization

CHAPTER 4

Compressed Gases

DOT Hazard Class 2 is composed of gases that are under pressure (see Figure 4.1). The hazard is the pressure, which can cause violent container failure when an accident occurs. Pressure containers can be dangerous under accident conditions; ambient temperature changes, flame impingement, and damage to containers can cause boiling liquid expanding vapor explosions (BLEVEs). Containers can also rocket under high pressure, creating an impact hazard when valves are knocked off. Class 2 is divided into three subclasses: 2.1 gases are flammable, 2.2 gases are nonflammable, and 2.3 gases are poisons. Each compressed-gas category presents its own special hazards, in addition to the hazard of being under pressure, in a specially designed and regulated pressure container. Container pressures range from 5 psi to as much as 6000 psi (see Figure 4.4). The higher the pressure, the more substantial the container must be constructed to contain the pressure; the higher the pressure, the greater the danger when the pressure is released or the container fails.

Gases may also have other hazards, such as flammability, toxicity, and reactivity. Gases may be heavier or lighter than air; however, most gases are heavier. Seven lighter-than-air gases along with some common compressed gases are listed in Figure 4.1.

Gases may also be liquefied. Liquefaction takes advantage of a gas's ability to be liquefied by pressure or cooling, or a combination of both. Liquefied gases have a large liquid-to-gas expansion ratio. This allows a larger quantity of gas to be shipped as a liquid compared to a gas. It is much more economical to ship a compressed gas as a liquid.

Three terms are important in understanding the process of the liquefaction of gases by pressure: critical point, critical temperature, and critical pressure. Critical point is the point at which a gas will exist as a gas or as a liquid. When a gas is heated to its critical temperature at its critical pressure, it becomes a liquid (see Figure 4.2). Critical temperature is the maximum temperature at which a liquid (in this case, a liquefied gas) can be heated and still remain a liquid. For example, for butane, the critical temperature is 305°F. As more heat is added, more of the liquid vaporizes. At the critical temperature, no amount of pressure can keep the liquid

Common Class 2 Gases

Lighter than Air		Heavier than Air	
Helium	He	Argon	Ar
Acetylene*	C_2H_2	Propane*	C_3H_8
Hydrogen*	H_2	Butane*	C_4H_{10}
Ammonia*	NH_3	Chlorine	Cl_2
Methane*	CH_4	Phosgene	$COCl_2$
Nitrogen	N_2	Hydrogen Sulfide*	H_2S
Ethylene*	C_2H_4	Butadiene*	C_4H_6

*Flammable Gases

Figure 4.1

Critical Temperature and Pressure

Gas	Boiling Point °F	Critical Temp °F	Critical Pressure (PSI)
Ammonia	–28	266	1691
Butane	31	306	555
Carbon Dioxide	–110	88	1073
Hydrogen	–422	–390	294
Nitrogen	–320	–231	485
Oxygen	–297	–180	735
Propane	–44	206	617

Figure 4.2

from turning into a gas. Critical pressure is the maximum pressure required to liquefy a gas that has been cooled to a temperature below its critical temperature. The critical pressure for butane is 525 psi. In order to liquefy any gas, it must be cooled to or below its critical temperature. For example, the critical temperature of butane is 305°F; at 305°F, it must be pressurized to 525 psi to become a liquefied compressed gas.

Because liquefied gases have large liquid-to-gas expansion ratios, this also increases the hazard of the gas during an emergency (see Figure 4.3). A small amount of a liquid leaking from a container can form a large gas cloud. Larger leaks will produce larger vapor clouds. This increases the danger of flammability if an ignition source is present, and of asphyxiation or toxicity when vapor clouds form. Some liquefied gases, such as propane and butane, are ambient-temperature liquids. They are shipped and stored in uninsulated tanks, and the temperature of the liquid inside is close to the ambient air temperature.

Cryogenic liquefied gases, such as hydrogen, nitrogen, oxygen, and argon are extremely cold liquids. The definition of a cryogenic liquid is any liquid with a boiling point below –130°F. They will be shipped and stored in insulated containers. Sometimes, they are stored under pressure. The liquids inside containers will have temperatures ranging from –130°F to –452°F below zero. Cryogenics are discussed further in the Nonflammable Gases section at the end of this chapter.

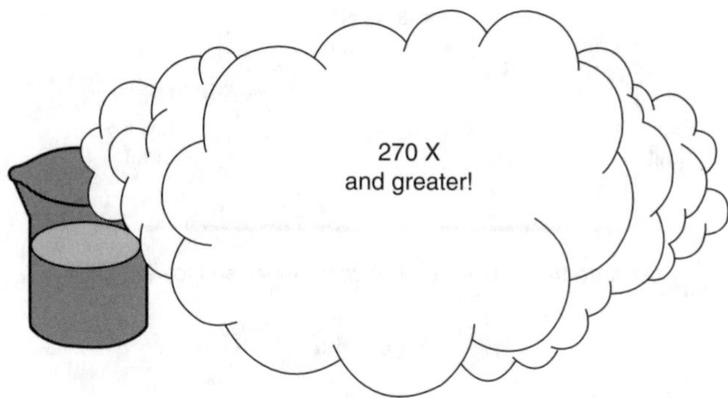

Figure 4.3

Container Pressures

Atmospheric Pressure	0–5 psi
Low Pressure	5–100 psi
High Pressure	100–3000 psi
Ultra-high Pressure	3000–6000 psi

Figure 4.4

When ambient temperatures are cold, the liquefied compressed gases will also be cold. Water from the booster tanks on fire apparatus can be 70°F or higher. Thus, water, if applied to a tank surface to cool it during a fire, could actually be heating the liquid in the tank rather than cooling it. Cryogenic liquids are already colder than the water at any temperature, and the water will act as a super-heated material, causing the cryogenic to heat up and vaporize faster. This difference in temperatures can cause problems for responders unless the containers are handled properly. Care should be taken when applying water to "cool" containers.

FLAMMABLE GASES

Class 2.1 compressed gases are flammable. Flammable gases may be shipped and stored as liquefied gases, cryogenic liquids, or compressed gases. The DOT defines a flammable gas as "any material which is a gas at 68°F or less and 14.7 psi of pressure or above, which is ignitable when in a mixture of 13% or less, by volume, with air, or has a flammable range with air, of at least 12 percentage points, regardless of the lower limit."

Flammable Range

Flammable range is the point at which there is a proper oxygen and fuel mixture present for combustion to occur. Flammable range is represented by a scale, numbered from

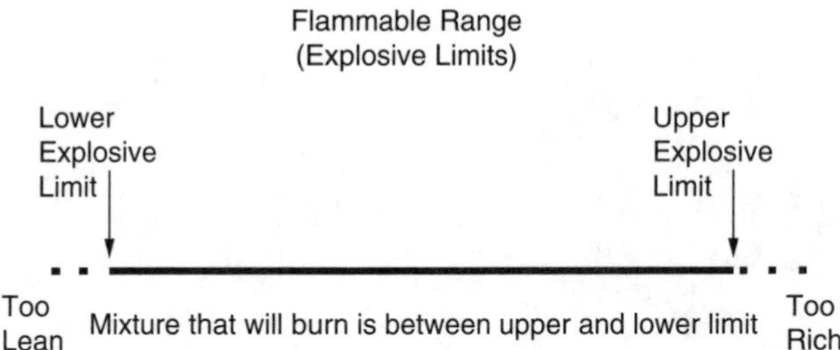

Figure 4.5

Flammable Ranges of Common Materials

Acetylene	2.5–80%
Ethyl ether	1.85–48%
Methyl alcohol	6–36.5%
Formaldehyde	7–73%
Propane	2.4–9.5%
Ammonia	16–25%
Methane	5–15%

Figure 4.6

0 to 100% (see Figure 4.5). Two common terms used to express where the mixture is located within the range are the upper explosive limit (UEL) and lower explosive limit (LEL). When a flammable gas/air mixture is above the UEL, it is considered too rich to burn, which means there is plenty of fuel for combustion to occur, but not enough oxygen. When a flammable gas/air mixture is below its LEL, there is plenty of oxygen for combustion to occur, but not enough fuel. For combustion to take place, a flammable gas and air mixture must be between its upper and lower explosive limits.

Different flammable gases have different flammable ranges (see Figure 4.6). Most common hydrocarbon fuel gases have ranges from 1 to 13%. Some flammable materials have wide flammable ranges, such as hydrogen and acetylene. The only way to determine if a flammable gas is within its flammable range is to use monitoring instruments. These instruments check for a percentage of the LEL. The rule of thumb, according to the EPA, is when you reach 10% of the LEL; it becomes too dangerous for personnel to proceed any further.

Vapor Density

Another important physical characteristic of gases is vapor density. Vapor density is the relationship between the molecular weight of a gas and the air. If you separate component gases in air into nitrogen, oxygen, and argon, and the rare gases helium, neon, krypton, and xenon, you can measure the weight of a molecule of air. (see Figure 4.7) The molecular weight of an average air molecule is 29 atomic mass units

Contents of Air

Substance	% by Volume
Nitrogen, N_2	78.08
Oxygen, O_2	20.95
Argon, Ar	0.93
Carbon dioxide CO_2	0.033
Neon, Ne	0.0018
Helium, He	0.00052
Methane, CH_4	0.0002
Krypton, Kr	0.00011
Nitrogen (I) oxide, N_2O	0.00005
Hydrogen, H_2	0.00005

Figure 4.7

(AMU). Therefore, gases that have a molecular weight of less than 29 AMU will be lighter than air, and those with a weight greater than 29 will be heavier. For example, the gas methane has one carbon atom, which weighs 12, and 4 hydrogen atoms, which weigh 1 each. Adding them together gives you a molecular weight for methane of 16, which is less than 29, so methane is lighter than air.

When determining vapor pressure from reference books, air is given a value weight of 1 when compared to other gases; any gas that has a vapor density greater than 1 will be heavier than air. Propane has a vapor density of 1.56; therefore, propane is heavier than air. If a gas has a vapor density less than 1, it will be lighter than air. Natural gas (methane) has a vapor density of 0.554; therefore, natural gas is lighter than air. An explosion in a structure that causes damage near the foundation usually indicates a heavier-than-air gas, such as propane. Because propane is heavier than air, it stays low to the ground and goes into basements and confined spaces. A natural gas explosion will cause damage in the upper part of the structure because natural gas is lighter than air and will rise to the upper parts of the structure.

DOT has established criteria for a material to be classified as a flammable gas in transportation. First of all, its LEL must be below 13%. Some materials have wide flammable ranges that make them much more dangerous than materials with narrow ranges. The DOT states that if a material has a flammable range greater than 12 percentage points, regardless of the lower limit, the gas is also classified as flammable.

Propane and butane are two common flammable liquefied compressed gases. Propane has a flammable range of 2.4 to 9.5% in air, and butane 1.9 to 8.5% in air. Propane and butane have boiling points of –44°F and 31°F, respectively. Both materials are above their boiling points under ambient temperature conditions in many parts of the country year-round. This makes the materials highly dangerous when a leak or fire occurs, especially if there is a fire with flame impingement on the container. The vapor density of propane is 1.56, and butane is 2.07. Thus, both propane and butane vapors are heavier than air.

Propane and butane are shipped on the highway in MC/DOT 331 uninsulated containers. On the rail, they are shipped in insulated, pressurized tank cars. The railcars are insulated to keep flame impingement away from the tank surface to extend the time it would take for tank failure to occur. Propane and butane are stored in uninsulated bulk pressure containers shaped much like the MC/DOT 331. They may also be found in varying sizes of portable pressure containers. Because the gases are already above their boiling points, flame impingement, radiant heat transfer, or increases in ambient temperature can cause the materials to boil faster. Faster boiling causes an increase in pressure within the container. Even though the containers are specially designed to withstand pressure and have relief valves provided to release excess pressure, there are limits to the pressure they can tolerate. If the pressure buildup in the container exceeds the ability of the tank to hold the pressure or of the relief valve to relieve the pressure, the container will fail, resulting in a BLEVE.

Flammable Gas Elements

Hydrogen, H_2, is a nonmetallic, diatomic, elemental flammable gas. Hydrogen is one of the most flammable materials known. It burns clean, without smoke or visible flame. The only way to detect a hydrogen fire is from the radiant heat. Hydrogen may be found in transportation or storage as a cryogenic liquid (refrigerated liquid), a compressed gas, or a liquefied compressed gas. The flammable range of hydrogen is wide, at 4 to 75%; its boiling point is –423°F; and its ignition temperature is 1075°F. Hydrogen gas is slightly soluble in water and is noncorrosive. It is an asphyxiant gas and can displace the oxygen in the air or in a confined space. There are traces of hydrogen in the atmosphere, and it is abundant in the sun and stars. In fact, it is the most abundant element in the universe. Hydrogen has a vapor density of 0.069, therefore, it is lighter than air. As a cryogenic liquid, hydrogen has a four-digit UN identification number of 1066; as a compressed gas, its four-digit UN identification number is 1049. The NFPA 704 classification for hydrogen is health 3, flammability 4, and reactivity 0. It is usually shipped in insulated containers, insulated MC 338 tank trucks, tank cars, and tube trailers. Hydrogen is used in the production of ammonia, hydrogenation of vegetable oils, fuel for nuclear engines, hydrofining of petroleum, and cryogenic research.

HYDROCARBON FAMILIES

Three of the hydrocarbon families of hazardous materials contain compounds, which are gases: alkanes, alkenes, and alkynes. Hydrocarbon families are compounds made up of carbon and hydrogen. Carbon has a unique ability to bond with itself almost indefinitely. It forms long chained compounds, with varying numbers of carbon atoms and hydrogen atoms filling the remaining bonds. Names of the compounds are based upon the number of carbons bonded together in the compound and the type of bond between the carbons. The number of carbons in a compound is indicated in the name of the compound by a specific prefix (see Chapter 2, Figure 2.60). All three of the families use the same prefixes to indicate the number of carbons. For example, a one-carbon prefix is “meth,” two carbons “eth,” three carbons “prop,” and four carbons “but.” The hypothetical number of carbons and prefixes is endless.

For the purposes of this book, only the prefixes from 1- to 10-carbon chains will be considered. Those compounds that have more than four carbons are usually liquids and will be discussed in Chapter 5, Flammable Liquids. A system for naming compounds with more than 10 carbons is located in the Appendix, page 416 under the IUPAC System.

Alkanes

Alkanes are the first hydrocarbon family to be presented. The alkane family has only single bonds between the carbons. The smallest alkane contains one carbon. The prefix for a one-carbon compound is "meth." The ending for the alkane family indicating all single bonds is "ane." Therefore, a one-carbon alkane is called methane, with a molecular formula of CH_4. A two-carbon compound has the prefix "eth" and the ending "ane," thus, the compound is called ethane, with a molecular formula of C_2H_6. A three-carbon compound has the prefix "prop" and the ending "ane." The compound is called propane, with a molecular formula of C_3H_8.

Liquefied petroleum gases (LPG), which include propane, butane, isobutane propylene, and mixtures, were ranked twenty-sixth in number of incidents in all modes of transportation during 1995, having been involved in 123 transportation incidents. These gases accounted for the largest number of serious injuries of all hazardous materials incidents and were second in the number of deaths. LPG is consistently one of the highest-volume hazardous materials shipped by rail. In 1994, LPG was ranked number two, with over 150,000 tank carloads shipped. LPG shipments accounted for 22% of all tank car shipments by rail and 6% of the leaks.

A four-carbon compound is named butane, with the molecular formula of C_4H_{10}. Structures, molecular formulas, and some physical characteristics of these alkane-compressed gas compounds are shown in Figure 4.8. Notice the differences in boiling points and flammable ranges. As the carbon content increases, the boiling point of the compounds increases. In addition to carbon content, polarity and branching of compounds will affect physical characteristics. This concept will be discussed in more detail in Chapter 5. For now, just be aware that there are relationships between the physical characteristics of flammable gases, which include ignition temperature, heat output, vapor content, and vapor pressure.

Alkane hydrocarbons occur naturally, and their major hazard is flammability. Gases may also act as asphyxiants by displacing oxygen in the air. Alkanes are considered saturated hydrocarbons because all of the single bonds are full. No other elements can be added to the compound without physically removing one or more of the hydrogen atoms attached to the carbons. When hydrogen atoms are removed, radicals are formed, these radicals are used to make hydrocarbon derivatives. The type of hydrocarbon family can be estimated from the molecular formula of the compound by looking at the ratio of carbons to hydrogen atoms. In the case of alkanes, there are twice as many hydrogen atoms as carbons, as well as two hydrogen atoms. For example, ethane has two carbons: $2 \times 2 = 4 + 2 = 6$, so there are two carbons and six hydrogen atoms in ethane. Butane has four carbons: $4 \times 2 = 8 + 2 = 10$, so there are four carbons and 10 hydrogen atoms in butane. The molecular formula for butane is C_4H_{10} (see Figure 4.8).

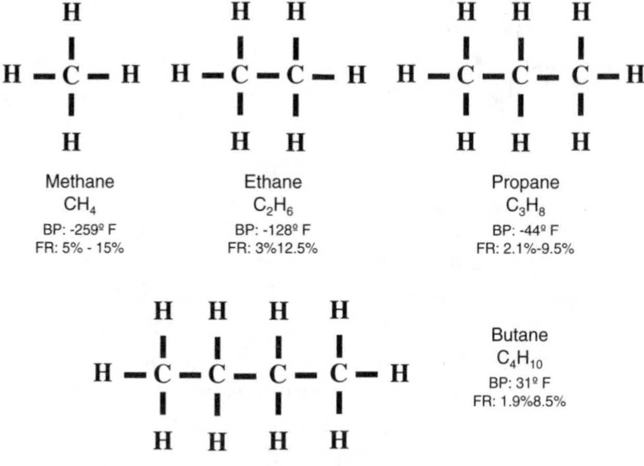

Figure 4.8

Isomers

Hydrocarbon structures may be altered so that their physical characteristics make them more economically valuable. One such alteration is called branching, or isomers. An **isomer** is a hydrocarbon that has the same molecular formula, i.e., the same number of hydrogen atoms and carbons, but a different structural form. The molecular formula for butane is C_4H_{10}; the molecular formula for the isomer of butane would be *i*-C_4H_{10} (Figure 4.9). The formula stays the same, but the structure

Fixed-facility tube bank with hydrogen gas.

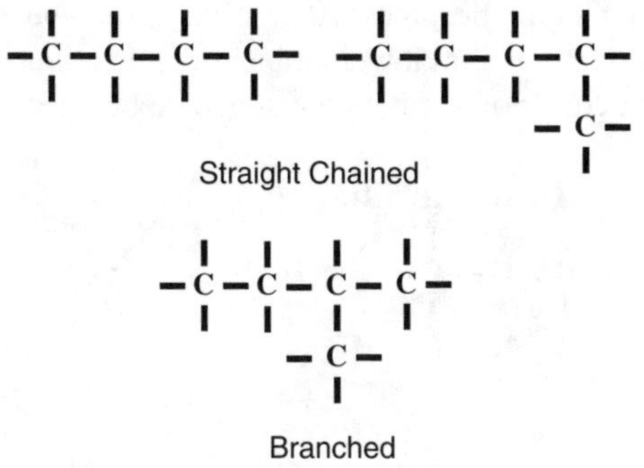

```
 |   |   |   |        |   |   |   |
-C - C - C - C-      -C - C - C - C-
 |   |   |   |        |   |   |   |
                                -C-
                                  |
```

Straight Chained

```
 |   |   |   |
-C - C - C - C-
 |   |   |   |
       -C-
         |
```

Branched

Figure 4.9

is different. In order to determine the difference, a prefix has to be added to the molecular formula and the name. When a structure is without a branch, it is sometimes referred to as normal, or straight-chained, i.e., all the carbons are connected together, end-to-end, in a chain. In Figure 4.9, in both examples all of the carbons are connected together end-to-end. In the second example, the end carbon has been placed below the chain, but is still connected end-to-end with the rest of the carbons. So the second example is not a branch or an isomer, even though it might appear to have a "branch." The fact is it is just the same as the first example, except for the arrangement of the carbons in the chain. To have a branched hydrocarbon, the branch cannot be on the end carbons, but must be on the carbons between the ends; see the third example.

The straight-chained hydrocarbon is sometimes referred to as the "normal" configuration. When a material is listed in a reference book, you may see a small "n-" in front of the name or molecular formula; this indicates that it is the "normal," or straight-chained, form of the compound. If the material is the isomer, or branched form, a small "i" will be placed in front of the molecular formula and the prefix "iso" in front of the name. For example, butane can be normal, which would be written as normal butane, or n-butane. The branched form of butane would be written isobutane. The molecular formula would be C_4H_{10} or n-C_4H_{10} for the normal, and *i*-C_4H_{10} for the branched compound. Normal is not used all of the time with the normal form of the compound; it is found mostly in reference books and on laboratory containers. If no prefixes appear in the name, it is understood to be the normal form. However, "iso" or the small "i" must be used to designate the branched form of the compound. Without this designation, there is no way to determine the branching from the name or the molecular formula.

The physical effect that branching has on a hydrocarbon is the lowering of the boiling point of that material. For example, butane is a liquefied compressed gas and has a boiling point of 31°F. One of the primary uses of butane is as a fuel for household and industrial purposes. Propane and butane tanks are usually located outdoors, and the liquefied gas takes on whatever the ambient temperature happens to be. In many parts of the country, the ambient temperature is below 31°F much of the winter. Because of the low ambient temperatures, butane would not be above its boiling point, and therefore would not be producing enough vapor to be used as a fuel. However, by changing the structure of butane and making it the branched isobutane, the boiling point becomes 10°F. Thus, with the lower boiling point, isobutane can be used as a fuel at lower ambient temperatures. The following is an illustration of the structures of normal butane and isobutane:

```
    H   H   H   H
    |   |   |   |
H — C — C — C — C — H
    |   |   |   |
    H   H   H   H
```

Normal Butane
C_4H_{10}
BP: 31º F

```
        H
        |   H   H
    H — C   |   |
      /   \ |   |
     H      C — C — H
    H — C /     |
      / |       H
     H  H
```

Iso-Butane
C_4H_{10}
BP: 11º F

Figure 4.10

One way of determining if the structure is branched or straight-chained is to try to draw a line through all of the carbons connected together in a chain without lifting the pencil or having to backtrack to reach another carbon. In the following examples, the structure on the left is normal, or straight-chained, because the line can be drawn through all of the carbons without backtracking. The structure on the right, however, requires lifting the pencil or backtracking to draw the line to the branched carbon. Therefore, the compound on the right is the isomer, or branched, compound.

```
                          — C
                              \
— C — C — C — C —              C — C —
                              /
                          — C
```

Straight Chained — Branched

Figure 4.11

Branching can also occur in the alkene and alkyne families. In order for branching to occur in hydrocarbons, there must be at least a four-carbon compound. Propane cannot be branched until the hydrocarbon derivatives, that is, elements other than carbon and hydrogen, are added to the structure of the compound. Other types of branching will be discussed in the Hydrocarbon Derivatives section of Chapter 5.

Alkenes

The alkene family has one or more double bonds between the carbons in the chain. Alkene compounds do not occur naturally; they are manmade. They are considered unsaturated because the double bond can be broken by heat or oxygen. The double bond is actually out-of-plane electrons between the carbons. The charges on the four

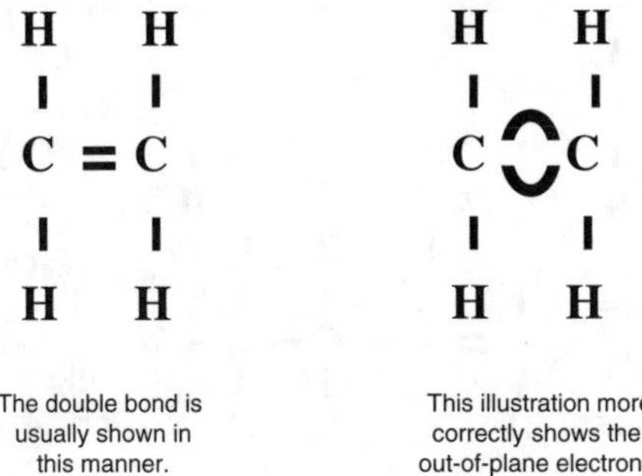

Figure 4.12

electrons between the carbons are negative. Charges that are the same repel each other. This forces the electrons out of plane and makes them vulnerable to the oxygen from the air. The illustration above shows the double bonds between two carbons (see Figure 4.12). The structure on the left is how the double bond is usually represented in a structure. The structure on the right shows how the out-of-plane electrons really appear.

When double bonds are broken, heat is created and other elements, including atmospheric oxygen, attach to the compound. The major hazard of the alkenes is flammability. Some of the compounds may be toxic or irritants, and some are suspected of being carcinogenic. Double-bonded compounds are usually unstable and reactive. Oxygen from the air can react with the double bonds and break them, creating heat and forming other compounds. Alkenes may also be used to make hydrocarbon derivatives. The same prefixes for determining the number of carbons are used for the alkenes as for the alkanes. However, since there must be at least one double bond between two carbons to have an alkene, there are no single-carbon alkenes. The smallest alkene would be two carbons. The prefix for a two-carbon compound is "eth." The ending that indicates at least one double bond is "ene." Therefore, a two-carbon alkene is called ethene, with the molecular formula of C_2H_4. Many times, there is more than one way to name chemical compounds. In the case of the alkene family, sometimes a "yl" is inserted between the prefix and the ending "ene." Therefore, ethene may sometimes be called ethylene. A three-carbon compound has the prefix "prop" and the ending "ene." The compound is called propene, or propylene, with the molecular formula C_3H_6. A four-carbon compound would have the prefix "but" and the ending "ene." The compound is named butene, or butylene, with the molecular formula of C_4H_8. The structures for these alkene compounds are shown in Figure 4.13.

Some compounds of alkenes have more than one double bond in the structure. Naming of the prefix for the number of carbons is the same as with the other alkenes. The ending "ene" is still used to indicate double bonds in the alkene family. There is, however, a prefix used to indicate the number of double bonds in the compound. The prefix "di" is inserted before the "ene" in the name to indicate two double bonds. For example, butene with two double bonds is called butadiene. The prefix "tri" is inserted to indicate three double bonds. Hexene, with three double bonds, is called hexatriene. There must be at least four carbons before two double bonds can

```
H   H
|   |
C = C
|   |
H   H
```

Ethene
Ethylene
C_2H_4
BP: -103.9 C
FR: 3% - 36%

```
H   H   H   H
|   |   |   |
C = C - C - C - H
|       |   |
H       H   H
```

Butene
Butylene
One of several Liquified Petroleum Gases

```
H   H   H
|   |   |
C = C - C - H
|       |
H       H
```

Propene
Propylene
C_3H_6
BP: 147.7 C
FR: 2% - 11%

Figure 4.13

be present. Two double bonds next to each other are highly unstable and will not hold together long. Following are examples of two and three double bonds in compounds.

```
H   H   H   H
|   |   |   |
C = C - C = C
|           |
H           H
```

Butadiene

```
H   H   H   H   H   H
|   |   |   |   |   |
C = C - C = C - C = C
|                   |
H                   H
```

Hexatriene

Figure 4.14

When trying to estimate the hydrocarbon family from the molecular formula, there is a ratio that indicates the alkenes. With the alkene family, there are twice as many hydrogen atoms as carbons. Propene has three carbons: $3 \times 2 = 6$, so there are three carbon atoms and six hydrogen atoms in propene (C_3H_6). In the case of alkenes that have more than one double bond, the ratio does not work. It is better to look at a molecular formula and draw out the structure rather than to try to guess from the ratio of the molecular formula. While the ratios work most of the time, there are exceptions, such as the two and three double-bonded compounds.

Alkynes

The alkyne hydrocarbons have at least one triple bond between the carbons in the chain. The ending for the alkyne family is "yne." Alkynes are unsaturated, with the

Fixed-facility propane tank. Note condensation on side, indicating liquid level.

triple bonds being reactive to heat and the oxygen in the air. As with the alkene family, there are no one-carbon alkynes. The most commercially valuable alkyne is the two-carbon compound. The prefix for two carbons is "eth," so the chemical name for the two-carbon, triple-bond compound is ethyne, with the molecular formula of C_2H_2. This is probably the only alkyne that emergency responders will ever encounter. Ethyne, however, is known by the commercial name **acetylene.** This is a tradename, so it is not derived from any of the naming rules of the hydrocarbon families. Acetylene is a highly flammable, colorless gas with a flammable range of 2.5 to 80%. Pure acetylene is odorless; however, the ordinary commercial purity has a distinct garlic-like odor. Acetylene can be liquefied and solidified; however, both forms are highly unstable. The vapor density is 0.91, so it is slightly lighter than air. Acetylene is produced when water is reacted with calcium carbide and other binary carbide salts. When these salts come in contact with water, acetylene gas is released. Acetylene is unstable and burns rich, with a smoky flame. The material may burn within its container. In fact, acetylene is so unstable that it can detonate under pressure. It is dissolved in a solvent, such as acetone, to keep it stable within a specially designed container. The container has a honeycomb mesh of ceramic material inside to help keep the acetylene dissolved in the acetone.

Acetylene is nontoxic and has no chronic harmful effects, even in high concentrations. In fact, it has been used as an anesthetic. Like most gases, acetylene can be a simple asphyxiant if present in high-enough concentrations that displace the oxygen in the air. The LEL of acetylene is reached well before asphyxiation can occur, and the danger of explosion is reached before any other health hazard is present. When fighting fires involving acetylene containers, the fire should be extinguished before closing the valve to the container. This is because the acetylene has such a wide flammable range that it can burn inside the container. Acetylene is incompatible with bromine, chlorine, fluorine, copper, silver, mercury, and their compounds. Acetylene has a four-digit UN identification number of 1001. The NFPA 704 designation is health 1, flammability 4, and reactivity 3. Reactivity is reduced to 2 when the acetylene is dissolved in acetone.

There are other alkyne compounds, but they do not have much commercial value, and will not be commonly encountered. A three-carbon compound with one triple bond has "prop" as a prefix for three carbons and is called propyne, with a molecular formula of C_3H_4. It is listed in the *Condensed Chemical Dictionary* as propyne, but you are referred to **methylacetylene** for information. It is listed as a dangerous fire risk, and it is toxic by inhalation. Propyne is used as a specialty fuel and as a chemical intermediate. A four-carbon alkyne has the prefix "but," and the compound is called butyne, with the molecular formula of C_4H_6. The chemical listing is under the name **ethylacetylene**, and it is designated as a dangerous fire risk. It is also used as a specialty fuel and as a chemical intermediate. The following shows the structures for ethyne, propyne, and butyne:

H — C ≡ C — H

Ethyne
C_2H_2

```
     H
     |
H —  C — C ≡ C — H
     |
     H
```

Propyne
C_3H_4

```
     H               H
     |               |
H —  C — C ≡ C — C — H
     |               |
     H               H
```

Butyne
C_4H_6

Figure 4.15

While there are no commercially valuable two or three triple-bonded compounds, the same rules for naming them would apply as in the alkenes: the prefixes "di" for two and "tri" for three would be used. There is a ratio of carbon atoms to hydrogen atoms that can be used to identify the compound from the molecular formula. With the alkyne family, there are twice as many hydrogen atoms as carbons: –2. Ethyne has two carbons: $2 \times 2 = 4 - 2 = 2$. So there are two carbon atoms and two hydrogen atoms in the compound ethyne, with a molecular formula of C_2H_2.

Hydrocarbon Derivatives

Hydrocarbon-derivative compounds do not occur naturally. They are manmade from hydrocarbon compounds, as discussed earlier, with some additional elements added. Hydrocarbon derivatives belong to families just as the hydrocarbons. In order to make hydrocarbon derivatives, hydrogen needs to be removed from the alkane family. Alkene hydrocarbons have one or more double bonds that can be broken and other elements added. They may or may not have hydrogen removed. Elements commonly added to hydrocarbon compounds to create hydrocarbon derivatives include oxygen, nitrogen, fluorine, chlorine, bromine, and iodine. Together with the hydrocarbons, these elements make up over 50% of all hazardous materials.

There are some hydrocarbon-derivative functional groups that have flammable gas compounds in their families (see Chapter 2, Figure 2.63). Alkyl halides are listed with toxicity as a primary hazard. However, there are some flammable alkyl halides. Vinyl chloride and methyl chloride are alkyl halides. Vinyl chloride and methyl chloride are extremely flammable gases. The amines are also primarily toxic as a group; there are, however, some flammable amine gases. Methylamine, dimethylamine, ethylamine, and propylamine are all flammable gases. There are a few ether flammable gases, most of which do not use the trivial naming system and may not be recognized as ethers. For example, propylene oxide is an ether compound. Methyl ether may be found as a compressed gas or a liquid. In the aldehyde family, most compounds are liquids, except for the one-carbon aldehyde, formaldehyde. Hydrocarbon-derivative functional groups will be discussed in detail in Chapter 5, Flammable Liquids.

Methyl chloride, CH_3Cl, is an alkyl-halide hydrocarbon derivative. It is a colorless compressed gas or liquid with a faintly sweet, ether-like odor. It is a dangerous fire risk, with a flammable range of 10.7 to 17% in air. The critical temperature is approximately 225°F, and the critical pressure is 970 psi. It is slightly soluble in water. The vapor density is 1.8, which is heavier than air. The boiling point is −11°F, and the flash point is 32°F. The ignition temperature is 1170°F. It is a narcotic, producing psychogenic effects. The TLV is 50 ppm in air. The four-digit UN identification number is 1063. The NFPA 704 designation is health 1, flammability 4, and reactivity 0. The primary uses are as a catalyst in low-temperature polymerization, as a refrigerant, as a low-temperature solvent, as an herbicide, and as a topical anesthetic. The structure for methyl chloride is shown in the following illustration:

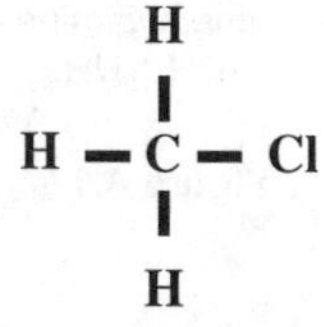

Methyl Chloride
CH_3Cl

Figure 4.16

Dimethylamine, $(CH_3)_2NH$, an amine hydrocarbon derivative, is a gas with an ammonia-like odor. It is a dangerous fire risk, with a flammable range of 2.8 to 14% in air. It is insoluble in water. The vapor density is 1.55, which is heavier than air. The boiling point is 44°F, and the ignition temperature is 806°F. Dimethylamine is an irritant, with a TLV of 10 ppm in air. The four-digit UN identification number is 1032. The NFPA 704 designation is health 3, flammability 4, and reactivity 0. The

Highway transportation tube trailer with compressed hydrogen gas.

primary uses are in electroplating and as gasoline stabilizers, pharmaceuticals, missile fuels, pesticides, and rocket propellants. The structure for dimethylamine is shown in the following illustration:

$$\begin{array}{ccccccccc} & & H & & H & & H & & \\ & & | & & | & & | & & \\ H & - & C & - & N & - & C & - & H \\ & & | & & & & | & & \\ & & H & & & & H & & \end{array}$$

Dimethylamine
$(CH_3)_2NH$

Figure 4.17

COMMON HYDROCARBON DERIVATIVES

Ethylene, C_2H_4 (ethene), has a boiling point of −155°F and is a dangerous fire and explosion risk. The flammable range is fairly wide, with an LEL of 3% and a UEL of 36%. The vapor density is 0.975, which is slightly lighter than air. The critical pressure is 744 psi, and the critical temperature is 9.5°C. Ethylene is not toxic, but can be an asphyxiant gas. The UN designation number for ethylene is 1962 as a compressed gas. The NFPA 704 designation is health 3, flammability 4, and reactivity 2. As a cryogenic liquid, the UN designation number is 1038, and the NFPA 704 designation is health 1, flammability 4, and reactivity 2. It is usually shipped in steel

MC 331 used to transport liquefied compressed gases.

pressurized cylinders and tank barges. Ethylene is used in the production of other chemicals, as a refrigerant, in the welding and cutting of metals, as an anesthetic, and in orchard sprays to accelerate fruit ripening. The structure for ethylene is shown in the Alkene section of this chapter.

Propylene, C_3H_6 (propene), has a boiling point of –53°F. The flammable range of propylene is 2 to 11%. The vapor density is 1.46, which is heavier than air. The four-digit UN identification number is 1077. The NFPA 704 designation is health 1, flammability 4, and reactivity 1. It is not toxic, but can be an asphyxiant gas by displacing the oxygen in the air. It is usually shipped as a pressurized liquid in cylinders, tank cars, and tank barges. The structure for propylene is shown in the Alkene section of this chapter.

Vinyl chloride, C_2H_3Cl, is a compressed gas that is easily liquefied. Vinyl chloride is the most important vinyl monomer and may polymerize if exposed to heat. It has an ether-like odor, and phenol is added as an inhibitor during shipment and storage. It is highly flammable, with a flash point of –108°F, and a boiling point of 7°F. The flammable range is 3.6 to 33% in air, with an ignition temperature of 882°F. Vinyl chloride is insoluble in water and has a specific gravity of 0.91, which is lighter than water. The vapor density is 2.16, which is heavier than air. It is toxic by inhalation, ingestion, and skin absorption. Vinyl chloride is a known human carcinogen. The TLV is 5 ppm in air. The four-digit UN identification number is 1086. The NFPA 704 designation is health 2, flammability 4, and reactivity 2; uninhibited, the values would be higher for reactivity. The primary uses are in making polyvinyl chloride and as an additive in plastics. The structure for vinyl chloride is shown in Figure 4.18.

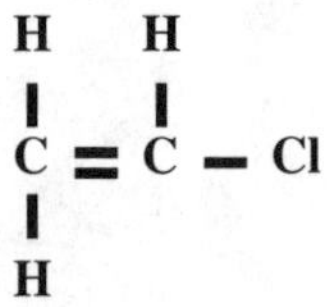

```
H   H
|   |
C = C – Cl
|
H
```

Vinyl Chloride
C_2H_3Cl

Figure 4.18

Formaldehyde, HCHO, is an aldehyde hydrocarbon derivative. Formaldehyde is a gas with a strong, pungent odor, and it readily polymerizes. Commercially, it is offered as a 37 to 50% solution, which may contain up to 15% methanol to inhibit polymerization. Boiling points for solutions range from 206° to 212°F. Commercial solutions have the trade name **formalin**. Formaldehyde is flammable, with a wide flammable range of 7 to 73% in air. The boiling point is –3°F, and the flash point is 185°F. The ignition temperature for the gas is 572°F. It is water-soluble. The vapor density is 1, which is the same as air. Formaldehyde is also toxic by inhalation, a strong irritant, and a carcinogen. The TLV is 1 ppm in air. Nonflammable solutions are Class 9 Miscellaneous Hazardous Materials, with a four-digit UN identification number of 2209. The NFPA 704 designation is health 3, flammability 2, and reactivity 0. Flammable solutions are Class 3 Flammable Liquids, with a four-digit UN identification number of 1198. The NFPA 704 designation is health 3, flammability 4, and reactivity 0. The following structure is for formaldehyde:

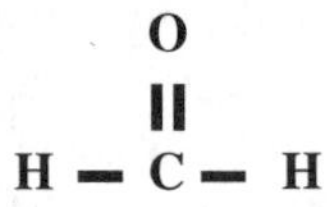

```
    O
    ||
H – C – H
```

Formaldehyde
HCHO

Figure 4.19

Butadiene, C_4H_6, has a boiling point of 24°F and a flammable range of 2 to 11%. The vapor density is 1.93 psi, which means it is much heavier than air. It is highly flammable and may polymerize. Butadiene may form explosive peroxides in contact with air. It has a four-digit UN identification number of 1010. The NFPA 704 designation is health 2, flammability 4, and reactivity 2. Butadiene must be inhibited during transportation and storage. It is usually shipped in steel pressurized cylinders, tank cars, and tank barges. The structure is shown in the Alkene section of this chapter.

INCIDENTS

There have been numerous incidents over the years involving pressurized containers and flammable gases. Two of the most recent incidents have resulted in the deaths of four firefighters, two in Carthage, Illinois, and two in Albert City, Iowa. Both incidents were similar in that they involved storage tanks on farms, one a commercial farm, and the other a private farm. Because the Albert City explosion involved a commercial facility, the incident was investigated by the U.S. Chemical Safety and Hazard Investigation Board. During the Albert City incident, the tank was engulfed in flames due to leaking propane underneath the tank. It was this flame impingement that resulted in the BLEVE that killed the firefighters. According to the investigative report, the firefighters were positioned too close to the burning propane tank when it exploded. They were under the impression they would be protected during an explosion if they stayed away from the ends of the tank. Three major factors resulting in the deaths were noted by the safety board's investigation.

- Protection of above-ground piping was inadequate.
- The diameter of the pipe downstream from an excess-flow valve was too narrow, which prevented the valve from functioning properly.
- Firefighter training for responding to BLEVEs was inadequate.

In Kingman, Arizona, a 33,500-gal railroad tank car containing liquefied petroleum gas was being offloaded at a gas distribution plant. As the liquid lines were attached to the tank, a leak was detected. During attempts by workers to tighten the fittings to stop the leak, a fire occurred. Both workers were severely burned, and one later died from injuries. Shortly after the fire department arrived, the liquid line failed and flame impingement began on the vapor space of the tank car. Nineteen minutes after the flames began contacting the surface of the tank car, a BLEVE occurred. A large section of the 20-ton tank was propelled 1200 feet by the explosion. Railroad conductor Hank Graham, who took many of the now-famous Kingman photographs, had the hair burned off his arms as he took pictures. Twelve Kingman firefighters were killed in the explosion. Only one firefighter from Kingman who was close to the scene survived. Ninety-five spectators on nearby Highway 66 were injured by the blast. The explosion set fire to lumber stored nearby, a tire company, and other businesses within 900 feet of the blast.

A quite different type of incident occurred in Waverly, Tennesee, involving two 28,000-gal propane tank cars. A 24-car train derailment occurred just six blocks from the town square in this town of 6,000 residents. The derailment created a mass of piled railcars that damaged at least one of the tank cars of propane. Initially, 50 to 100 people were evacuated within one-quarter mile of the accident site as a precaution, although no leaks, fires, or explosions occurred when the train derailed. The accident occurred around 5:10 a.m. on a cold winter day. There were approximately 2 in of snow on the ground, it was cloudy, and the temperature had been below freezing for several days.

The task of cleaning up the crash site began about mid-day on Wednesday. By Friday, the situation was felt to be well under control, security was relaxed, and people were allowed to return to their homes. Spectators and nonessential personnel were allowed into the immediate area of the derailment. Workmen smoked and used acetylene cutting torches at will in the process of cleaning up the site. A tank truck was brought in to offload the damaged propane tank cars. Friday afternoon, the

Kingman, Arizona: railcar fire just before explosion. Note flame impingement on vapor space of tank. Courtesy of Hank Graham.

Kingman, Arizona: propane explosion. Note pieces of tank car propelled by the explosion. Courtesy of Hank Graham.

temperature began to rise, causing an increase in pressure inside the damaged tank. Normally, the pressure-relief valve would function to release pressure created by increases in ambient temperature. However, in this instance, the damaged portion

Waverly, Tennessee: propane tank car UTLX 83013 before BLEVE. Courtesy of Waverly Fire Department.

of the tank was weaker than the relief-valve pressure setting. At approximately 3:00 p.m., the tank could no longer withstand the increased pressure and opened up, releasing into the environment liquid propane, which instantly vaporized. The vapors quickly found an ignition source, and flames shot 1000 feet into the air. The giant fireball was visible for 30 miles around Waverly. The resulting explosion and fire killed 16 people, including the Waverly fire and police chiefs and five firefighters. Fifty-four people were injured, many severely burned.

Crescent City, Illinois, was the site of still another accident involving propane in transportation. Sixteen cars of an eastbound freight train derailed in the center of town at approximately 6:40 a.m., including 10 cars each containing 34,000 gal of liquid propane. Two additional propane tanks remained on the tracks. During the derailment, one of the propane tank cars was punctured by the coupler of another car, causing a leak that ignited almost immediately. Flames reached several hundred feet into the air. A nearby house and business were set on fire from the radiant heat, injuring several residents. Relief valves on the other tank cars began to open as the pressure built up from the surrounding fires. The first explosion (BLEVE) occurred around 7:33 a.m., almost one hour after the derailment. In that first blast, several firefighters and bystanders were injured and some fire equipment was damaged. Additional explosions occurred at 9:20, 9:30, 9:45, 9:55, and 10:10 a.m. Parts of tank cars were propelled all over town, setting fires and damaging structures. However, there were no injuries to civilians from the explosions, because of a quick evacuation after the derailment; 73 firefighters, police officers, and press personnel were injured by the explosions, but there were no fatalities. Twenty-four living quarters were destroyed by fire and three homes destroyed by “flying” tank cars;

Waverly, Tennessee: propane tank car UTLX 83013 after BLEVE. Courtesy of Waverly Fire Department.

Crescent City, Illinois: fireball from BLEVE of propane tank. Courtesy of Irma Hill.

Weyauwega, Wisconsin: train derailment burning feed mill from propane fires. Photograph by Robert Ehrenberg of the Weyauwega Fire Department.

numerous other homes received damage. Eighteen businesses were destroyed. The remaining propane tanks were allowed to burn, which took some 56 hours after the derailment.

The city of Weyauwega, Wisconsin, population 1700, was evacuated for 20 days following a train derailment in the middle of town on March 4, 1996, at 5:55 a.m. Thirty-four cars were derailed, including seven containing LPG, seven containing propane, and two containing sodium hydroxide. The resulting fires from leaking propane damaged a feed mill and a storage building. There were no explosions as a result of the derailment and resulting fires. The incident commander, Jim Baehnman, Assistant Chief of the Weyauwega Fire Department, said, "The tone of the response from the beginning was not time-driven, but rather safety-driven." This may have very well accounted for the fact that not a single death or injury occurred as a direct result of this incident.

When dealing with emergencies involving pressurized containers and flammable gases, great caution should be taken. Flame impingement on the vapor space of a container is a "no-win" situation. If a BLEVE is going to occur, it is just a matter of time. To try to fight a fire under those conditions is to play Russian roulette. The NFPA *Fire Protection Guide* says that BLEVE times range from 8 to 30 min, with the average time being 15 min. There is usually no way to know how long the flame impingement has been going on prior to the fire department arrival, and no way to know exactly when the BLEVE will occur. If the only threat is to the emergency responders, there is little reason to risk their lives needlessly. If the impingement is on the liquid space, the liquid will absorb the heat for a period of time, and will boil faster as it does. There will be an increase in pressure within the tank as the liquid boils faster. This can still be a dangerous situation if not handled properly. Conditions involving the tank must be monitored constantly for changes, including liquid level, pressure increases, and signs of tank failure (see Figure 4.20).

Precautions should be taken to ensure personnel safety when fighting fires involving flammable gases. Flammable gas fires should not be extinguished until the source

Albert City, Iowa, propane explosion killed two firefighters; tank was blown into a chicken house.

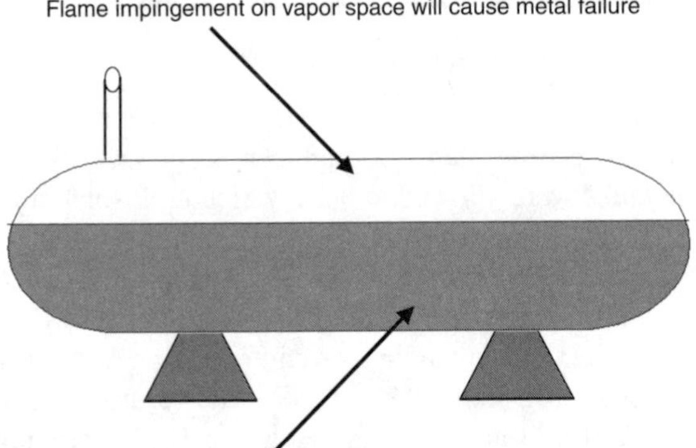

Figure 4.20

of the gas has been shut off. It is much safer to have the gas on fire and know where it is than to have the gas leaking and going where it wants to go. Flammable gases are more dangerous than flammable liquids. Emergency responders should do a risk/benefit analysis before determining tactics for dealing with flammable gas emergencies. This analysis should be based upon the physical and chemical characteristics of the gas, the container characteristics, ambient conditions, and the life-safety factor of the public and the firefighters. If the risk/benefit analysis determines that the only lives in danger are those of the firefighters, then they should initiate unmanned hose streams and withdraw to a safe location and let the incident take its course. Property can be replaced; firefighter lives cannot. Lessons learned from the previously mentioned incidents should be used to develop standard operating procedures.

NONFLAMMABLE COMPRESSED GASES

Class 2.2 includes gases that are nonflammable and nonpoisonous. Gases in this subclass can be compressed, liquefied, cryogenic, or gases in solution. Although these gases are nonflammable, the containers can still BLEVE, under flame impingement conditions on the tank or from a fire involving other materials. BLEVE can also occur from a damaged or weakened container or from an overpressure of the container caused by overfilling. Increases in pressure caused by increases in ambient temperature can also cause container failure.

The DOT definition of a nonflammable gas is "a material that exerts in the packaging an absolute pressure of 41 psi or greater at 20°C, and does not meet the definitions of Division 2.1 or 2.3." If the pressure in a container is less than 41 psi, a gas does not belong in this category. Many cryogenic materials are shipped at atmospheric pressure and so are not considered compressed gases. Cryogenics are not considered a DOT hazard class as a group except when under pressure or regulated by some other hazard class. If cryogenic gases are shipped above 41 psi, they are considered a compressed gas. Cryogenics may be required to be placarded if they have other hazards, such as flammable, poison, or oxidizer. If cryogenics do not have another hazard, they are not required to be placarded under DOT regulations. Compressed gases that are shipped as liquefied gases, such as cryogenics, exhibit other hazards not indicated by the placard. Liquefied refrigerated gases, such as cryogenics, are extremely cold materials; boiling points are –130°F or greater (see Figure 4.21). Liquid helium has a boiling point of –452°F; it is the coldest material known. It is also the only material on earth that never exists as a solid under normal temperatures and pressures; it only exists as a cryogenic liquid or as a gas.

Gases are processed into cryogenic liquids by a combination of pressurization, cooling, and ultimate release of pressure. Therefore, cryogenics do not require pressure to keep them in the liquid state, unless they will be in the container for a long period of time; then they are pressurized. Cryogenics are kept cold by the temperature of the liquid and the insulated containers. The cryogenic liquefaction process begins when gases are placed into a large processing container. They are pressurized to 1500 psi. The process of pressurizing a gas causes the molecules to move faster, causing more collisions with each other and the walls of the container. This causes heat to be generated. An example of this heating process occurs while filling an SCBA bottle — the top feels hot. Continuing with the liquefaction process, once the pressure of 1500 psi is reached, the material is cooled to 32°F using ice water. When the gas is cooled, the pressure is once again increased, this time up to 2000 psi, with an accompanying increase in temperature. The gas is then cooled to –40°F using liquid ammonia. As the gas is cooled, all of the pressure is released. The resulting temperature decrease turns the cryogenic gas into a liquid.

Boiling Points of Cryogenic Liquids

Helium	–452°F	Air	–318°F
Neon	–411°F	Fluorine	–307°F
Nitrogen	–321°F	Hydrogen	–423°F
Argon	–303°F	Methane	–257°F
Oxygen	–297°F	Nitric Oxide	–241°F
Krypton	–244°F	CO	–312°F
Xenon	–162°F	NF_3	–200°F

Figure 4.21

Several gases found on the Periodic Table are extracted from the air and turned into cryogenic liquids. These include neon, argon, krypton, xenon, oxygen, and nitrogen. All but oxygen are considered inert, i.e., they are nontoxic, nonflammable, and nonreactive. To extract these gases from the air, the air is first turned into a cryogenic liquid. Then the liquid is run through a type of distillation process, where each component gas is drawn off as it reaches its boiling point. Gases are then liquefied by the same process mentioned previously. Some common cryogenic gases and their characteristics are listed in the following paragraphs.

Helium, He, is a gaseous, nonmetallic element from the noble gas family, family eight on the Periodic Table. Helium is colorless, odorless, and tasteless. It is nonflammable, nontoxic, and nonreactive. Helium has a boiling point of –452°F, and is slightly soluble in water. Even though helium is an inert gas, it can still displace oxygen and cause asphyxiation. Helium has a vapor density of 0.1785, which is lighter than air. It is derived from natural gas by liquefaction of all other components. Helium has a four-digit UN identification number of 1046 as a compressed gas and 1963 as a cryogenic liquid. Helium is used to pressurize rocket fuels, in welding, in inflation of weather and research balloons, in luminous signs, geological dating, lasers, and as a coolant for nuclear-fusion power plants.

Neon, Ne, is a gaseous, nonmetallic element from the noble gas family. It is colorless, odorless, and tasteless, and is present in the Earth's atmosphere at 0.0012% of normal air. It is nonflammable, nontoxic, and nonreactive, and does not form chemical compounds with any other chemicals. Neon is, however, an asphyxiant gas and will displace oxygen in the air. The boiling point of neon is –410°F, and it is slightly soluble in water. Neon has a vapor density of 0.6964, which is lighter than air. The four-digit UN identification number is 1065, when compressed, and 1913 as a cryogenic liquid. Its primary uses are in luminescent electric tubes and photoelectric bulbs. It is also used in high-voltage indicators, lasers (liquid), and cryogenic research.

Argon, Ar, is a gaseous, nonmetallic element of family eight. It is present in the Earth's atmosphere to 0.94%, by volume. It is colorless, odorless, and tasteless. Argon does not combine with any other chemicals to form compounds. It has a boiling point of –302°F and it is slightly soluble in water. Argon has a vapor density of 1.38, which makes it heavier than air. The four-digit UN identification number is 1006, as a compressed gas, and 1951 as a cryogenic liquid. Argon is used as an inert shield in arc welding, in electric and specialized lightbulbs (neon, fluorescent, and sodium vapor), in Geiger-counter tubes, and in lasers.

Krypton, Kr, is a gaseous, nonmetallic element of family eight. It is present in the Earth's atmosphere to 0.000108%, by volume. It is a colorless and odorless gas. Krypton is nonflammable, nontoxic, and nonreactive. It is, however, an asphyxiant gas and can displace oxygen in the air. At cryogenic temperatures, krypton exists as a white, crystalline substance with a melting point of 116°K. The boiling point of krypton is –243°F. Krypton is known to combine with fluorine at liquid nitrogen temperature by means of electric discharges or ionizing radiation to form KrF_2 or KrF_4. These materials decompose at room temperature. Krypton is slightly water-soluble. The vapor density is 2.818, which is heavier than air. The four-digit UN

Fixed-facility cryogenic liquid oxygen container with heat exchanger outside of a hospital.

identification number is 1056, for the compressed gas, and 1970 for the cryogenic liquid. Krypton is used in incandescent bulbs, fluorescent light tubes, lasers, and high-speed photography.

Xenon, **Xe**, is a gaseous, nonmetallic element from family eight. It is a colorless, odorless gas or liquid. It is nonflammable and nontoxic at standard temperatures and pressures, but is an asphyxiant and will displace oxygen in the air. The boiling point is −162°F, and the vapor density is 05.987, which is heavier than air. Xenon is chemically unreactive; however, it is not completely inert. The four-digit UN identification number is 2036 for the compressed gas and 2591 for the cryogenic liquid. Xenon is used in luminescent tubes, flash lamps in photography, lasers, and as an anesthesia.

Xenon compounds. Xenon combines with fluorine through a process of mixing the gases, heating in a nickel vessel to 400°C, and cooling. The resulting compound is xenon tetrafluoride, XeF_4, composed of large, colorless crystals. Compounds of xenon difluoride, XeF_2, and hexafluoride, XeF_6, can also be formed in a similar manner. The hexafluoride compound melts to a yellow liquid at 122°F and boils at 168°F. Xenon and fluorine compounds will also combine with oxygen to form oxytetrafluoride, $XeOF_4$, which is a volatile liquid at room temperature. Compounds formed with fluorine must be protected from moisture to prevent the formation of xenon trioxide, XeO_3, which is a dangerous explosive when dried out. The solution of xenon trioxide is a stable, weak acid that is a strong oxidizing agent.

Expansion Ratios for Cryogenics

Argon	841/1
Ethane	487/1
Fluorine	981/1
Helium	754/1
Hydrogen	840/1
LNG*	637/1
Nitrogen	697/1
Oxygen	862/1

*Liquefied Natural Gas

Figure 4.22

Cryogenic liquids have large expansion ratios, some as much as 900 or more to 1 (see Figure 4.22). Because of this expansion ratio, if the cryogenic liquid is flammable or toxic, these hazards are intensified because of the potential for large gas cloud production from a small amount of liquid. As the size of a leak increases, so does the size of the vapor cloud.

Even though some cryogenic liquids do not require placards, these materials can still pose a serious danger to responders. Gases formed by warming of cryogenic liquids can displace oxygen in the air, which can harm responders by asphyxiation. Normal atmospheric oxygen content is about 21%. When the oxygen in the lungs and ultimately the blood is reduced, unoxygenated blood reaches the brain, and the brain shuts down. It may only be a few seconds between the first breath and collapse. Being quite cold, cryogenic liquids can cause frostbite and solidification of body parts. When the parts thaw out, the tissue is irreparably damaged.

Cryogenic liquids are shipped and stored in special containers. On the highway, the MC 338 tanker is used to transport cryogenic liquids. The tank is usually not pressurized, but is heavily insulated to keep the liquids cold. A heat exchanger is located underneath the belly of the tank truck to facilitate the offloading of product as a gas. Railcars are also specially designed to keep the cryogenic liquids cold inside the containers to minimize the boiling off of the gas. Fixed storage containers of cryogenic liquids are usually tall, small-diameter tanks. These are insulated and resemble large vacuum bottles that keep the liquid cold. Cryogenic storage containers are also under pressure to keep the material liquefied. A heat exchanger is used to turn the liquid back into a gas for use. It is composed of a series of metal tubes with fins around the outside. As the liquid runs through the tubes, it is warmed and turns into a gas. Other gases, such as hydrogen, are liquefied, sometimes made into cryogenics, and placarded as flammable gases. Liquid oxygen is placarded as an oxidizer or nonflammable compressed gas.

There are some “foolers” in the nonflammable compressed gas subclass. Anhydrous ammonia, for example, is regulated by the DOT as a nonflammable compressed gas. The United States is the only country in the world that placards anhydrous ammonia in this manner. Such placarding of anhydrous ammonia as a nonflammable gas is a result of lobbying efforts by the agricultural fertilizer industry. In most other parts of the world, anhydrous ammonia is placarded as a poison gas, not to mention

Trailer used to haul anhydrous ammonia to be used as a fertilizer in farming.

that it is also flammable under certain conditions. If anhydrous ammonia leaks inside a building or in a confined space, it may very well be found within its flammable range and burn if an ignition source is present. If anhydrous ammonia met the DOT definition of a flammable gas, which is a LEL of less than 13% or a flammable range of greater than 12 percentage points, it would be placarded as such. It does not, however, meet the DOT definition; it has a LEL of 16% and a flammable range of 16 to 25%. Therefore, it is placarded as a nonflammable gas even though, in reality, it should be placarded by its most severe hazard, which is a poison gas.

Hydrocarbon Derivatives

A few hydrocarbon derivatives from the alkyl-halide family are 2.2 nonflammable compressed gases. This illustrates the wide range of hazards of the alkyl halides as a group. Some are flammable, some are toxic, and some are nonflammable and nontoxic. They can still act as asphyxiants and displace the oxygen in the air. It is important to remember that the primary hazard of the alkyl halides is toxicity. Some of them are also flammable; therefore, all must be assumed to be toxic and flammable until the individual chemical is researched and the actual hazards are determined. It is interesting to note that while the DOT lists tetrafluoromethane as a nonflammable, nonpoisonous gas, the *Condensed Chemical Dictionary* lists the compound as toxic by inhalation. The *NIOSH Pocket Guide to Chemical Hazards* does not list the compound. The best source of information about this compound and others may be the MSDS (material safety data sheet). Examples of nonflammable Class 2.2 alkyl halides are tetrafluoromethane and trifluoromethane.

Tetrafluoromethane, CF_4, also known as carbon tetrafluoride and fluorocarbon 14, is a colorless gas that is slightly soluble in water. It is nonflammable, but is listed as toxic by inhalation in the *Condensed Chemical Dictionary.* The four-digit UN identification number is 3159. The primary uses are as a refrigerant and gaseous insulator. The structure is illustrated in the following example:

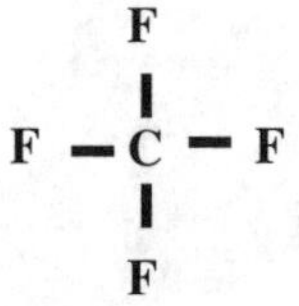

```
     F
     |
F — C — F
     |
     F
```

Tetrafluoromethane
CF_4

Figure 4.23

Trifluoromethane, CHF_3, also known as fluoroform, propellant 23, and refrigerant 23, is a colorless gas that is nonflammable. There are no hazards listed for trifluoromethane. It may be an asphyxiant gas and displace oxygen in the air and in confined spaces. The four-digit UN identification number is 2035 for the compressed gas and 3136 for the cryogenic liquid. The primary uses for trifluoromethane are as a refrigerant, as a direct coolant for infrared detector cells, and as a blowing agent for urethane foams. The structure is shown in the following illustration:

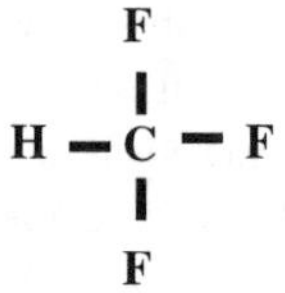

```
     F
     |
H — C — F
     |
     F
```

Trifluoromethane
CHF_3

Figure 4.24

Nonflammable Gas Compunds

Nitrogen, N_2, is colorless, odorless, and tasteless and makes up 78% of the air in the Earth's atmosphere. Nitrogen has a boiling point of –320°F and is slightly soluble in water. Nitrogen is nonflammable and nontoxic. It may, however, displace oxygen and be an asphyxiant gas. The vapor density of nitrogen is 0.96737, which makes it slightly lighter than air. The four-digit UN identification number is 1066 as a compressed gas and 1977 as a cryogenic liquid. As a cryogenic liquid, the NFPA 704 designation is health 3, flammability 0, and reactivity 0. Nitrogen is used in the production of ammonia, cyanides, and explosives; as an inert purging agent; and as a component in fertilizers. It is usually shipped in insulated containers, insulated MC 338 tank trucks, and tank cars.

MC 338 used for the transportation of cryogenic liquids.

Dewier container of cryogenic nitrogen.

Oxygen, O_2, like nitrogen, is a nonmetallic elemental gas. Oxygen, although nontoxic, is highly reactive with hydrocarbon-based materials and is an oxidizer. Oxygen makes up approximately 21% of the air breathed. The boiling point of oxygen is –297°F. It is nonflammable, but supports combustion. Oxygen can explode when exposed to heat or organic materials. The vapor density of oxygen is 1.105, which makes it slightly heavier than air. Oxygen is incompatible with oils, grease, hydrogen, flammable liquids, solids, and gases. The four-digit UN identification number for oxygen is 1072 as a compressed gas and 1073 as a cryogenic liquid. The NFPA 704 designation for liquid oxygen is health 3, flammability 0, and reactivity 0. Liquid oxygen is shipped in Dewier flasks and MC 338 tank trucks. It may also be encountered in cryogenic railcars. Liquid oxygen in contact with an asphalt surface, such as a parking lot or highway, can create a contact explosive; dropping an object, driving, or even walking on the area can cause an explosion to occur.

Anhydrous ammonia, NH_3, seeks water, "anhydrous" meaning "without water." This can be particularly dangerous to responders because ammonia can seek water in the eyes, lungs, and other moist parts of the body. Ammonia is toxic, with a TLV of 25 ppm; the inhalation of concentrated fumes can be fatal above 2000 ppm. Ammonia odor is detectable at 1 to 50 ppm. It has a boiling point of –28°F and a flammable range of 16 to 25% in air, although it does not meet the DOT definition of a flammable gas. The ignition temperature is 1204°F. The vapor density of ammonia is 0.6819, which makes ammonia gas lighter than air. It is also water-soluble. Hose streams can be used to control vapor clouds of ammonia gas. The runoff created, however, is ammonium hydroxide, which is corrosive, so the runoff should be contained.

Ammonia is incompatible with mercury, hydrogen fluoride, calcium hypochlorite, chlorine, and bromine. The four-digit UN identification number is 1005, and the NFPA 704 designation is health 3, flammability 1, and reactivity 0. Ammonia may be shipped as a cryogenic liquid or a liquefied compressed gas. Ammonia is used as an agricultural fertilizer and as a coolant in cold-storage buildings and food lockers. It is usually shipped in MC 331 tank trucks, railcars, barges, and steel cylinders. Anhydrous ammonia is the second most-released chemical from fixed facilities on the EPA listing, with 3586 accidents resulting in the release of over 19 million lbs of ammonia. Anhydrous ammonia is consistently among the seven highest-volume hazardous materials shipped on the railroads.

Mild exposure to anhydrous ammonia can cause irritation to eye, nose, and lung tissues. When NH_3, is mixed with moisture in the lungs, it causes severe irritation. Ammonium hydroxide is actually produced in the lungs. Prolonged breathing can cause suffocation.

The human eye is a complex organ made up of nerves, veins, and cells. The front of the human eye is covered by membranes, which resist exposure to dust and dirt. However, these cannot keep out anhydrous ammonia, because the entire eye is about 80% water. A shot of ammonia under pressure can cause extensive, almost immediate, damage to the eye. The ammonia extracts the fluid and destroys eye cells and tissue in minutes. If you get a shot of anhydrous ammonia in your eye, the first few seconds are crucial. Immediately flush the eyes with copious amounts of water. If wearing contact lenses, remove them. Your eyes will fight to stay closed because of the extreme pain, but they must be held open so the water can flush out the ammonia. Continue to flush the eyes for at least 15 minutes. Get professional medical help as soon as possible to prevent permanent damage. If water is not available, fruit juice or cool coffee can be used to flush the eyes. Remove contaminated clothing and thoroughly wash the skin.

Clothing frozen to skin by liquid ammonia can be loosened with liberal application of water. Wet clothing and body thoroughly, and then remove the clothing. Leave burns exposed to the air and do not cover with clothing or dressings. Immediately after first-aid treatment with water, get the burn victim to a physician. Do not apply salves, ointments, or oils, as these cause ammonia to burn deeper. Let a physician determine the proper medical treatment. Remove the victim to an area free from fumes if an accident occurs. If the patient is overcome by ammonia fumes and stops breathing, get him or her to fresh air and give artificial respiration. The patient should be placed in a reclining position with head and shoulders elevated. Basic life support should be administered if needed. Oxygen has been found useful in treating victims who have inhaled ammonia fumes. Administer 100% oxygen at atmospheric pressure.

Any person who has been burned or overcome by ammonia should be placed under a physician's care as soon as possible. Begin irrigation with water immediately. The rescuer should use fresh water, if possible. If the incident is a farm accident, there is a requirement for water tanks for irrigation of the eyes and body on the anhydrous ammonia tank. Open water in the vicinity of an anhydrous ammonia leak may have picked up enough NH_3 to be a caustic aqua-ammonia solution. This could aggravate the damage if used in the eyes or for washing burns. The victim should be kept warm, especially to minimize shock. If the nose and throat are affected, irrigate them with water continuously for at least 15 minutes. Take care not to cause the victim to choke. If the patient can swallow, encourage drinking lots of some type of citrus drink, such as lemonade or fruit juice. The acidity will counteract some of the effect of the anhydrous ammonia.

Ammonia and propane are often shipped in the same type of container. In the late winter and early spring, the containers are purged of the propane and used for ammonia. In the late summer and early fall, the containers are purged of ammonia and used for propane. One of the primary uses of ammonia is as a fertilizer. The tanks used for farm application of ammonia can also be used for propane in the winter for heating and grain-drying purposes. Ammonia will attack copper, zinc, and their alloys of brass and bronze. Propane valves, fittings, and piping are often made of these materials. If the valves and other fixtures are not changed to steel before being used for ammonia, or if the ammonia is not completely purged from the containers before propane use, serious accidents can occur. Ammonia can damage fittings and cause leaks of the highly flammable propane, which may result in fires. There have also been situations in which propane has been transported in ammonia containers and the placards for ammonia have been left on the containers. This can present a serious hazard for emergency responders.

Heat exchanger underneath MC 338 allows the cryogenic liquid to be offloaded as a gas.

Carbon dioxide, **CO_2**, is colorless and odorless. It can also be a solid (dry ice), which will undergo sublimation and turn back into carbon dioxide gas. Carbon dioxide may also be encountered as a cryogenic liquid. CO_2 is miscible with water, and is nonflammable and nontoxic, but can be an asphyxiant and displace oxygen. In 1993, two workers were killed aboard a cargo ship when a carbon dioxide fire-extinguishing system discharged. The oxygen in the area was displaced by the carbon dioxide, and the men were asphyxiated. Carbon dioxide has a vapor density of 1.53, which is heavier than air. It has a four-digit UN number of 2187 as a cryogenic and 1013 as a compressed gas. The NFPA 704 designation is health 3, flammability 0, and reactivity 0. It is used primarily in carbonated beverages and fire-extinguishing systems.

INCIDENTS

According to NFPA studies on ammonia incidents between 1929 and 1969, there were 36 incidents in which released ammonia gas was ignited; 28 resulted in a combustion explosion. All of the explosions occurred indoors.

On January 18, 2002, a Canadian Pacific freight train derailed outside Minot, North Dakota. Five of the cars carried anhydrous ammonia. Leaking ammonia killed one person and sent dozens of others to hospitals for treatment. Ten of those seeking treatment were admitted to the hospital. Some local residents were evacuated, while others were asked to shelter in place. Civil defense sirens and local radio and television stations alerted residents.

In Shreveport, Louisiana, one firefighter was killed and one badly burned in a fire involving anhydrous ammonia. A leak developed inside a cold-storage plant.

The firefighters donned Level A chemical protective clothing and went inside to try to stop the leak. Something caused a spark, and the anhydrous ammonia caught fire.

In Verdigris, Oklahoma, a tank car was being filled with anhydrous ammonia. It was unknown that there was a weakened place on the tank car, which gave way from the pressure of the ammonia and resulted in a BLEVE. There was no fire, just a vapor cloud that traveled downwind, defoliating trees and turning other vegetation brown. One worker who was filling the tank car was killed in the incident.

In Delaware County, Pennsylvania, ammonia was being removed from an abandoned cold-storage facility when a leak occurred. Firefighters were exposed to ammonia and complained of irritation and burning of the face and other exposed skin surfaces. Several were transported to local hospitals for treatment after going through decontamination at the scene.

In Ortanna, Pennsylvania, two workers were killed while doing routine maintenance on an ammonia system in a cold-storage building used for storing fruit. Both of the workers were members of the local volunteer fire company, and one was the assistant chief. Several firefighters were injured by the ammonia vapors while trying to rescue the workers.

Anhydrous ammonia is a common, but dangerous, material when not properly handled. Firefighters should wear full Level A chemical protective clothing when exposed to ammonia vapors. Be aware, however, that ammonia is also flammable, and Level A protective clothing provides no thermal protection. When responding to incidents involving ammonia, use caution. Firefighter turnouts do not provide adequate protection from ammonia vapors. If there are victims exposed to ammonia for any length of time, the chance of rescue is slim. Do not expose unprotected rescue personnel to ammonia vapor. Victims can be decontaminated using emergency decon procedures by first responders. This will reduce the impact of NH_3 on victims.

POISON GASES

Subclass 2.3 materials are an inhalation hazard, and some may also be absorbed through the skin. Firefighters are exposed to certain types of toxic materials whenever they fight a fire. These toxic materials are by-products of the combustion process. Toxic fire gases include carbon monoxide (CO), hydrogen chloride (HCl), hydrogen cyanide (HCN), sulfur dioxide (SO_2), nitrogen dioxide (NO_2), ammonia (NH_3), hydrogen sulfide (H_2S), and phosgene ($COCl_2$). These toxic materials kill thousands of persons each year. There are, however, poisons in every community that are potentially more dangerous, toxic materials capable of killing tens of thousands of people in a matter of minutes.

The DOT definition of poison gas is "a material that is a gas at 68°F or less at 14.7 psi and is so toxic to humans as to pose a hazard to health during transportation, or in the absence of adequate data on human toxicity, is presumed to be toxic to humans because when tested on laboratory animals it has an LC_{50} value of not more than 5000 ml/m^3." These materials are considered so toxic that, when transported, the vehicle must be placarded regardless of the quantity. The potential exists for 2.3 materials to affect large populations by creating toxic gas clouds. In order to understand the

Ton containers of chlorine.

toxic effects of poisons, it is necessary to know some toxicological terminology. One thing to remember about toxicological data is that little human data are available. The data available are the result of tests on laboratory animals. Toxicity for humans is really nothing more than an educated guess. Most of the terms mentioned here are applied to workplace exposures. Acceptable exposures in many cases are 8 hours a day, 40 hours a week. Concentrations encountered on the scene of an incident will be much higher than any ordinary workplace exposure, but for a shorter period of time.

TLVTWA is the threshold limit value–time-weighted average concentration for a normal 8-hour workday and a 40-hour workweek, to which nearly all workers may be repeatedly exposed, day after day, without adverse effect.

IDLH is immediately dangerous to life and health. IDLH determines the highest concentration that a person can be exposed to for a maximum of 30 min and still escape without any irreversible health effects.

LC_{50} is the lethal concentration by inhalation for 50% of the laboratory animals tested.

STEL is the short-term exposure limit, defined as a 15-min TWA exposure, which should not be exceeded at any time during a workday, even if the 8-hour TWA is within the TLV-TWA. Exposures above the TLV-TWA up to the STEL should not be longer than 15 min and should not occur more than four times a day. There should be at least 60 min between successive exposures in this range.

Concentration is the amount of one substance found in a given volume of another substance. Depending on the materials involved, there are many different ways of expressing concentration. Two of the most common ways are ppm (parts per million) and milligrams per kilogram (mg/kg). Toxicology will be discussed further in Chapter 8. Examples of some 2.3 poison gases are fluorine, chlorine, carbon monoxide, hydrogen sulfide, phosgene, phosphine, and chloropicrin-methyl bromide mixture.

Carbon monoxide, CO, is an odorless, colorless, tasteless gas that is toxic by inhalation, with a TLV of 50 ppm. A 1% concentration is lethal to adults if inhaled for 1 min. Carbon monoxide binds to the blood hemoglobin 220 times tighter than oxygen. The more carbon monoxide that binds to the blood, the less oxygen can be carried. Carbon monoxide will prevent oxygen from being taken into the blood, thus causing a type of chemical asphyxiation. In addition to its primary hazard of toxicity, it is also highly flammable and is a dangerous fire and explosion risk. The boiling point is –313°F, and the ignition temperature is 1292°F. Carbon monoxide has a flammable range from 12 to 75%. Its vapor density is 0.967, which is slightly less than that of air. It is slightly soluble in water. The four-digit UN identification number is 1016 as a compressed gas and 9202 as a cryogenic liquid. The NFPA 704 designation for carbon monoxide is health 3, flammability 4, and reactivity 0. The primary uses are in the synthesis of organic compounds, such as aldehydes, acrylates, alcohols, and in metallurgy.

Hydrogen sulfide, H_2S, is a colorless gas with an odor like rotten egg. It is the only common material that can halt respiration. It is toxic by inhalation with a TLV of 10 ppm. The minimal perceptible odor is found at concentrations of 0.13 ppm; at 4.60 ppm, the odor is moderate; at 10 ppm, tearing begins; at 27 ppm, there is a strong, unpleasant, but not intolerable, odor. When TLV reaches 100 ppm, coughing begins, eye irritation occurs, and loss of sense of smell begins after 2 to 15 min. Marked eye irritation occurs at 200 to 300 ppm, and respiratory irritation after 1 hour of exposure. Loss of consciousness takes place at 500 to 700 ppm, with the possibility of death in 30 min to 1 hour. Concentrations of 700 to 1000 ppm cause unconsciousness, cessation of respiration, and death. Instant unconsciousness occurs at 1000 to 2000 ppm, with cessation of respiration and death in a few minutes. Death may occur even if the victim is removed to fresh air at once. Hydrogen sulfide is highly flammable and a dangerous fire and explosion risk. The boiling point is –76°F. The flammable range is 4.3 to 46%. The ignition temperature is 500°F. Its vapor density is 1.189, which is heavier than air. Hydrogen sulfide is soluble in water. It is incompatible with oxidizing gases and fuming nitric acid. The four-digit UN identification number is 1053. The NFPA 704 designation for hydrogen sulfide is health 4, flammability 4, and reactivity 0. It is used in the purification of hydrochloric and sulfuric acids and is a source of hydrogen and sulfur. Hydrogen sulfide is usually shipped in steel pressure cylinders.

Fluorine, F_2, is a nonmetallic elemental gas from the halogens, which is family seven on the Periodic Table. It is the most electronegative and powerful oxidizing agent known. It reacts vigorously with most oxidizible substances at room temperature, frequently causing combustion. Fluoride compounds form with all elements except helium, neon, and argon. Fluorine is a pale yellow gas with a pungent odor. It is nonflammable, but will support combustion because it is an oxidizer. The boiling point is –307°F. The vapor density is 1.31, which is heavier than air. Fluorine is water-reactive. The primary hazard is toxicity; fluorine is toxic by inhalation and extremely irritating to tissues. The TLV is 1 ppm, and the IDLH is 25 ppm in air. Fluorine is incompatible with and should be isolated from everything! The four-digit

Pressure railcar of chlorine.

UN identification number is 1045. The NFPA 704 designation is health 4, flammability 0, and reactivity 4. The white section at the bottom of the diamond contains a W with a slash through it, indicating water reactivity. Because of the strong reactivity with other materials, it is shipped in special steel containers. The primary uses are in the production of metallic and other fluorides, fluorocarbons, fluoridation of drinking water, and in toothpaste.

Boron trifluoride, **BF_3**, is a colorless gas with a vapor density of 2.34, which is heavier than air. It is water-soluble and does not support combustion. It is also water-reactive, toxic by inhalation, and corrosive to skin and tissue. The TLV is 1 ppm, and the IDLH is 100 ppm in air. The boiling point is –148°F. The four-digit UN identification number is 1008. The NFPA 704 designation is health 4, flammability 0, and reactivity 1. The primary uses are as a catalyst in organic synthesis, in instruments for measuring neutron intensity, in soldering fluxes, and in gas brazing.

Dichlorosilane, **H_2SiCl_2**, is a pyrophoric, water-reactive gas. It is flammable, with a wide flammable range of 4.1 to 99% in air. The boiling point is 47°F, and the flash point is –35°F. The ignition temperature is 136°F. The vapor density is 3.48, which is heavier than air. It is immiscible in water and highly water-reactive. Contact with water releases hydrogen chloride gas. It is toxic by inhalation and skin absorption. Hydrogen chloride causes severe eye and skin burns and is irritating to the skin, eyes, and respiratory system. The four-digit UN identification number is 2189. The NFPA 704 designation is health 4, flammability 4, and reactivity 2. The white area at the bottom of the diamond contains a W with a slash through it, indicating water reactivity. It is shipped in carbon steel cylinders.

Phosgene, **$COCl_2$**, is a clear to colorless gas or fuming liquid, with a strong stifling or musty hay-type odor. It is slightly soluble in water. The vapor density is 3.41, which is heavier than air. Phosgene is a strong irritant to the eyes, is highly toxic by inhalation, and may be fatal if inhaled. The TLV is 0.1 ppm, and the IDLH is 2 ppm in air. The boiling point is 46°F, and it is noncombustible. When carbon tetrachloride comes in contact with a hot surface, phosgene gas is evolved, which is one of the main reasons that carbon tetrachloride fire extinguishers are no longer approved. The four-digit UN identification number is 1076. The NFPA 704 designation is health 4, flammability 0, and reactivity 1. It is shipped in steel cylinders, special tank cars, and tank trucks. The primary uses are in organic synthesis, including isocyanates, polyurethane, and polycarbonate resins; in carbamate, organic carbonates, and chloroformates pesticides; and in herbicides. The structure for phosgene follows.

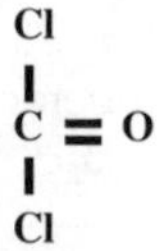

Cl
|
C = O
|
Cl

Phosgene
$COCl_2$

Figure 4.25

Phosphine, **PH_3**, a nonmetallic compound, is the gas evolved when binary phosphide salts come in contact with water. It is colorless, with a disagreeable, garlic-like or decaying fish odor. It is toxic by inhalation and is a strong irritant. It has a TLV of 0.3 ppm and an IDLH of 200 ppm in air. It is also highly flammable (pyrophoric) and will spontaneously ignite in air. The flammable range is extremely wide, at 1.6 to 98% in air. It is slightly soluble in cold water. The vapor density is 1.17, which is heavier than air. The four-digit UN identification number is 2199. The NFPA 704 designation is health 4, flammability 4, and reactivity 2. It is shipped in steel cylinders. The primary uses are in organic compounds, as a polymerization initiator, and as a synthetic dye. The structure is shown below.

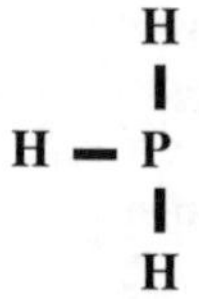

H
|
H — P
|
H

Phosphine
PH_3

Figure 4.26

Diborane, B_2H_6, is a colorless gas with a nauseating, sweet odor. It decomposes in water and is highly reactive with oxidizing materials, including chlorine. It is toxic by inhalation and a strong irritant, with a TLV of 0.1 ppm in air. The IDLH is 40 ppm. In addition to being toxic, diborane is also a dangerous fire risk. It is pyrophoric and will ignite upon exposure to air. The boiling point is −135°F and the flammable range is 0.8 to 88% in air. The ignition temperature is 100° to 140°F, and the flash point is 130°F. Diborane will react violently with halogenated fire-extinguishing agents, such as the halons. The four-digit UN identification number is 1911. The NFPA 704 designation is health 4, flammability 4, and reactivity 3. The white section of the diamond has a W with a slash through it, indicating water reactivity. The primary uses are as a polymerization catalyst, as fuel for air-breathing

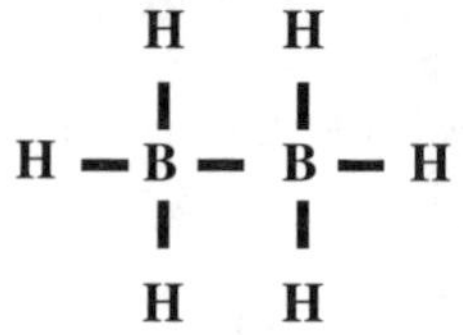

```
     H    H
     |    |
H — B — B — H
     |    |
     H    H
```

Diborane
B_2H_6

Figure 4.27

engines and rockets, as a reducing agent, and as a doping agent for p-type semiconductors. The structure is shown in Figure 4.27.

Poison gases may be encountered as gases, liquefied gases, or cryogenics. The placard will indicate poison gas; it will not tell you the material has been liquefied or turned into a cryogenic liquid. The container type will help determine the physical state of the materials. If there is a four-digit identification number on the placard, it may provide physical state information when looked up in the *Emergency Response Guide Book.*

Chlorine, **Cl_2**, an elemental gas, is one of the most common poison gases transported and stored. Chlorine does not occur freely in nature. Chlorine is derived from the minerals halite (rock salt), sylvite, and carnallite, and is found as a chloride ion in seawater. Chlorine is nonflammable. Its vapor density is about 2.45, which makes it heavier than air. Chlorine is toxic, with a TLV of 0.5 ppm in air, and is also a strong oxidizer. Chlorine will behave much the same way as oxygen in accelerating combustion during a fire. It is also corrosive. The specific gravity of chlorine is 1.56, which makes it heavier than water, and it is only slightly soluble in cold water. Chlorine is incompatible with ammonia, petroleum gases, acetylene, butane, butadiene, hydrogen, sodium, benzene, and finely divided metals. The four-digit UN identification number is 1017. The NFPA 704 designation for chlorine is health 3, flammability 0, and reactivity 0. It may be encountered in 150-lb cylinders, 1-ton containers, and tank car quantities. Chlorine is nonflammable, but a BLEVE is possible if the container is exposed to flame, because it is a liquefied gas. However, the likelihood of container failure is low. The NFPA has never recorded an incident of a BLEVE involving chlorine.

Because chlorine is so common, its hazards are sometimes taken for granted. There was a time when firefighters handled chlorine leaks with turnouts and SCBAs (self-contained breathing apparatus); that is no longer an acceptable practice. Poison gases pose a large threat not only to the public, but also to emergency responders. To properly protect responders, full Level A chemical protective clothing and SCBAs must be worn for protection. Chlorine is used as a swimming pool chlorinator, a water-treatment chemical, and for many other industrial uses. Chlorine is the fourth most-released chemical from fixed facilities on the EPA listing, with 2099 accidents resulting in the release of over 84 million lbs of chlorine. Chlorine is also consistently among the seven highest-volume hazardous materials shipped on the railroads,

Level A chemical protection is necessary when dealing with poison gases.

accounting for over 49,000 carloads in 1994. Chlorine is not listed among the top 50 transportation incidents; however, it is number 16 in terms of total injuries. Chlorine is not often involved in transportation incidents, but when it is, it causes serious injuries.

Ethylene oxide, $\mathbf{CH_2OCH_2}$, is a colorless gas at room temperature. It is miscible with water and has a specific gravity of 0.9, which is lighter than water. It is an irritant to the skin and eyes, with a TLV of 1 ppm in air. Ethylene oxide is a suspected human carcinogen. In addition to toxicity, it is highly flammable, with a wide flammable range of 3 to 100% in air. The flash point is –20°F, and the boiling point is 120°F. The ignition temperature is 1058°F. The vapor density is 1.5, which is heavier than air. The four-digit UN identification number is 1040. The NFPA 704 designation is health 3, flammability 4, and reactivity 3. The primary uses are in the manufacture of ethylene glycol and acrylonitrile, as a fumigant, and as a rocket propellant. The structure is shown below.

```
    H   H
    |   |
H — C — C — H
     \ /
      O
```

Ethylene Oxide
CH_2OCH_2

Figure 4.28

INCIDENTS

In Atlanta, Georgia, a small pressurized cylinder fell from a truck in the garage of the Hilton Hotel. The resulting leak of chlorine, a 2.3 poison gas, sent 33 people to the hospital, including six firefighters and four police officers.

Eight people were killed and 88 others were injured as the result of leaking chlorine from a railroad tank car during a derailment in Youngstown, Florida. The liquid chlorine car ruptured, releasing a toxic cloud of chlorine. Chlorine is 2.5 times heavier than air, stays close to the ground, and has an expansion ratio of 460:1, which means that 1 gal of liquid chlorine will vaporize into 460 gal of chlorine gas. The chlorine gas settled into a low area on a nearby highway. As cars passed through the chlorine cloud, they stalled. Drivers were overcome by the chlorine and eight of them died. A few breaths of chlorine at 1000 ppm concentration can be fatal. The concentrations at the accident scene were estimated by environmental personnel to be 10,000 to 100,000 ppm.

SUMMARY

Particular attention must be paid to compressed gases in emergency response situations. Compressed gases present responders with multiple hazards, including poisons, flammables, oxidizers, cryogenics, and the hazard of the pressure in the container. If the container fails or opens up, it can become a projectile or throw pieces of the container over a mile from the incident scene. Learn to recognize pressure containers and be cautious when there is flame impingement on a pressure container.

REVIEW QUESTIONS CHAPTER 4

1. Which of the following is true of cryogenic liquids?
   A. They have wide flammable ranges
   B. They are extremely cold materials
   C. They have large expansion ratios
   D. All of the above
2. Hazard Class 2 is composed of compressed gases that may have which of the following hazards?
   A. Flammability
   B. Elevated temperature
   C. Sublimation
   D. Volatility
3. The flammable range of a 2.1 compressed gas occurs at which of the following locations?
   A. Below the lower limit
   B. Within the first 12%
   C. Above the upper limit
   D. Between the upper and lower limits

4. A compressed gas with a vapor density greater then 1 will have what relationship with air?
   A. Heavier
   B. Equal weight
   C. Lighter
   D. Will mix with air
5. Which of the following hydrocarbon compounds has at least one double bond?
   A. Aromatic
   B. Alkyne
   C. Alkane
   D. Alkene
6. Which of the following hydrocarbon compounds is considered saturated?
   A. Alkene
   B. Alkane
   C. Alkyne
   D. Aromatic
7. Acetylene belongs to which of the following hydrocarbon families?
   A. Alkyne
   B. Aromatic
   C. Alkene
   D. Alkane
8. Provide names and formulae for the compounds represented by the following structures:

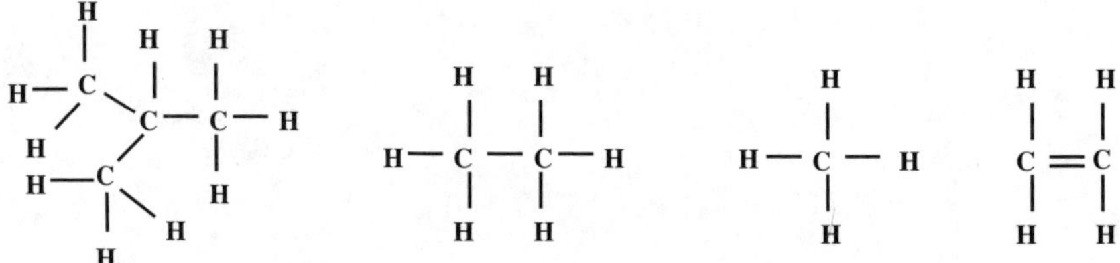

Figure 4.29

9. Indicate whether the following compounds are alkanes, alkenes, or alkynes.

   CH_4 C_2H_2 C_3H_6 C_2H_6 C_3H_4

10. Match the following with the appropriate hydrocarbon family.
   a. Alkane Saturated
   b. Alkene Unsaturated
   c. Alkyne

CHAPTER 5

Flammable Liquids

Class 3 materials are liquids that are flammable or combustible. Flammable liquids cause more fires than flammable gases because they are more abundant. Vapors of many flammable liquids are heavier than air. Most flammable liquids have a specific gravity of less than 1, so they float in water. They may also be incompatible with ammonium nitrate, chromic acid, hydrogen peroxide, sodium peroxide, nitric acid, and the halogens. According to the DOT, flammable liquids "have a flash point of not more than 141°F, or [are] any material in a liquid phase with a flash point at or above 100°F, that is intentionally heated and offered for transportation or transported at or above its flash point in bulk packaging." There is an exception to this definition that involves flammable liquids with a flash point between 100° and 140°F. Those liquids may be reclassified as combustible liquids, and at the option of the shipper, may be placarded flammable, combustible, or fuel oil. Even though the DOT wanted all liquids up to 140°F to be placarded flammable, this exception was made because of public comments, particularly from the fuel oil industry. Combustible liquids are defined as "materials that do not meet the definition of any other hazard class specified in the DOT flammable liquid regulations and have flash points above 141°F and below 200°F."

The NFPA uses a classification system for flammable and combustible liquids in fixed storage facilities (see Figure 5.1). This system is part of the consensus standard NFPA 30, the Flammable and Combustible Liquids Code. The NFPA system further divides the flammable and combustible liquid categories into subdivisions based upon the flash points and boiling points of the liquids. NFPA's classification system does not apply to transportation of hazardous materials, since DOT regulations supersede NFPA 30. Examples of liquids in the various classification categories are listed in Figure 5.2.

As previously mentioned, all of the DOT hazard classes identify only the most severe hazard of that material. All of the classes have hidden hazards that are both chemical and physical in nature. Flammability is not the only hazard associated with Class 3 flammable liquids. They may also be poisonous or corrosive. For general purposes, there are no UN/DOT subclasses of flammable liquids. Emergency

Flammable liquids must be at their flash point before combustion can occur if an ignition source is present and the material is within its flammable range.

NFPA 30
Flammable and Combustible Liquid Classification

Flammable
Class IA = FP < 73°F - BP < 100°F
Class IB = FP < 73°F - BP > 100°F
Class IC = FP > 73°F - BP < 100°F

Combustible
Class II = FP > 100°F < 140°F
Class IIIA = FP > 140°F < 200°F
Class IIIB = FP > 200°F

Figure 5.1

NFPA Flammable and Combustible Liquids

Class I	Class II	Class III
Acetone	Acetic Acid	Benzyl Chloride
Benzene	Fuel Oil	Corn Oil
Carbon Disulfide	Kerosene	Linseed Oil
Gasoline	Decane	Nitro Benzene
Methanol	Pentanol	Parathion

Figure 5.2

responders should realize that all materials with red placards will burn under certain conditions. Appropriate precautions should be taken when dealing with flammable liquids. The dividing line for flammable and combustible liquids is 140°F in the DOT regulations, and 100°F in the NFPA standard. Those liquids with flash points below 100° and 140°F, respectively, are considered flammable; those above are considered combustible. The problem with classifying flammable and combustible liquids in emergency response situations is ambient temperature and radiant heat.

Ambient temperatures near or above 100°F are common in many parts of the country. The radiant heat from the sun or an exposure fire can reach temperatures well above 100°F. Many of the liquids classified as combustible have flash points at or near 100°F. Surfaces such as roadways may have temperatures well above the flash points of combustible liquids. When a combustible liquid is spilled on a surface, it may be heated above its flash point and the combustible liquid may act like a flammable liquid.

Because of the uncertainty of the potential flammability of combustible liquids, they will be referred to as flammable liquids throughout this book. It is highly recommended that they be treated the same way on the incident scene for the purpose of responder and public safety. Flash point will be discussed in detail later in this chapter. It is important, however, to note at this point that the flash point temperature is the most critical factor in determining if a flammable liquid will burn. The most important precautions are to control ignition sources at the incident scene and keep personnel from contacting the liquid without proper protective clothing.

According to DOT and EPA statistics, flammable liquids are involved in over 52% of all hazardous materials incidents. This should not be surprising since flammable liquids are used as motor fuels for highway vehicles, railroad locomotives, marine vessels, and aircraft. Additionally, many flammable liquids are used to heat homes and businesses. Effective handling of flammable liquids at an incident scene requires that emergency responders have a basic understanding of the physical characteristics of flammable liquids.

EFFECTS OF TEMPERATURE ON FLAMMABLE LIQUIDS

Many of the physical characteristics of flammable liquids involve temperature. It is important to understand that there are different temperature scales listed in reference books; make sure which scale is used when materials are researched. There is a big difference between the temperatures of the Fahrenheit and Centigrade (Celsius) scales. The Fahrenheit scale is familiar to most emergency responders because it is the temperature used most commonly in the United States. The Centigrade scale is used predominantly throughout the rest of the world. It is also used within the scientific and technical community of the United States. Reference books used by emergency responders to obtain information on hazardous materials may also use the Centigrade scale. Shown in the following examples are temperature-conversion formulas used to convert Centigrade to Fahrenheit and Fahrenheit to Centigrade (see Figure 5.3).

Another temperature scale that may be encountered on a less frequent basis is the Kelvin scale, also known as absolute temperature. This scale is used principally in theoretical physics and chemistry and in some engineering calculations. Absolute temperatures are expressed either in degrees Kelvin or in degrees Rankine, corresponding respectively to the Centigrade and Fahrenheit scales. Temperatures in

Temperature Conversion Formulas

$$\text{From C° to F°: } F° = \frac{9 \times C°}{5} + 32$$

Example:

$$\text{C Temp} = 40° = \frac{9 \times 40}{5} + 32 = \frac{360}{5} = 72 + 32 = 104°F$$

$$\text{From F° to C°: } C° = \frac{5(F° - 32)}{9}$$

Example:

$$\text{F Temp} = 104° = \frac{5(104 - 32)}{9} = \frac{5 \times 72}{9} = \frac{360}{9} = 40°C$$

Figure 5.3

Kelvin are obtained by adding 273 degrees to the Centigrade temperature (if above 0°C), or subtracting the Centigrade temperature from 273 (if below 0°C). Degrees Rankine are obtained by subtracting 460 from the Fahrenheit temperature. Absolute zero is the temperature at which the volume of a perfect gas theoretically becomes zero and all thermal motion ceases, which occurs at –273.13°C, or –459.4°F.

BOILING POINT

The boiling point of a flammable liquid is the first physical characteristic that will be discussed. It is a physical characteristic that is affected by the temperature of the liquid and atmospheric pressure. Boiling point is defined as "the temperature at which the vapor pressure of a liquid equals the atmospheric pressure of the air." Atmospheric pressure is 14.7 psi at sea level. Liquids naturally want to become gases; it is atmospheric pressure that keeps a liquid from becoming a gas at normal temperatures and pressures. Atmospheric pressure decreases as altitude increases. The higher the altitude, the lower the boiling point of any given material (see Figure 5.4). For example, water boils at 212°F at sea level. In Denver, Colorado, the altitude is 1 mile (5280 ft) above sea level; water boils at approximately 203°F. At Pikes Peak, Colorado, the altitude is more than 14,000 feet above sea level, and

Boiling Point of Water and Atmospheric Pressure (psi) vs. Altitude

Altitude	Atmospheric Pressure	Boiling Point
Sea level	14.7 psi	212°F
3300 ft	13.03 psi	203°F
5280 (1 mile)	12.26 psi	201°F
10,000 ft	10.17 psi	194°F
13,000 ft	9.00 psi	188°F

(Values are approximate)

Figure 5.4

water boils at approximately 186°F. Atmospheric pressure decreases as altitude increases because the air is thinner at higher altitudes.

Atmospheric pressure is always pushing down on the surface of a liquid in an open tank or in a spill. With atmospheric pressure pushing down on a liquid, there is not much vapor moving away from the surface of the liquid. There is a direct relationship between the boiling point of a liquid and the amount of vapor present in a spill or in a container. Inside a container, increased vapor will increase the pressure in the tank. In an open spill, the vapor content will be greater. If a flammable liquid is above its boiling point, more vapor will be produced. The more heat that is applied to the container or spill, the more vapor that is produced, the higher the pressure in the container or the further the vapor will travel away from a spill.

If a flammable liquid is in a closed container, the vapor pressure will increase inside the container as the temperature of the liquid increases. This increase in temperature can come from many different sources. Increases in ambient temperature, radiant heat from the sun, or a nearby fire can increase the vapor pressure in a container. As the pressure increases in a container, it will reach the setting on the pressure-relief valve and the relief valve will function. If this pressure increase occurs in a container that does not have a relief valve, the container may rupture. Rupture may also occur in a container with a relief valve if the pressure rises too fast for the relief valve to vent the material into the air, or if the relief valve is not working properly. In either case, the rupture may be violent, with a fireball and flying pieces of tank that can travel over a mile from the blast site. This phenomenon is referred to as a boiling liquid expanding vapor explosion (BLEVE).

Factors Affecting Boiling Point

Molecular Weight

A number of outside factors can help determine the relative boiling point of a flammable liquid. They include molecular weight, polarity, and branching. The first consideration in determining boiling point is the molecular weight of a compound. Each element in a compound has an atomic weight, listed on the Periodic Table. Most flammable liquids are made up of hydrogen, carbon, and a few other elements. They can be organized into families called hydrocarbons and hydrocarbon derivatives.

Hydrocarbons are made up of just carbon and hydrogen. Carbon has an atomic weight of 12, so each carbon atom in a compound weighs 12 AMU. Hydrogen has an atomic weight of 1, so each hydrogen atom in a compound will weigh 1 AMU. Methane is the smallest hydrocarbon, made up of one carbon atom and four hydrogen atoms. The hydrocarbon compound methane has a weight of 16 AMU. In the compound butane, there are four carbon atoms, each weighing 12 AMU (which equals 48) and 10 hydrogen atoms, each weighing 1 AMU (which equals 10). Butane therefore weighs 58 AMU. The molecular weight of a compound is the sum of all the weights of all the atoms. The more carbons in a hydrocarbon compound, the heavier it will be. The heavier it is, the more energy it will take to get the liquid to boil and overcome atmospheric pressure. Therefore, the heavier a compound is, the

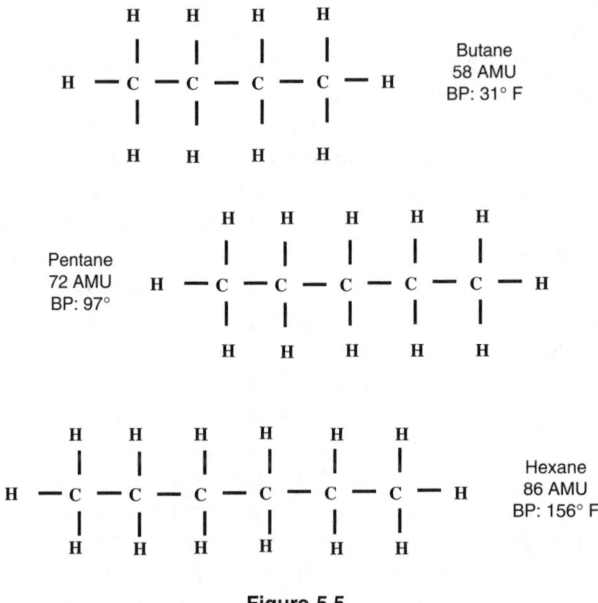

Figure 5.5

higher the boiling point it will have. In Figure 5.5 there are three hydrocarbon compounds. Compare the molecular weights of each and look at their boiling points. You can readily see the relationship between heavy compounds and high boiling points.

Polarity

Hydrocarbon derivatives are compounds with other elements in addition to hydrogen and carbon. Weight will still determine boiling points when comparing hydrocarbon derivatives within the same family. However, when comparing different families, the concept of polarity has to be considered with some of the compounds. Polarity is the second factor that affects boiling point. There is a rule in chemistry that says, "Like materials dissolve like materials." Materials that are polar will mix together. Another term used when materials mix together is miscibility. If a material is miscible, it will mix with another. If a material is immiscible, it will not mix with another.

All polar compounds are alike in terms of being polar. Therefore, polar compounds are soluble in polar compounds. One of the main reasons polarity is discussed in terms of emergency response is because of the foam used for fighting flammable liquid fires. Two general types of foam are used to extinguish flammable liquid fires: hydrocarbon foam and polar-solvent foam, which is sometimes referred to as alcohol-type foam. The reason that different types of foam are necessary for flammable liquid fires is polarity.

Water is a polar compound. Because "like dissolves like," water is miscible with most polar solvents. The main ingredient of firefighting foam is water, along with foam concentrate and air. If regular foam is put on a polar-solvent liquid, such as

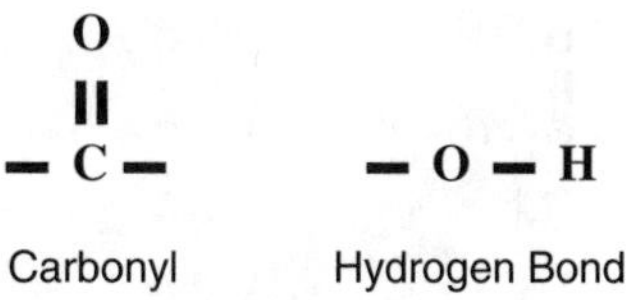

```
  O
  ||
— C —        — O — H
```

Carbonyl Hydrogen Bond

Figure 5.6

alcohol, the water will be removed from the foam by the polar solvent and the foam blanket will break down. To effectively extinguish fires with polar solvents, it is necessary to use polar-solvent or alcohol-type foam.

Polarity is presented here to discuss the types of flammable liquids that are polar solvents and the effect that polarity has on them. Polar solvents, such as alcohols, aldehydes, organic acids, and ketones, require special foam to extinguish fires. Choosing the right type of foam is part of the process of effectively managing a flammable liquid incident. Polar liquids tend to have higher boiling points than nonpolar liquids. Polarity is said to have the effect of raising the boiling point of a liquid. There are two types of structure that represent polarity in hydrocarbon-derivative compounds: the carbonyl structure and hydrogen bonding (see Figure 5.6).

A carbonyl contains a carbon-to-oxygen double bond. Most double bonds are reactive; however, with the carbonyl family, except for aldehyde, the double bond is protected by hydrocarbon radicals on either side of the carbonyl. This prevents oxygen from getting to the bond and breaking it.

Double bonds are stable within the carbonyl families, except for aldehydes. The carbonyl structure is shown in the following illustration (see Figure 5.7), first by itself, and then in the hydrocarbon functional groups ketone, aldehyde, organic acid, and ester. All four compounds are polar because of the carbonyl, and the organic acid also has a hydrogen bond. The amount of polarity is generally the same between the ketone, ester, and aldehyde. The polarity of an acid is higher because of the hydrogen bond and the carbonyl in the same compound, like a double dose of polarity.

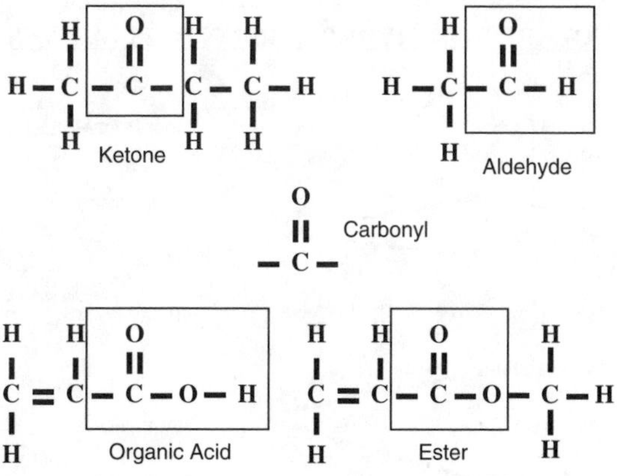

```
    H   O   H   H             H   O
    |   ||  |   |             |   ||
H — C — C — C — C — H     H — C — C — H
    |       |   |             |
    H       H   H             H
      Ketone                    Aldehyde

              O
              ||   Carbonyl
            — C —

H   H   O                 H   H   O       H
|   |   ||                |   |   ||      |
C = C — C — O — H         C = C — C — O — C — H
|                         |               |
H    Organic Acid         H     Ester     H
```

Figure 5.7

```
    H   H   H              O                   H                   H   O   H
    |   |   |              ||                  |                   |   ||  |
H - C - C - C - H      H - C - O - H       H - C - O - H       H - C - C - C - H
    |   |   |                                  |                   |       |
    H   H   H                                  H                   H       H
```

Propane, 44 AMU, BP: -40° F, Non-Polar

Formic Acid, 46 AMU, BP: 170° F, Super Duper Polar

Ethyl Alcohol, 46 AMU, BP: 170° F, Super Polar

Dimethyl Ketone, 56 AMU, BP: 133° F, Polar

Figure 5.8

When oxygen and hydrogen covalently bond, the bond is polar. Water has a hydrogen-oxygen bond that gives water polarity. Water is a liquid between 32° and 212°F and has a molecular weight of 18 AMU. The molecular weight of an average air molecule is 29 AMU. Therefore, the water molecule is lighter than air! Water should be a gas at normal temperatures and pressures, but it is a liquid due to polarity. Even though water has a molecular weight of 18 AMU, it has a boiling point of 212°F! If water was not polar, there would be no life on earth as we know it.

The hydrogen bond is also a polar bond. There are two hydrocarbon-derivative families presented in this book that have hydrogen bonding: alcohols and organic acids. The organic acid also has a carbonyl bond. This double polarity makes organic acid the most polar material among the hydrocarbon derivatives. Many materials exhibit some degree of polarity. However, only the polarity of carbonyl and hydroxyl groups of hydrocarbon derivatives will be discussed here. All other materials mentioned will be considered nonpolar. The hydrocarbon and hydrocarbon-derivative compounds are listed in Figure 5.9. They are ranked in order of descending degree of polarity. Organic acids are the most polar of all the materials listed and will be referred to as "super-duper polar." Alcohols are the second most-polar compounds and they will be referred to as "super polar." The remaining carbonyls are polar, with the exception of amines, which are "slightly polar." All other compounds discussed will be considered nonpolar. In the structures in Figure 5.8, compare the polarity of the compounds listed and notice the effect that polarity has on boiling point.

POLARITY OF HYDROCARBON DERIVATIVES

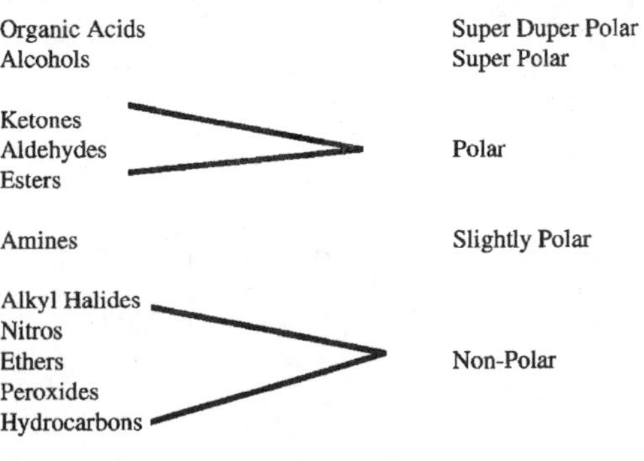

Figure 5.9

Formic acid has the highest boiling point even though it weighs less than dimethyl ketone. Methyl alcohol has the next-highest boiling point; it also weighs less than dimethyl ketone. The weights of methyl alcohol and ethane are about the same, yet the difference in boiling point is 275°F.

To fully understand the concept of polarity, we must revisit the structure of the atom presented in Chapter 2. The nucleus contains positive protons, and the energy levels outside the nucleus contain negative electrons. There are normally an equal number of negative electrons and positive protons in an element and the elements of a compound. Hydrogen has only one electron in its outer energy level. Oxygen has 16 electrons in its outer energy levels. As hydrogen and oxygen bond together, oxygen has a tendency to draw the one electron from hydrogen toward the oxygen side of the covalent bond. This exposes the positive nucleus of hydrogen, creating a slightly positive side to the hydrogen end of the molecule. Oxygen has drawn the electron from hydrogen toward its nucleus. In doing so, oxygen has more negative electrons near the nucleus than positive protons inside. This creates a slightly negative field around the oxygen side of the molecule. This concept is represented in the following illustration:

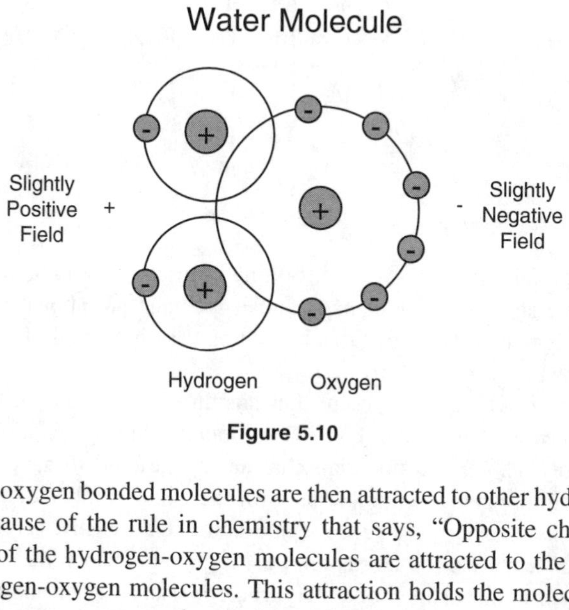

Water Molecule

Slightly Positive Field +

\- Slightly Negative Field

Hydrogen Oxygen

Figure 5.10

Hydrogen-oxygen bonded molecules are then attracted to other hydrogen-oxygen molecules because of the rule in chemistry that says, "Opposite charges attract." Positive ends of the hydrogen-oxygen molecules are attracted to the negative ends of other hydrogen-oxygen molecules. This attraction holds the molecules together, so that it takes more energy to break them apart and cause the compound to boil. This attraction of molecules is illustrated in Figure 5.11.

Branching

The last factor that affects the boiling point of a flammable liquid is branching. (Branching was discussed in Chapter 4 under Isomers in the Hydrocarbon section.) Branched compounds are all manmade. Since branching does not occur naturally, it is done for a particular purpose, usually because there is an economic value. When

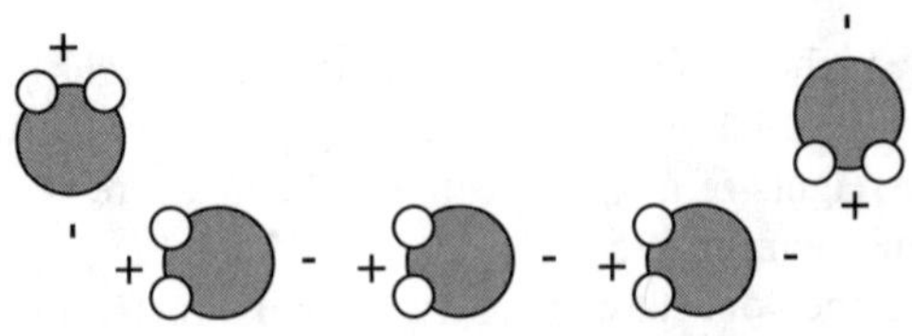

Water molecules shown with a polar attraction holding them together.

Figure 5.11

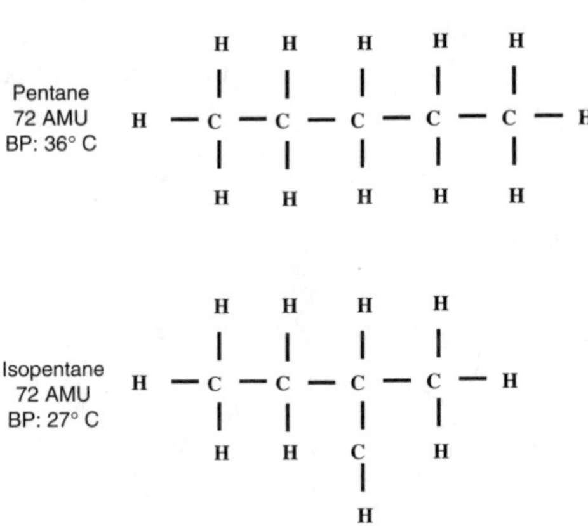

Figure 5.12

a compound is branched, it has a lower boiling point than the unbranched liquid. In Figure 5.12, the structures for pentane and isopentane are shown with the corresponding boiling points. The effect that branching has on the boiling point of a liquid is clearly visible.

There is another type of structure that has an effect on the boiling point of a liquid, called a cyclic compound. Cyclic compounds with five, six, or seven carbon atoms are highly stable, and materials that are cyclic tend to have higher boiling points than straight-chained compounds with the same number of carbons. Those with less than five carbons are reactive and come apart easily. Cyclic compounds with more than seven carbon atoms tend to fragment. The cyclic compounds will be discussed later in this chapter under Hydrocarbons and the Aromatic Compounds. In the following example, hexane is compared to cyclohexane (see Figure 5.13). Notice the difference that the cyclic structure has on the boiling point.

FLASH POINT

The next and probably the most important physical characteristic of a flammable liquid is its flash point. Flash point is the most important information for emergency

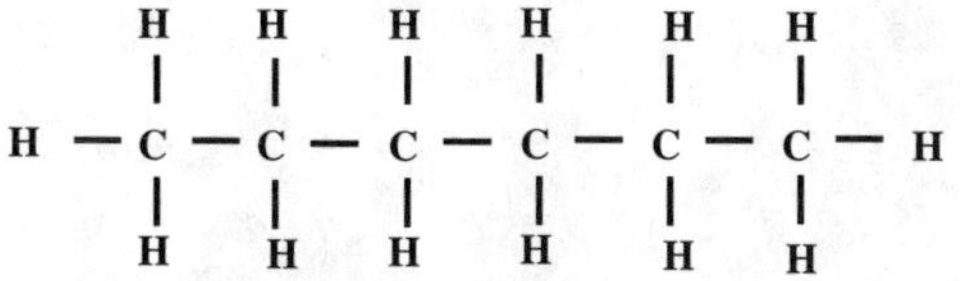

Hexane
C_6H_{14}
BP: 156° F

Cyclo Hexane
C_6H_6
BP: 179° F

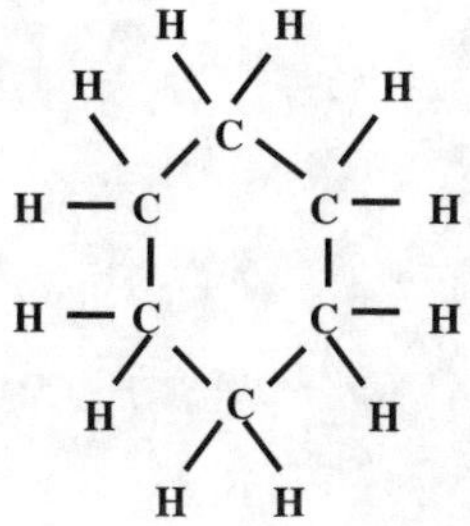

Figure 5.13

responders to know about a flammable liquid. Flash point, more than any other characteristic, helps to define the flammability hazard of a liquid. If a flammable liquid is not at its flash point temperature, **it will not burn**. Flash point is defined as "the minimum temperature to which a liquid must be heated to produce enough vapor to allow a vapor flash to occur (if an ignition source is present)." After all, it is the vapor that burns, not the liquid, so the amount of vapor present is critical in determining whether the vapor will burn.

Flash point is a measurement of the temperature of the liquid. Therefore, even if the ambient temperature is not at the flash point temperature, the liquid may have been heated to its flash point by some external heat source. For example, the radiant heat of the sun, heat from a fire, or heat from a chemical process may heat the liquid to its flash point. If an ignition source is present, ignition can and probably will occur. In order for ignition to occur if a flammable liquid is at its flash point temperature, there has to be an ignition source present that has a temperature at or above the ignition temperature of the vapor.

There are three basic methods in which ignition temperatures can be reached, or, to put it another way, there are three types of ignition sources. They are external, external-internal (autoignition), and internal (spontaneous combustion). External ignition sources produce heat that enters the vapor or liquid itself and transfers their heat energy directly to the flammable material. Examples of external ignition sources are open flames, sparks (electrical, static, or frictional), and heated objects. Sparks are capable of developing temperatures ranging from 2000° to 6000°F. External-internal (autoignition) sources heat the vapor or liquid through an indirect method. Three types of indirect methods are radiant-heat transfer, convection-heat transfer, and combustion-heat transfer. From any of these sources, heat is transferred until ignition temperature is reached and ignition occurs, without the presence of any

Underflow dams used to stop the flow of flammable liquids that are lighter than water.

open flame or spark. Spontaneous ignition is the final type of ignition source. In this case, the material itself produces heat sufficient to reach its own ignition temperature. This can occur in two ways: as a result of the biological processes of some microorganisms and slow oxidation. Biological processes are usually not associated with flammable liquids, but rather with organic materials, such as hay and straw.

The activity of biological organisms within the material generates heat, and the heat is confined by the materials, until the ignition temperature is reached and ignition occurs. Slow oxidation is a chemical reaction. Chemical reactions may produce heat. If the heat is insulated from dissipating to the outside of the material, it will continue to build up. As the heat builds, the material is heated from within. The process continues until the ignition temperature of the material is reached and ignition occurs.

Reference books, which are used to research chemical characteristics in hazardous materials emergency response, may show different flash point values. There are two different tests used to determine the flash point of a liquid. They are known as the open-cup and closed-cup test apparatus. The differences in the testing procedures often produce somewhat different flash point temperatures. Open-cup flash point tests try to simulate conditions of a flammable liquid in the open, such as a spill from a container to the ground. Open-cup tests usually result in a higher flash point temperature for the same flammable liquid than the closed-cup method. The flash point of a liquid varies with the oxygen content of the air, pressure, purity of the liquid, and the method of testing. If reference books give conflicting flash point temperatures, **use the lowest flash point value given**.

Flash point should not be confused with fire point. Fire point is the temperature at which the liquid is heated to produce enough vapor for ignition and sustained combustion to occur. Fire point temperature is 1 to 3 degrees above the flashpoint temperature. The fact that the fire point is so close to the flash point really does not give it much significance to emergency responders: if a liquid is at its flash point, prepare for a fire.

The author is familiar with only one occasion in which there was a vapor flash from vapors being ignited and no fire occurred after the vapors burned off. In that situation, there may not have been a fire because all of the vapor was burned off and there was nothing else to burn. A gasoline tanker was filling the underground storage tanks of a service station early one morning. The vent pipes for the underground tanks were not up to code and were just 3 ft above the ground. (NFPA 30, The Flammable and Combustible Liquids Code, requires that vents be a minimum of 3 ft above the roofline of the nearest building.) As a result of the improper vents and the vapors being heavier than air, the vapors collected near the ground. When a soda machine compressor turned on near the vapor spill, it provided an ignition source and a vapor flash occurred. Other than a surprised deliveryman, there was not any injury or damage from the flash fire. Do not be concerned about fire point; it is the flash point that is the important temperature to look up and be aware of at the scene of an incident.

There is a direct, parallel relationship between boiling point and flash point temperatures. Generally speaking, a liquid that has a low boiling point temperature will have a low flash point temperature. If a flammable liquid has a high boiling point, it will have a high flash point.

VAPOR PRESSURE

Boiling point is also related to vapor pressure and vapor content, although the relationship is opposite in nature. Materials with low boiling points and flash points will have high vapor pressure and high vapor content. The vapor pressure of a liquid is defined in the *Condensed Chemical Dictionary* as "the characteristic at any given temperature of a vapor in equilibrium with its liquid or solid form. This pressure is often expressed in millimeters of mercury, mm Hg."

In simple terms, vapor pressure is the pressure exerted by the liquid against atmospheric pressure. When the pressure of the vapor is greater than atmospheric pressure, vapor will spread beyond an open container or an open spill. If liquid is in a container, vapor pressure is the pressure exerted by the liquid vapors on the container. For example, when a gas has been liquefied, the only thing keeping it a liquid is the pressure in the tank; the liquid is already above its boiling point. The pressure inside the container is the atmospheric pressure in that container. It can be much higher than outside atmospheric pressure. For example, if the pressure in the tank is 50 psi, the atmospheric pressure in that tank is 50 psi regardless of conditions outside the tank.

Physical Relationships

BP = Boiling Point
FP = Flash Point
IT = Ignition Temp

VP = Vapor Pressure
VC = Vapor Content
HO = Heat Output

HIGH

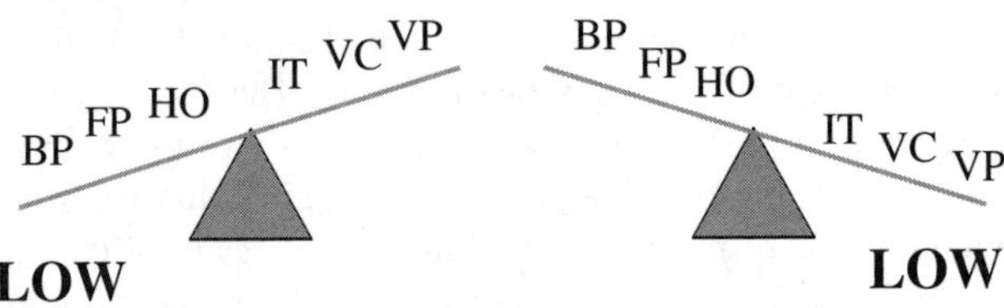

Figure 5.14

VAPOR CONTENT

Vapor content is the amount of vapor that is present in a spill or open container. The lower the boiling point and the flash point of a liquid, the more vapor there will be. The parallel relationship of boiling point and flash point is comparable to the opposite relationship of vapor pressure and vapor content shown in the diagram, using a seesaw to illustrate the up-and-down and opposing relationship (see Figure 5.14).

As shown in the previous illustration, when the boiling point and flash point are low, vapor content and vapor pressure are high. When boiling point and flash point are high, vapor pressure and vapor content are low. The lower the boiling point or flash point of a liquid, the higher the vapor content at a spill and the higher the vapor pressure inside a container. If a liquid is above its boiling point temperature, there is likely to be more vapor moving farther away from a spill. If the liquid is below its boiling point temperature, there will be some vapor above the surface of the liquid, but it will not travel far. If the flammable liquid in a container is below its boiling point, the vapor content and vapor pressure in the container will be low. If the flammable liquid in a container is above its boiling point, the vapor content and vapor pressure in the container will be high.

VAPOR DENSITY

Vapor density is a physical characteristic that affects the travel of vapor; it is the weight of a vapor compared to the weight of air (see Figure 5.15). Vapor density is usually determined in the reference books by dividing the molecular weight of a compound by 29, which is the assumed molecular weight of air. Air is given a weight

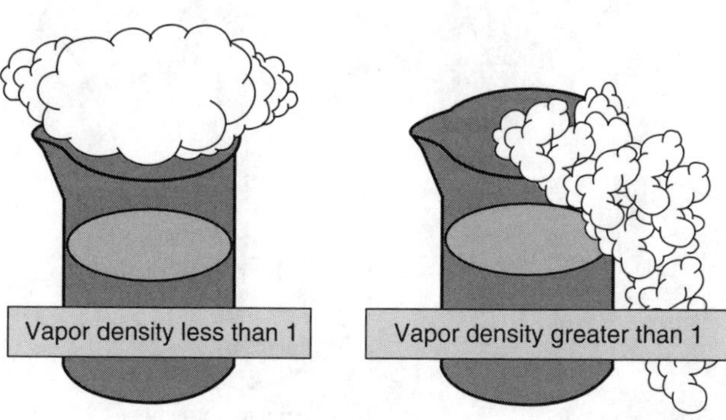

Figure 5.15

value of 1, which is used to compare the vapor density of a material. If the vapor of a material has a density greater than 1, it is considered heavier than air. Heavier-than-air vapor will lie low to the ground and collect in confined spaces and basements. This can cause problems because many ignition sources are in basements, such as hot-water heaters and furnace pilot lights. If the vapor density is less than 1, the vapor is considered to be lighter than air, so it will move up and travel farther from the spill.

Another term associated with vapor is volatility. It is the tendency of a solid or a liquid to pass into the vapor state easily. This usually occurs with liquids that have low boiling points. A volatile liquid or solid will produce significant amounts of vapor at normal temperatures, creating an additional flammability hazard. The vapor produced by a volatile liquid is affected by wind, vapor pressure, temperature, and surface area. Temperature always causes an increase in vapor pressure and vapor content in an incident. The more vapor pressure in a container, the greater the chance of container failure. The more vapor content, the farther the vapor may travel away from a spill.

SPECIFIC GRAVITY

Specific gravity is to water what vapor density is to air. Specific gravity is the relationship of the weight of a liquid to water or another liquid (see Figure 5.16). Like air, water is given a weight value of 1. If a flammable liquid has a specific gravity greater than 1, it is heavier than water and will sink to the bottom in a water spill. If a flammable liquid has a specific gravity less than 1, it will float on top of the water. The specific gravity of a flammable liquid is important in a water spill because it will determine what tactics are necessary to contain the spill. Specific gravity is the theory behind the construction of overflow and underflow dams, which are used to stop the flow of hazardous materials in water spills. Overflow dams are constructed for liquids heavier than water. The liquid sinks to the bottom of the

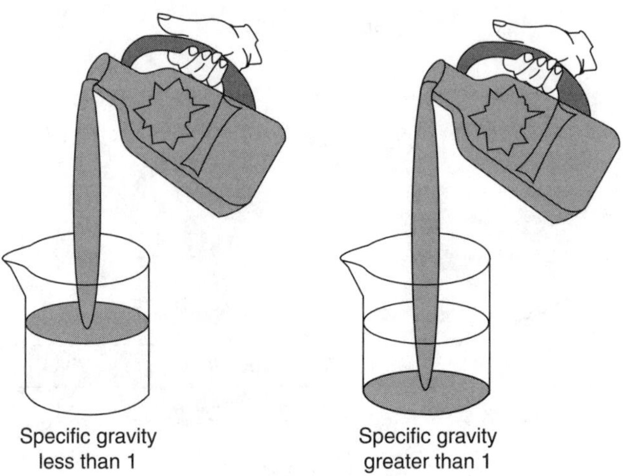

Figure 5.16

water, and the water flows over the dam, while the hazardous liquid is stopped by the dam. Underflow dams are constructed for liquids lighter than water. The liquid floats on the surface of the water and is stopped by the top of the dam, while the water continues to flow through a pipe at the bottom of the dam.

A term often associated with hazardous materials and water is miscibility. If a chemical is miscible with water, it will mix with water, which could make clean up difficult. If the chemical is not miscible with water, it will form a separate layer. The layer will form on top or on the bottom of the water, depending on the specific gravity of the liquid. Most flammable liquids are lighter than water and immiscible, so they float on the surface.

POLYMERIZATION AND PLASTICS

Flammable liquids may undergo a chemical reaction called polymerization, in which a large number of simple molecules, called monomers, combine to form long-chained molecule called a polymer. This process is used under controlled conditions to create plastics (see Figure 5.17). Alkene hydrocarbon compounds and hydrocarbon derivatives, such as aldehydes, alkyl halides, and esters, and the aromatic hydrocarbon styrene may undergo polymerization. There are other monomers that are flammable and can polymerize, but their primary hazard is poison. Monomers can be flammable liquids, flammable gases, and poisons.

When a monomer, such as styrene, is transported or stored, an inhibitor is included in solution to keep the styrene from polymerizing. An inhibitor, usually an organic compound, retards or stops an unwanted polymerization reaction. If an accident should occur, this inhibitor can become separated from the monomer and a runaway polymerization may occur. Phenol — a deadly poison — is used as an inhibitor for vinyl chloride. Dibutylamine is used as an inhibitor for butadiene.

Types of Plastics

Thermoplastics	Thermosets
ABS	Polyurethane
Acrylics	Amino Resins
Nylons	Epoxy Resins
Polycarbonate	Phenolic Resins
Polyesters	Polyesters
Polyethylene	
Polypropylene	
Polystyrene	
Polyurethane	
Polyvinyl Chloride	

Figure 5.17

During a normal chemical reaction to create a particular polymer from a monomer, a catalyst is used to control the reaction. A catalyst is any substance that in a small amount noticeably affects the rate of a chemical reaction, without itself being consumed or undergoing a chemical change. For example, phosphoric acid is used as a catalyst in some polymerization reactions. Once an uncontrolled polymerization starts at an incident scene, it will not be stopped until it has completed its reaction, no matter what responders may try to do. If the polymerization occurs inside a tank, the tank may rupture violently. If a container of a monomer is exposed to fire, it is important to keep the container cool. Heat from an exposure fire may start the polymerization reaction. In Figure 5.18, the monomer vinyl chloride is shown along with the process of polymerization of the vinyl chloride molecules.

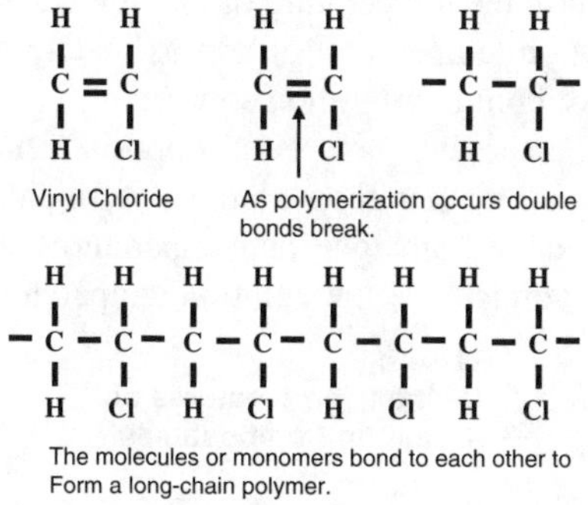

Vinyl Chloride

As polymerization occurs double bonds break.

The molecules or monomers bond to each other to Form a long-chain polymer.

Figure 5.18

This has been an abbreviated explanation of polymers and plastics, which are fairly complicated subjects. An entire book could be written on them. The most important thing for responders to understand about monomers and polymers is to be able to recognize them and the danger they present in the uncontrolled conditions of the incident scene.

IGNITION TEMPERATURE

An often misunderstood physical characteristic associated with flammable liquids is ignition temperature. The definition of ignition temperature (also known as autoignition temperature) is "the minimum temperature to which a material must be heated to cause autoignition without an ignition source." In other words, the material autoignites by being heated to its ignition temperature. For example, consider a pan of cooking oil on a stove. Cooking oils are animal or vegetable oils. They are combustible liquids with high boiling points and flash points. If the heat is turned up too high on a stove with a pan of cooking oil, the oil may catch fire. Many kitchen fires occur because of cooking oils or grease being overheated. The reason for this is ignition temperature. Liquids that have high boiling and flash points have low ignition temperatures. Corn oil, commonly used as cooking oil, has an ignition temperature of 460°F. If corn oil is heated on a stove to 460°F, it will autoignite. Gasoline has an ignition temperature of approximately 800°F, depending on the blend. If gasoline were placed on a stove without an ignition source, the stove would not produce a temperature high enough to autoignite the gasoline. It would just boil away into vapor.

This can be a real "fooler" on the incident scene, particularly when dealing with combustible liquids where the boiling points and flash points are high. Responders sometimes become complacent when dealing with combustible liquids. They think that because the boiling and flash points are high, the danger of fire is low. If, however, ignition sources are not controlled, what little vapor is present above the liquid can ignite if the temperature of the ignition source is above the ignition temperature of the liquid (see Figure 5.19). For example, a lighted cigarette, with no drafts, has a surface temperature of about 550°F, and Number 1 fuel oil has an ignition temperature of 444° F. A cigarette can be an ignition source for a combustible liquid because of the low ignition temperatures. A cigarette cannot be an ignition source for gasoline because the temperature of the lighted cigarette, without a draft, is below gasoline's ignition temperature of approximately 800°F. Figure 5.20 provides the temperatures of common ignition sources.

Referring back to the seesaw, there is an opposite relationship between boiling/flash point and ignition temperature. Flammable liquids that have low boiling points and flash points have high ignition temperatures. Liquids that have high boiling points and flash points have low ignition temperatures.

Ignition Temperatures of Common Combustibles

Wood	392°F
#1 Fuel oil	444°F
Paper	446°F
60 Octane gas	536°F
Acetylene gas	571°F
Wheat flour	748°F
Corn	752°F
Propane gas	871°F

Figure 5.19

Temperatures of Common Ignition Sources

Lighted cigarette, no drafts	550°F
Lighted cigarette, with drafts	1350°F
Struck match	2000°F+
Electric arc	2000°F+

Figure 5.20

Combustible liquids generally are more difficult to ignite than flammable liquids. However, once they are ignited, they have a much higher heat output than flammable liquids. Because of this high heat output, combustible liquid fires are much more difficult to extinguish than flammable liquid fires. There is a parallel relationship between boiling point and heat output. Materials that have low boiling points have low heat outputs. This relationship between boiling point, flash point, ignition temperature, vapor content, vapor pressure, and heat output is illustrated in Figure 5.14.

FLAMMABLE RANGE

The last physical characteristic to be presented is flammable range. Flammable range is defined as the percent of vapor in air necessary for combustion to occur and is referred to as the explosive limit. It is expressed on a scale from 0 to 100% (see Figure 5.21). There is an upper explosive limit (UEL) and a lower explosive limit (LEL); between the two, there is a proper mixture of vapor (fuel) and air for combustion to occur. Above the UEL, there is too much vapor and not enough air; in other words, the mixture is too rich to burn. Below the LEL, there is enough air, but too little vapor; therefore, the mixture is too lean to burn. Most flammable liquids have explosive limits between 1 and 12%. They have a narrow flammable range; all of the conditions must be just right for combustion to occur. The liquid must be at its flash point, the air–vapor mixture must be within its flammable range, and the

Flammable Range (Explosive Limits)

The percentage of fuel in the air that will burn

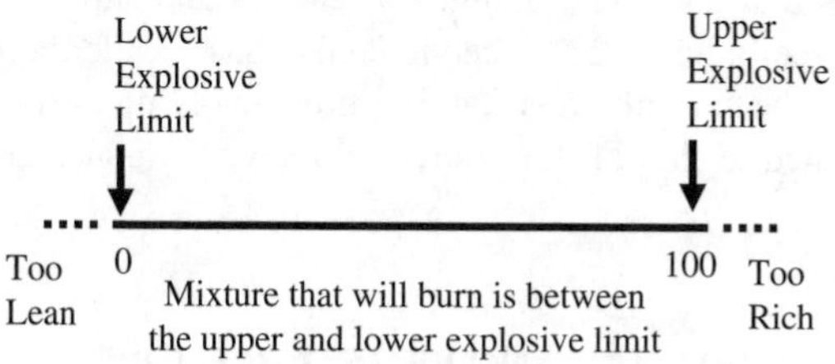

Figure 5.21

Typical Flammable Ranges of Flammable Liquid Families

Fuel family	1–8%
Aromatic hydrocarbons	1–7%
Ketones	2–12%
Esters	1–9%
Amines	2–14%
Alcohols	1–36%
Ethers	2–48%
Aldehydes	3–55%
Acetylene	2–85%

Figure 5.22

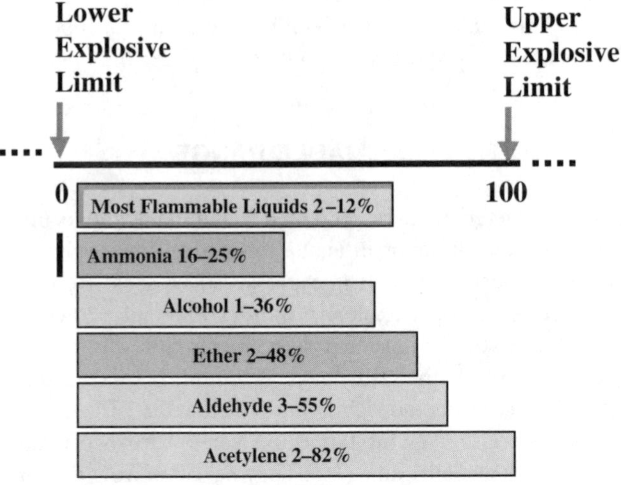

Figure 5.23

ignition-source temperature must be above the ignition temperature of the liquid. Figure 5.22 provides typical flammable ranges of flammable liquid families.

Wide flammable-range materials are dangerous because they can burn inside a container since they burn rich. Alcohols, ethers, and aldehydes are families of flammable liquids that have wide flammable ranges and should be addressed with extreme caution (see Figure 5.23). Acetylene also has a wide flammable range, and firefighters need to be careful when fighting fires involving acetylene tanks. Before any valves are turned to shut off the source of acetylene, make sure there is not fire inside the tank.

ANIMAL AND VEGETABLE OILS

Some combustible liquids, such as animal and vegetable oils, have a hidden hazard: they may burn spontaneously when improperly handled. They have high boiling and flash points, narrow flammable ranges, low ignition temperatures, and are nonpolar. Examples of these liquids are linseed oil, cottonseed oil, corn oil, soybean oil, lard, and margarine. These unsaturated materials can be dangerous when rags containing residue are not properly disposed of or they come in contact with other combustible materials.

There is a double bond in the chemical makeup of animal and vegetable oils that reacts with oxygen in the air. This reaction causes the breakage of the double bond, which creates heat. If the heat is allowed to build up in a pile of rags for example, spontaneous combustion will occur over a period of hours.

In Verdigris, Oklahoma, a fire occurred in an aircraft hangar at a small airport. The owner's living quarters were on the second level of the hangar. Workers had been polishing wooden parts of an airplane in the afternoon. The rags used to apply linseed oil were placed in a plastic container in a storage room in the hangar, just below the living quarters. Around 2 a.m., the rags with the linseed oil spontaneously ignited and the fire traveled up the wall into the living quarters. Fortunately, the owner had smoke detectors; the family was awakened, and the fire department was called promptly. The fire was quickly extinguished with a minimum of damage. The V-pattern on the wall led right back to the box where the linseed oil-soaked rags had been placed. There was little doubt what had happened; the confinement of the pile allowed the heat to build up as the double bonds were broken in the linseed oil, which combined with oxygen in the air, and spontaneous combustion occurred.

A fire occurred on February 23, 1991, at the One Meridian Plaza Building in Philadelphia, resulting in the deaths of three firefighters. The fire was started by spontaneous combustion in linseed oil-soaked rags that were improperly disposed

Closed floating-roof tank used to store polar-solvent flammable liquids, such as alcohols and ketones.

A fire at the One Meridian Plaza building in Philadelphia that killed three firefighters, caused by linseed oil-soaked rags that were not properly stored and disposed of after use.

of after use. The fire occurred on the 22nd floor of the 30-story building. PCB contamination from the fire made the building uninhabitable, and it is currently being torn down.

Ordinary petroleum products, such as motor oil, grease, diesel fuel, and gasoline, to name a few, do not have a double bond in their chemical makeup. Therefore, those materials **do not undergo spontaneous combustion**! This fact may come as a surprise to some people because the author knows there have been numerous fires blamed on soiled rags with those products on them. The fact is that those types of flammable liquids do not spontaneously ignite and cannot start to burn without some other ignition source.

FIRE-EXTINGUISHING AGENTS

The theory behind fire extinguishing was first represented by the fire triangle, and more recently by the fire tetrahedron. The triangle (see Figure 5.24) represents the three components that were thought to be necessary for fire to occur: heat, fuel, and oxygen. If any of the components were removed, the fire would go out. The current theory (see Figure 5.25), uses a four-sided geometric figure called a tetrahedron, representing the four components necessary for fire to occur: the original three, plus a chemical chain reaction. It is believed that fire is a chemical chain reaction; certain extinguishing agents work by interrupting this chemical chain reaction.

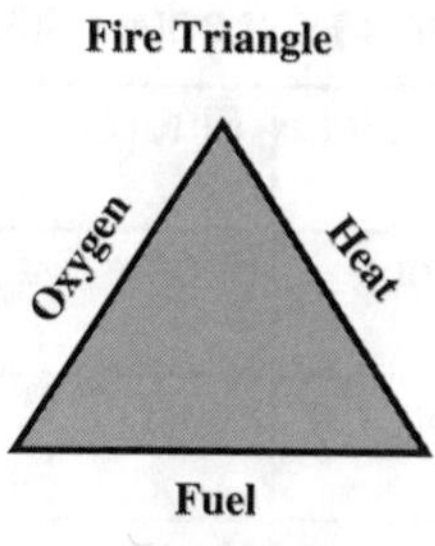

Figure 5.24

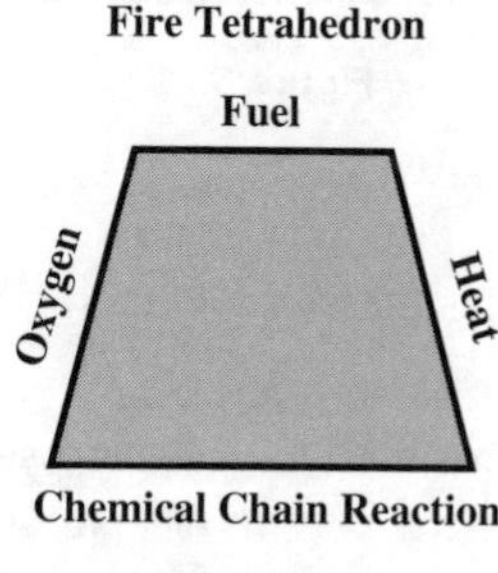

Figure 5.25

Extinguishing agents, such as foam, act to eliminate one of the four components. Foam excludes oxygen from the fuel by blanketing the surface of the liquid, and can also cool the material (removing heat). There are two general categories of firefighting foam: chemical and mechanical. Chemical foam was developed in the late 1800s. The foam bubble was produced by a chemical reaction of sodium bicarbonate powder, aluminum sulfate powder, and water. The reaction produces carbon dioxide, which is then encapsulated in the interior of a bubble. Chemical foams are expensive to produce and use, and create a rigid foam blanket that does not reseal well when disrupted. Presently, chemical foams have been replaced by more effective and economical mechanical foams.

There are three types of mechanical firefighting foams: protein, fluoroprotein, and aqueous film forming (AFFF). Different foams have different physical characteristics that affect their ability to form a foam blanket and extinguish a fire (see Figure 5.26). The type of foam selected will depend on the makeup of the fuel feeding the fire, the type of foam available, and the firefighting tactics chosen to extinguish the fire. Make sure that whatever extinguishing agent is used, it is the proper agent for the type of fire. Be sure there is enough agent on-scene to extinguish the fire before fire suppression efforts are started. Fires require a certain volume of water to effect extinguishment. Flammable liquid fires also require a certain volume of foam to extinguish the fire. If that amount of foam is not available, putting a lesser amount of foam on the fire is not going to extinguish it and is just a waste of time and foam.

FOAM TYPE COMPARISON

	Protein	**Fluoroprotein**	**AFFF**
Drain Time	Long	Shorter	Rapid
Viscosity	High	Low	Very Low
Flow Rate	Low	Moderate	Very Fast
Polar Compatibility	None	None	Varies by Supplier

Figure 5.26

Classes of Fires

Class A Ordinary Combustibles	Water All-Purpose Dry Chemical
Class B Flammable Liquids	All-Purpose Dry Chemical Foam Purple K™ Halon Carbon Dioxide Sodium Bicarbonate
Class C Electrical	All-Purpose Dry Chemical Carbon Dioxide Halon
Class D Combustible Metals	Dry Powder Graphite Sodium Chloride

Figure 5.27

In addition to foam, fires involving flammable liquids can be extinguished by using dry chemical, dry powder, halon, Purple K,® carbon dioxide, and water. Fire-extinguishing agents are rated according to the class of fire that they are effective in extinguishing (see Figure 5.27). Dry powder acts primarily to exclude atmospheric oxygen. Dry chemical, dry powder, halon, and Purple K® interrupt the chemical chain reaction. Dry-chemical fire extinguishers use sodium bicarbonate and monoammonium phosphate as agents. Purple K® extinguishers use potassium bicarbonate, which is where the purple color comes from. Dry-powder extinguishers use sodium chloride and graphite as agents. Water acts by cooling the fire or, in other words, removing the heat. Water may be ineffective against some flammable liquid fires. This precaution usually applies to materials with flash points below 100°F. When water is used, it should be applied in the form of a water spray. Halon and carbon dioxide displace the oxygen needed for the fire to burn, so if there is not enough oxygen, the fire goes out. Because of their effect on the ozone layer above the earth, halon fire-extinguishing agents are being phased out, along with other gaseous halogenated chemicals.

Carbon tetrachloride was once used as a fire-extinguishing agent. In fact, it was one of the first halons, halon 1040. It was discovered, however, that carbon tetrachloride has a hidden hazard. When carbon tetrachloride contacts fire or a hot surface, it gives off phosgene gas. As a result, carbon tetrachloride is no longer an approved extinguishing agent, but this does not mean that you will not see carbon tetrachloride extinguishers still in use. As recently as 1987, during routine fire inspections, they were found in use in a rural school, an apartment building laundry room, and in the basement of a drug store. They may also be found in antique stores and homes of collectors.

Soda-acid fire extinguishers are also obsolete and no longer approved in fire extinguishment. They are the familiar copper, brass, or stainless steel extinguishers that are now collectors' items. They have a screw-on cap that exposes a glass bottle of sulfuric acid when removed. The rest of the tank is filled with soda water. When the extinguisher is turned upside down, the acid and soda water mix, expelling the mixture through a rubber hose with a nozzle on the end. After 10 to 15 years of service, these extinguishers became dangerous. The pressure that builds up to expel the acid-soda water mix is around 100 psi; however, if the hose or nozzle is plugged or the hose is kinked, the pressure can reach 300 psi. Many of these extinguishers failed when used, causing injury to operators. As recently as the late 1980s, this type of extinguisher was found still in use in rural schools, old hotels, and main-street businesses. If these obsolete extinguishers are encountered, they should promptly be removed from service and properly disposed of.

HYDROCARBONS

The hydrocarbon families of flammable materials were introduced in Chapter 4 in the Flammable Gases section. The first family is the alkane, which has all single bonds and is considered saturated. The lighter alkanes (methane, ethane, propane, and butane) are all gases under normal conditions. All of these gases, however, may be encountered as liquefied gases. The alkane family is recognized by the "ane" suffix of the compound names and the single bonds between the carbons in the compounds. Starting with pentane, a five-carbon alkane, the remaining alkanes are flammable liquids. They are all naturally occurring and are byproducts of crude oil and include pentane, hexane, heptane, octane, nonane, and deckane. Other hydrocarbon compounds exist with more than 10 carbons, but the trivial naming system only works for the first 10. Other hydrocarbon compounds are named using the IUPAC system of naming. Those naming conventions are listed in the Appendix. As a group, pentane through decane have varying boiling points, flash points, narrow flammable ranges, high ignition temperatures (depending on the number of carbons in the compound), and are all nonpolar.

Pentane is a five-carbon alkane and has a boiling point of 97°F, a flash point of –40°F, a flammable range of 1.5 to 7.8%, and an ignition temperature of 500°F. As the carbon content of a compound increases, so do the boiling point and flash point. The structure and molecular formula for pentane are shown in Figure 5.28.

```
    H   H   H   H   H
    |   |   |   |   |
H - C - C - C - C - C - H
    |   |   |   |   |
    H   H   H   H   H
```

Pentane C_5H_{12}

Figure 5.28

Hexane is a six-carbon alkane and has a boiling point of 156°F, a flash point of –7°F, a flammable range of 1.1 to 7.5%, and an ignition temperature of 437°F. The structure and molecular formula for pentane are shown in Figure 5.29.

```
    H   H   H   H   H   H
    |   |   |   |   |   |
H - C - C - C - C - C - C - H
    |   |   |   |   |   |
    H   H   H   H   H   H
```

Hexane C_6H_{14}

Figure 5.29

Heptane is a seven-carbon alkane and has a boiling point of 98°C, a flash point of 25°F, a flammable range of 1.05 to 6.7%, and an ignition temperature of 433°F. The structure and molecular formula is shown in Figure 5.30.

```
    H   H   H   H   H   H   H
    |   |   |   |   |   |   |
H - C - C - C - C - C - C - C - H
    |   |   |   |   |   |   |
    H   H   H   H   H   H   H
```

Heptane C_7H_{16}

Figure 5.30

Octane is an eight-carbon alkane and has a boiling point of 125.6°C, a flash point of 56°F, a flammable range of 1.0 to 6.5%, and an ignition temperature of 428°F. The structure and molecular formula for pentane are shown in Figure 5.31.

```
    H   H   H   H   H   H   H   H
    |   |   |   |   |   |   |   |
H - C - C - C - C - C - C - C - C - H
    |   |   |   |   |   |   |   |
    H   H   H   H   H   H   H   H
```

Octane C_8H_{18}

Figure 5.31

Nonane is a nine-carbon alkane and has a boiling point of 150.7°C, a flash point of 86°F, a flammable range of 0.8 to 2.9%, and an ignition temperature of 403°F. The structure and molecular formula for pentane are shown in Figure 5.32.

```
    H   H   H   H   H   H   H   H   H
    |   |   |   |   |   |   |   |   |
H — C — C — C — C — C — C — C — C — C — H
    |   |   |   |   |   |   |   |   |
    H   H   H   H   H   H   H   H   H
```

Nonane C_9H_{20}

Figure 5.32

Decane is a 10-carbon alkane, and has a boiling point of 345°F, a flash point of 115°F, a flammable range of 0.8 to 5.5%, and an ignition temperature of 410°F. The structure and molecular formula are shown in Figure 5.33.

```
    H   H   H   H   H   H   H   H   H   H
    |   |   |   |   |   |   |   |   |   |
H — C — C — C — C — C — C — C — C — C — C — H
    |   |   |   |   |   |   |   |   |   |
    H   H   H   H   H   H   H   H   H   H
```

Decane $C_{10}H_{22}$

Figure 5.33

There is usually an opposite relationship between boiling point and ignition temperature. Compounds that have low boiling and flash points generally have high ignition temperatures. Compounds with high boiling and flash points, generally have low ignition temperatures.

While it is entirely possible to encounter these compounds in transportation and fixed facilities as individual chemicals, they will more likely be found in mixtures with other compounds. Mixtures do not involve chemical reactions or bonding. They are pure chemical compounds that have been mixed together to form a solution without losing their individual chemical makeup. However, the overall physical characteristics of the mixture will change based on the physical characteristics of the compounds' makeup. Boiling and flash points of the mixture will fall somewhere between the boiling and flash points of the components. In simple terms, a pure compound is octane, pentane, or isooctane. It is possible to draw structures or write formulas for these compounds. When pure compounds are mixed together, you cannot draw a structure or write a formula for the mixture. These mixtures include gasoline, diesel fuel, jet fuel, fuel oil, petroleum, ether, and others. Gasoline has a boiling point between 100° and 400°F, depending on the contents of the mixture; its flash point is –36° to –45°F; the flammable range is 1.4 to 7.6%; and the ignition temperature ranges from 536° to 853°F. Diesel fuel has a flash point range of 100° to 130°F. Hazardous materials reference books do not list a boiling point or ignition temperature for diesel fuel. Figure 5.34 shows the physical characteristics of some

Physical Characteristics of Some Common Fuels

	Gasoline	Kerosene	Fuel Oil #4	Jet Fuel
Boiling point	100–400°F	338–572°F		250°F
Flash point	–45°F	100–150°F	130°F	100°F
Ignition temp	536–853°F	444°F	505°F	435°F
Vapor density	3.0–4.0	4.5		1
Specific gravity	0.8	0.81	<1	0.8
LEL	1.4%	0.7%		0.6%
UEL	7.6%	5%		3.7%

Figure 5.34

common hydrocarbon fuels. Remember that the fuels are mixtures of hydrocarbons and other additives rather than pure compounds.

Physical characteristics of mixtures will vary, depending on the components of the mixture. Many times, mixtures are designed to have certain ranges of flash points so they will perform a particular function. If a high boiling and flash point liquid is mixed with a low boiling and flash point liquid, the boiling and flash point of the mixture will be somewhere between the two liquids that were mixed together. Mixtures will be discussed further in the Hydrocarbon Derivatives section.

Alkenes are the second hydrocarbon family. Alkenes have one or more double bonds between the carbons in the structure of the compound. The gases ethene (ethylene), propene (propylene), and butene (butylene) were discussed in Chapter 4. The first flammable liquid in the alkene family is pentene. Because it is an alkene, it has a double bond in the compound. Pentene has a boiling point of 86°F, a flash point of 0°F, a flammable range of 1.5 to 8.7%, and an ignition temperature of 527°F. The structure and formula for pentene is shown in Figure 5.35.

```
H   H   H   H   H
|   |   |   |   |
C = C - C - C - C - H
|       |   |   |
H       H   H   H
```

Pentene C_5H_{10}

Figure 5.35

Hexene has a boiling point of 146°F, a flash point of less than 20°F, and an ignition temperature of 487 F. The structure and molecular formula for hexane are shown in Figure 5.36.

```
H   H   H   H   H   H
|   |   |   |   |   |
C = C - C - C - C - C - H
|       |   |   |   |
H       H   H   H   H
```

Hexene C_6H_{12}

Figure 5.36

As with the alkanes, as the carbon content increases, so do the boiling point and flash point temperatures. There are alkene-family flammable liquids that have more than one double bond. **Pentadiene**, also known as 1,3,pentadiene and piperylene is a five-carbon compound with two double bonds. Pentadiene is a highly flammable liquid with an NFPA 704 designation for flammability of 4. It has a boiling point of –45°F and a flash point of 112°F. **Hexadiene** is a six-carbon liquid compound that is highly flammable with a boiling point of 147°F, a flash point of –6°F, and a flammable range of 2 to 6.1%. The structures and molecular formulas of pentadiene and hexadiene are illustrated in Figure 5.37.

```
H   H   H   H   H       H   H   H   H   H   H
|   |   |   |   |       |   |   |   |   |   |
C = C - C - C = C       C = C - C - C - C = C
|       |       |       |       |   |       |
H       H       H       H       H   H       H
```

Pentadiene C_5H_8

Hexadiene C_6H_{10}

Figure 5.37

Since there are no commercially valuable alkyne compounds that are likely to be encountered by emergency responders, the alkyne family will not be covered here. Acetylene, the alkyne gas, and the characteristics of the alkyne family were covered in Chapter 4.

The third and final hydrocarbon family to be discussed is known as the aromatic hydrocarbons, sometimes referred to as the BTX fraction (benzene, toluene, xylene). One additional aromatic beyond the BTX fraction is called styrene and will be covered as the fourth aromatic compound. Aromatics as a group are toxic and flammable. They have moderate boiling and flash points, narrow flammable ranges, high ignition temperatures, and are nonpolar.

Benzene is a known carcinogen, while toluene and xylene should be suspect. The parent member of the family is benzene, which has a molecular formula of C_6H_6. There is a ratio that can be used to recognize the benzene ring in a formula: there has to be a minimum of six carbons to have a benzene ring, and the number of carbon atoms to hydrogen atoms is almost in a 1:1 ratio. None of the other hydrocarbon families have that kind of carbon-to-hydrogen ratio. The structure of the compound is ringed with six carbons, sometimes referred to as the benzene ring.

Aromatics were thought at one time to be unsaturated because the structure was thought to have double bonds. (See the first-theory benzene structure in Figure 5.38.) The structure appeared to have three double bonds to satisfy the octet rule of bonding. However, in reality, aromatics do not behave like unsaturated compounds. They burn with incomplete combustion. They are unreactive, so it is theorized that instead of three double bonds, they have a unique structure where the six extra electrons are in a state of resonance within the benzene ring. They are not attached to any one of the carbons, but rather go from one to another at a speed faster than the speed of light, much the same way a rotor works inside a distributor in an automobile. The

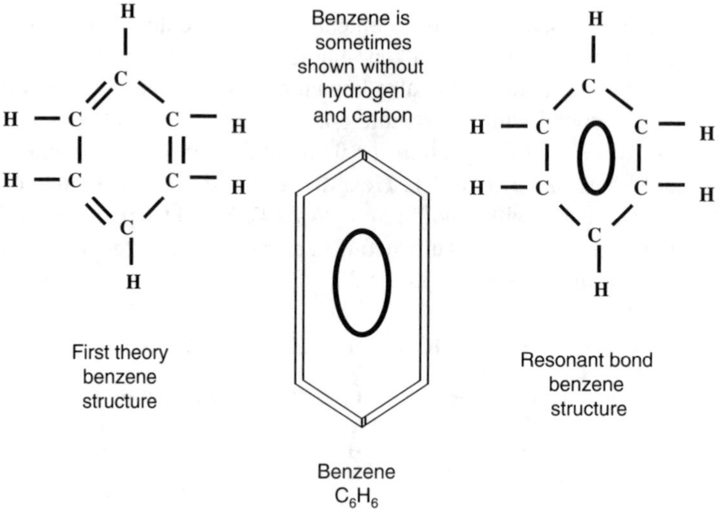

First theory benzene structure

Benzene is sometimes shown without hydrogen and carbon

Benzene C_6H_6

Resonant bond benzene structure

Figure 5.38

bond in the aromatic hydrocarbons is called a resonant bond. Resonant bonding is represented by a large circle located within the benzene ring.

The benzene ring is the backbone of three other aromatic compounds that will be presented here. The first is toluene, with the molecular formula $C_6H_5CH_3$. Sometimes, all of the carbon and hydrogen in the compound are combined and the formula is shown as C_7H_8. Toluene is a benzene ring with one hydrogen atom removed to attach a methyl radical CH_3. Xylene is the next aromatic, with a molecular formula of $CH_3C_6H_4CH_3$. If all of the carbon and hydrogen were added together, the formula would be C_8H_{10}. The benzene ring is the backbone for xylene with two hydrogen atoms removed. Two methyl radicals replace two hydrogen atoms in the ring to make xylene. The structures of benzene, toluene, and xylene are shown in Figure 5.39.

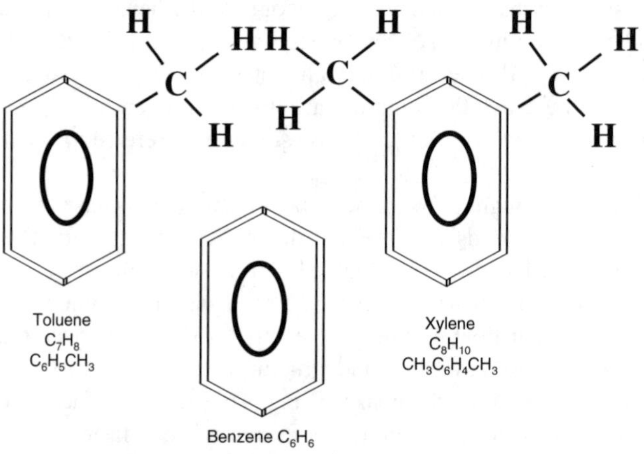

Toluene C_7H_8 $C_6H_5CH_3$

Benzene C_6H_6

Xylene C_8H_{10} $CH_3C_6H_4CH_3$

Figure 5.39

The positioning of the methyl radicals on the xylene ring is of particular importance. There are names for the different positions around the benzene ring. If the methyl radicals are placed on the top and first side position, it is referred to as the "ortho" position; "ortho" refers to "straight ahead." The ortho formula and structure of xylene are shown in Figure 5.40.

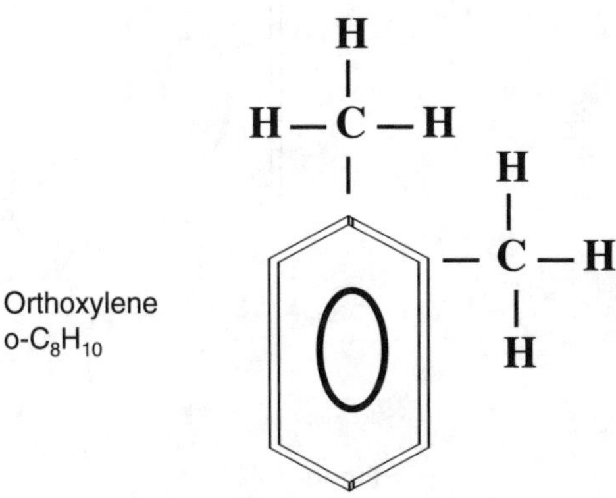

Figure 5.40

If the methyl radicals are placed on the top and on the second side position, it is referred to as the "meta" position; "meta" translates to "beyond." The structure and formula for meta xylene are shown in Figure 5.41.

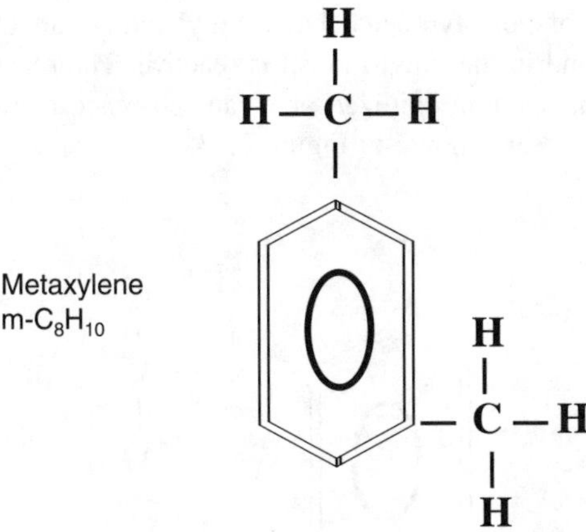

Figure 5.41

If the methyl radicals are on the top and bottom of the ring, it is referred to as the "para" position; "para" means "opposite." The structure and molecular formula for paraxylene are shown in Figure 5.42.

The effects that positioning have on a compound are changes in toxicity and some physical characteristics, such as melting point. The changes to boiling point,

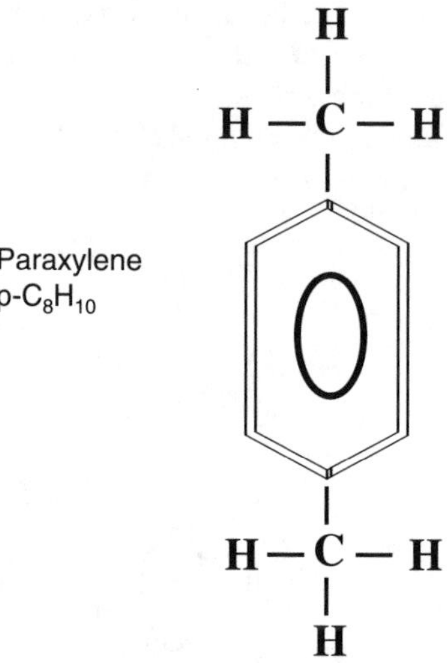

Figure 5.42

flash point, and ignition temperature are insignificant for these isomers of xylene. It is important when looking up these materials that you make sure which xylene is involved in an incident. The "para," "meta," and "ortho" structural isomers of xylene are often used in the making of pesticides.

The final aromatic compound we will discuss is styrene. It is, however, quite different from the other members of the aromatic family. Styrene is a monomer used in the manufacture of polystyrene. It has a vinyl radical attached to the benzene ring. The double bond in the vinyl radical is reactive. The reaction can occur with the oxygen in the air, with an oxidizer, or it can self-react in storage. The structure and molecular formula are shown in Figure 5.43.

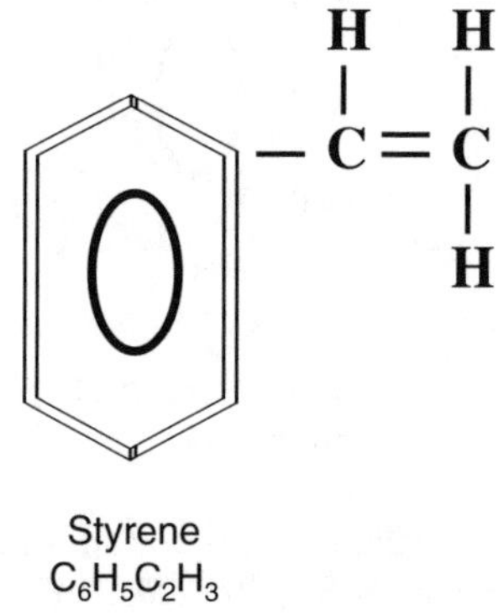

Figure 5.43

Benzene is the parent member of the aromatic hydrocarbon family. It is a colorless to light yellow liquid with a characteristic aromatic odor. Benzene is nonpolar and burns with incomplete combustion, producing a smoky fire. The flammable range is 1.5 to 8% in air. Benzene is also toxic, with a TLV of 10 ppm, and is a known carcinogen. Concentrations of 8000 ppm for 30 to 60 min are fatal. Its boiling point is 176°F, and the flash point is 12°F. The ignition temperature is 928°F. Its vapor density is 2.8, making it heavier than air. Small fires involving benzene should be fought with dry-chemical foams and large fires with hydrocarbon foams. Water may be ineffective and, if used, should be applied gently to the liquid surface. Benzene is immiscible in water and has a specific gravity of 0.9. It is lighter than water and will float on the surface. The four-digit UN identification number for benzene is 1114. The NFPA 704 classification is health 2, flammability 3, and reactivity 0. Benzene reacts with oxidizing materials and should be stored away from them in fixed facilities. It is shipped in 55-gal drums, highway tank trucks, railcars, and barges. Benzene is used in the manufacture of many other chemicals and as a solvent.

Toluene is an aromatic hydrocarbon. It is a colorless liquid with an aromatic odor. It is nonpolar and immiscible in water. Toluene is a dangerous fire risk, with a flammable range of 1.27 to 7% in air. The boiling point is 231°F, with a flash point of 40°F. The ignition temperature of toluene is 896°F. Small fires may be extinguished with dry-chemical foam and large fires with hydrocarbon-type foam. Water may be ineffective and should be applied gently to the surface of the liquid if used. In addition to flammability, toluene is toxic by ingestion, inhalation, and skin absorption, with a TLV of 100 ppm in air. Its vapor density is 3.1, which is heavier than air. The specific gravity is 0.9, so it is lighter than water and will float on the surface. The four-digit UN identification number is 1294. The NFPA 704 designation is health 2, flammability 3, and reactivity 0. It is shipped in 55-gal drums, tank trucks, railcars, and barges.

Xylene and **para-Xylene**, sometimes referred to as dimethylbenzene and xylol, is an aromatic hydrocarbon. The *Condensed Chemical Dictionary* refers to xylene as "a commercial mixture of the three isomers: "ortho," "meta," and "para." Xylene is a clear liquid that is nonpolar and immiscible in water. It is a moderate fire risk, with a flammable range of 0.9 to 7% in air. The boiling point is between 281° and 292°F, depending on the mixture. The flash point ranges from 81° to 90°F, and the ignition temperature ranges from 867° to 984°F. Small fires may be extinguished with dry-chemical foam and large fires with hydrocarbon-type foam. Water may be ineffective and should be applied gently to the surface of the liquid if used. In addition to flammability, xylene is toxic by ingestion and inhalation, with a TLV of 100 ppm in air. The vapor density is 3.7; therefore, it is heavier than air. The specific gravity is 0.9, so the xylenes will float on top of water. The four-digit UN identification number is 1307. The NFPA 704 designation for xylene is health 2, flammability 3, and reactivity 0. It is shipped in 55-gal drums, tank trucks, railcars, and barges.

Styrene is a colorless, oily, liquid aromatic hydrocarbon with a characteristic odor. It is sometimes called vinylbenzene or phenylethylene. Styrene is a monomer and must be inhibited during transportation and storage to prevent polymerization. It is a moderate fire risk, with a flammable range of 1.1 to 6.1%. The boiling point is 295°F, with a flash point of 88°F and an ignition temperature of 914°F. Small fires may be extinguished with dry-chemical foam and large fires with hydrocarbon-type foam. Water may be ineffective and should be applied gently to the surface of the liquid if used. In addition to flammability, styrene is toxic by ingestion and inhalation, with a TLV of 50 ppm in air. The vapor density is 3.6, which is heavier than air. Styrene is nonpolar and it is immiscible in water, with a specific gravity of 0.9. The four-digit UN identification number for styrene monomer, inhibited, is 2055. The NFPA 704 designation is health 2, flammability 3, and reactivity 2. It is shipped in 55-gal drums, tank trucks, railcars, and barges. When stored, it should be kept away from oxidizers.

HYDROCARBON DERIVATIVES

There are seven hydrocarbon-derivative families whose primary hazard is flammability (see Chapter 2, Figure 2.63): alkyl halide, amine, ether, alcohol, ketone, aldehyde, and ester. The alkyl halides, amines, and ethers are nonpolar. The ethers, alcohols, and aldehydes are polar and have wide flammable ranges. Some organic acids are flammable; inorganic acids do not burn. However, flammability is not the primary hazard of most organic acids. They will be discussed in detail in Chapter 10.

Some of the alkyl halides are also flammable, but their primary hazard is toxicity and they will be presented in Chapter 8. Hydrocarbon derivatives are manmade materials. They are made from hydrocarbons by removing hydrogen and adding some other element. The primary elements used in making hydrocarbon derivatives, in addition to carbon and hydrogen, are oxygen, nitrogen, chlorine, fluorine, bromine, and iodine. Once a hydrogen atom is removed from a hydrocarbon, the hydrocarbon becomes a radical. The same prefixes are used for single-bonded hydrocarbon radicals as for the hydrocarbons; however, a "yl" is added to the prefix, indicating that it is a radical. For example, a one-carbon radical is called "methyl," two carbons "ethyl," three carbons "propyl," etc. Remember that hydrogen has been removed, so the radical is not a complete compound. It must be attached to a hydrocarbon-derivative functional group to be complete (see Chapter 2, Figure 2.63). If a compound has more than one radical of the same type (except for ether), prefixes are used to indicate the number of radicals present. The prefixes are "di" for two, "tri" for three, and "tetra" for four. Radicals for single-bonded hydrocarbons are shown in Figure 5.44.

Hydrocarbon compounds with double bonds can also be made into radicals. In order to have a double bond, there must first be at least two carbons; there are no double-bonded radicals with only one carbon. Only two double-bonded hydrocarbon radicals are important here. They are two-carbon and three-carbon radicals with double bonds between the carbons. Since the prefixes for two and three carbons have been used up in the single-bonded compounds, the names for these double-bonded radicals are different from the others: a two-carbon compound with a double bond is called vinyl, which is actually a radical of ethene or ethylene; the three-carbon compound with one double bond is called "acryl," which is a radical of propene or propylene. The structures for the vinyl and acryl radicals are shown in Figure 5.45.

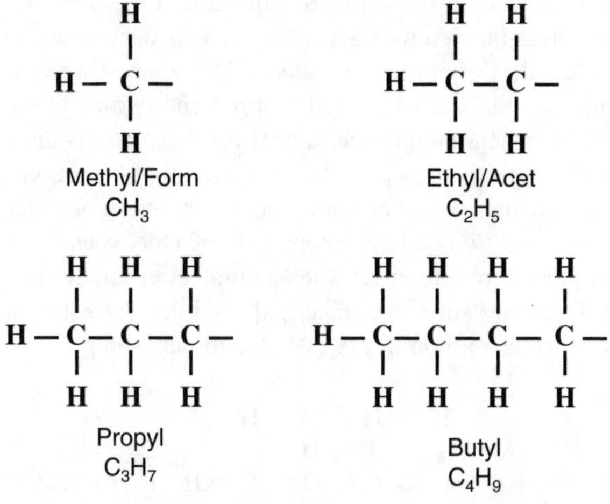

Figure 5.44

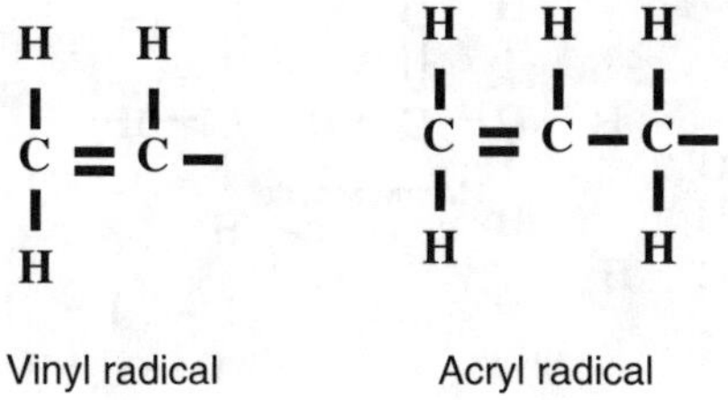

H H
C = C —
H

Vinyl radical

H H H
C = C — C—
H H

Acryl radical

Figure 5.45

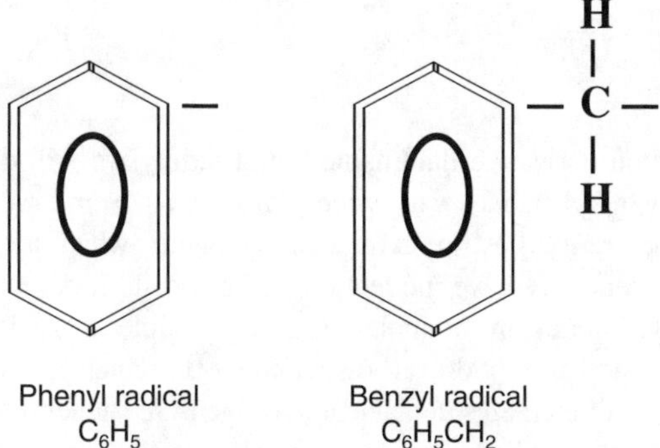

Figure 5.46

Hydrogen can be removed from aromatic compounds to make radicals. If one carbon is removed from benzene, the radical is called "phenyl." If one carbon is removed from toluene, the radical is called "benzyl." The structures and molecular formulas for the phenyl and benzyl radicals are shown in Figure 5.46.

With certain hydrocarbon-derivative functional groups, alternate names for one- and two-carbon, single-bonded radicals are used. This occurs when the radicals are used with the aldehydes, esters, and organic acids. A one-carbon radical for aldehydes, esters, and organic acids is called "form." The two-carbon radical is called "acet." Additionally, when naming the radical for these compounds, the carbon in the functional group is counted as part of the total number of carbons when choosing the prefix. The following examples show the structures, molecular formulas, and names for the one- and two-carbon compounds of aldehydes, esters, and organic acids. (Just a hint for future reference: when naming ester functional groups, nothing is named ester. There are some alternate naming rules for esters based on which radical is attached to the carbon in the ester functional group.)

Formaldehyde $HCHO$

Methyl acrylate $C_2H_3COOCH_3$

Formic acid $HCOOH$

Methyl acetate CH_3COOCH_3

Acetaldehyde CH_3CHO

Acetic acid CH_3COOH

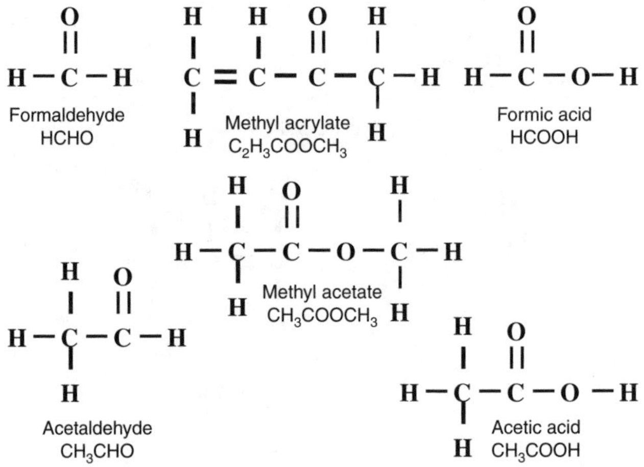

Figure 5.47

Alkyl Halide

The first hydrocarbon-derivative flammable liquid family we will discuss is the alkyl halide. As a family, alkyl halides vary widely in hazards. Some are flammable, some are toxic, and some are used as fire-extinguishing agents, which are not flammable or overly toxic. They generally have moderate boiling and flash points and high ignition temperatures. Alkyl halides are nonpolar. The alkyl halide group is represented by a single hydrocarbon radical or hydrocarbon backbone, to which is attached one or more of the halogen family of elements in place of hydrogen. The general formula is a radical and one of the halogens: **R-X**. Halogens are represented in the general formula by X, because it can be any of the halogens. Generally, fluorine, chlorine, bromine, and iodine may be found in the alkyl halides. They may be by themselves, have more than one atom of an element, or multiple atoms of different elements.

Ethylene dichloride is an alkyl halide hydrocarbon derivative. The primary hazard of the alkyl halides is toxicity. There are, however, some individual alkyl halides that are flammable and classified as flammable liquids. Ethylene dichloride is a colorless, oily liquid with a chloroform-like odor and a sweet taste. It is a dangerous fire risk, with a flammable range of 6 to 16% in air. The boiling point is 183°F, the flash point is 56°F, and the ignition temperature is 775°F. Small fires involving ethylene dichloride should be fought with dry-chemical foam and large fires with hydrocarbon foam. Water may be ineffective and, if used, should be applied gently to the surface of the liquid. Water is generally ineffective against flammable liquid fires where the liquid has a flash point below 100°F. The farther below 100°F the liquid's flash point is, the less effective water will be.

In addition to flammability, ethylene dichloride is toxic by ingestion, inhalation, and skin absorption; it is also a known carcinogen, with a TLV of 10 ppm in air. The vapor density is 3.4, which is heavier than air, so the vapors will stay close to the ground. The specific gravity is 1.3, which makes it heavier than water, so it will sink to the bottom. Alkyl halides are nonpolar, and ethylene dichloride is only slightly miscible in water. The four-digit UN identification number is 1184. The NFPA 704 designation for ethyl dichloride is health 2, flammability 3, and reactivity 0. Ethylene dichloride is shipped in metal cans, drums, tank trucks, railcars, and barges. It is usually packaged under nitrogen gas, which is an inert material. Ethylene dichloride is used in the production of vinyl chloride and trichloroethane. It is also used in metal degreasing, as a paint remover, a solvent, and a fumigant. The structure and molecular formula are shown in Figure 5.48.

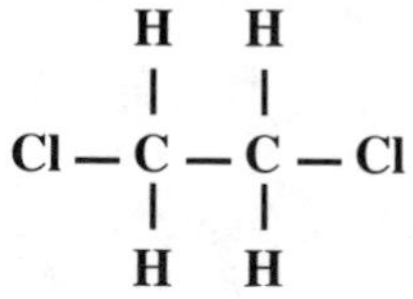

```
     H   H
     |   |
Cl — C — C — Cl
     |   |
     H   H
```

Ethylene dichloride
$C_2H_4Cl_2$

Figure 5.48

Amines

The next flammable liquid hydrocarbon-derivative family we will discuss is the amine. Generally, amines have low boiling points and flash points, narrow flammable ranges, and high ignition temperatures. In addition to being flammable, amines are toxic and irritants. They have a characteristic unpleasant odor, similar to the odor of the bowel or rotten flesh. The amines are considered slightly polar when compared to nonpolar materials. The amine functional group is represented by a single nitrogen atom surrounded by two or fewer hydrogen atoms. Nitrogen requires three bonds to satisfy the octet rule of bonding. The general formulas for the amines are **$R\text{-}NH_2$**, **R_2NH**, and **R_3N**. Nitrogen identifies the amine group, not the numbers of hydrogen attached to the nitrogen. There may be one, two, or three radicals connected to the nitrogen. $R\text{-}NH_2$ indicates that one radical and two hydrogen atoms are attached to the nitrogen. $R_2\text{-}NH$ indicates that two radicals and one hydrogen atom are attached to the nitrogen. $R_3\text{-}N$ indicates that three radicals and no hydrogen atoms are attached to the nitrogen. To name the amines, start with the smallest radical and proceed through however many more radicals there are, in order of size, and end with the word amine. The type of radicals attached to the nitrogen may be the same or different.

Propylamine is a colorless liquid that is slightly soluble in water. The specific gravity is 0.7, which is lighter than water. Propylamine is flammable, with a flammable range of 2 to 10% in air. The boiling point is 120°F, and the flash point is –35°F. The ignition temperature is 604°F. The vapor density is 2, which is heavier than air. In addition to being flammable, propylamine is corrosive and is a strong irritant to skin and tissue. The four-digit UN identification number is 1277. The NFPA 704 designation is health 3, flammability 3, and reactivity 0. It is shipped in glass bottles, cans, drums, and tank cars. The primary uses are as a chemical intermediate and as a lab reagent. The structure and molecular formula for propylamine are shown in Figure 5.49.

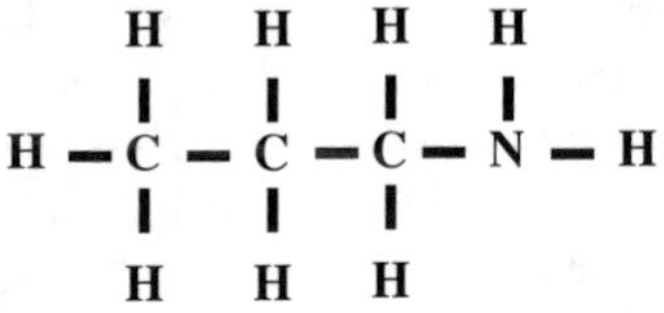

```
     H   H   H   H
     |   |   |   |
 H — C — C — C — N — H
     |   |   |
     H   H   H
```

Propylamine
$C_3H_7NH_2$

Figure 5.48

Butylamine is a colorless, volatile liquid with an amine-like odor. It is miscible with water and has a specific gravity of 0.8, which is lighter than water. It is a dangerous fire risk, with a flammable range of 1.7 to 9.8% in air. The flash point is 10°F, with a boiling point of 172°F. The ignition temperature is 594°F. The vapor density is 2.5, which is heavier than air. Butylamine is also a skin irritant, with a TLV ceiling of 5 ppm in air. The four-digit UN identification number is 1125. The NFPA 704 designation is health 3, flammability 3, and reactivity 0. It is shipped in glass bottles, cans, drums, and tank cars. The primary uses are in the manufacture of pharmaceuticals, insecticides, dyes, and rubber chemicals. The structure and molecular formula for butylamine are shown in Figure 5.50.

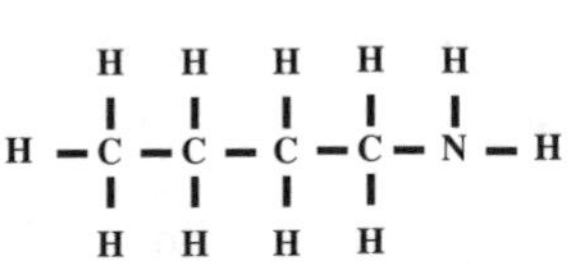

```
     H   H   H   H   H
     |   |   |   |   |
 H — C — C — C — C — N — H
     |   |   |   |
     H   H   H   H
```

Butylamine
$C_4H_9NH_2$

Figure 5.50

Isopropylamine is a colorless, volatile liquid. It is highly flammable, with a flammable range of 2 to 10.4% in air. The boiling point is 93°F, the flash point is

−15°F, and the ignition temperature is 756°F. It is miscible in water, with a specific gravity of 0.69, which is lighter than water. The vapor density is 2.04, which is heavier than air. In addition to flammability, isopropylamine is a strong irritant to tissue and has a TLV of 5 ppm in air. The four-digit UN identification number is 1221. The NFPA 704 designation for isopropylamine is health 3, flammability 4, and reactivity 0. The primary uses for isopropylamine are pharmaceuticals, dyes, insecticides, and as a dehairing agent. The structure and formula for isopropylamine are shown in Figure 5.51.

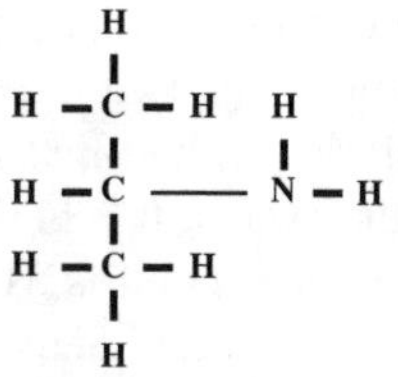

```
       H
       |
   H — C — H   H
       |       |
   H — C ———— N — H
       |
   H — C — H
       |
       H
```

Isopropylamine
$C_3H_7NH_2$

Figure 5.51

Ethers

The next flammable liquid hydrocarbon-derivative family is ether. The primary hazard of ether is flammability. Ethers have low boiling and flash points, low ignition temperatures, and are nonpolar. In addition to being flammable, ethers are anesthetic; they have wide flammable ranges, from 2 to 48% in air; and can form explosive peroxides as the ether ages. Ethers are nonpolar materials. When a container of ether is opened, oxygen from the air gets inside and bonds with the oxygen in the ether, forming an unstable peroxide. Heat, friction, or shock can cause the peroxide to explode. Oxygen has been known to permeate the soldered seam in a metal container even without the container being opened.

Ethers can become quite dangerous in storage. Most ether should not be stored longer than six months. If an aging container of ether is discovered, the nearest bomb squad should be called; the container should be treated as if it were a bomb. Ether is composed of a single oxygen atom with two hydrocarbon radicals, one on either side of the oxygen. It is expressed by the general formula **R-O-R**. Ether names do not always follow the trivial naming system. However, the formula will have a single oxygen atom, which indicates the ether family. The radicals on either side of the oxygen may be the same or they may be different. Ether compounds are named by identifying the two hydrocarbon radicals and ending with the word "ether." As a general rule, if the radicals are different, start with the smallest and name it, then name the second radical, and end with the word "ether." For example, the coupound methyl ethyl ether has a methyl radical and an ethyl radical on either side of the oxygen. The smallest radical is methyl, the second radical is ethyl, and the name ends with "ether."

There may be occasions when a compound is looked up in a reference book, but it is not listed by the trivial naming system. When this happens, just look it up using the other radical name, such as ethyl methyl ether. When the radicals on each side of the oxygen in the ether compound are the same, the compound is named with just the one radical name. No prefix is used to indicate two of the same radical. Because ether must have two radicals, and there is only one radical in the name, it is understood that there are two of the same radical. For example, methyl ether has two methyl radicals, one on each side of the oxygen. Ether is the only hydrocarbon derivative where the prefixe "di," indicating the number of radicals, is not used with the trivial naming system. However, even though uncommon, the "di" prefix is used and listed as a synonym for the common name of the ether compound. Therefore, it is not wrong to use "di" to identify the two radicals, it is just not common.

Ethyl methyl ether is a colorless liquid that is soluble in water. The specific gravity is 0.70, which is lighter than water. It is highly flammable, with a flammable range of 2 to 10.1% in air. The boiling point is 51°F, the flash point is –35°F, and the ignition temperature is 374°F. The vapor density is 2.07, which is heavier than air. In addition to flammability, ethyl methyl ether is an anesthetic and can form explosive peroxides as it ages. The four-digit UN identification number is 1039. The NFPA 704 designation is health 1, flammability 4, and reactivity 1. The primary use is in medicine as an anesthetic. The structure and molecular formula for ethyl methyl ether are shown in Figure 5.52.

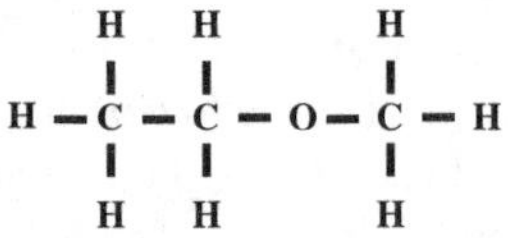

```
    H   H       H
    |   |       |
H — C — C — O — C — H
    |   |       |
    H   H       H
```

Ethyl Methyl Ether
$C_2H_5OCH_3$

Figure 5.52

Ethyl ether (diethyl ether) is a colorless, volatile, mobile liquid. It is slightly soluble in water, with a specific gravity of 0.7, which is lighter than water. It is a severe fire and explosion risk when exposed to heat or flame. The compound forms explosive peroxides from the oxygen in the air as it ages. The flammable range is wide, from 1.85 to 48% in air. The boiling point is 95°F, the flash point is –49°F, and the ignition temperature is 356°F. The vapor density is 2.6, which is heavier than air. In addition to flammability, it is an anesthetic, which causes central nervous system depression by inhalation and skin absorption, with a TLV of 400 ppm in air. The four-digit UN identification number is 1155. The NFPA 704 designation is health 1, flammability 4, and reactivity 1. The primary uses are in the manufacture of smokeless powder, as an industrial solvent, in analytical chemistry, and as an anesthetic. The structure and molecular formula for ethyl ether are shown in Figure 5.53.

Isopropyl ether (diisopropyl ether) is a colorless, volatile liquid, which is slightly soluble in water. The specific gravity is 0.7, which is lighter than water. It is highly flammable, with a wide flammable range of 1.4 to 21% in air. The boiling

```
    H   H       H   H
    |   |       |   |
H — C — C — O — C — C — H
    |   |       |   |
    H   H       H   H
```

Ethyl Ether
$C_2H_5OC_2H_5$

Figure 5.53

point is 156°F, the flash point is −18°F, and the ignition temperature is 830°F. The vapor density is 3.5, which is heavier than air. In addition to flammability, isopropyl ether is toxic by inhalation and a strong irritant, with a TLV of 250 ppm in air. The four-digit UN identification number is 1159. The NFPA 704 designation is health 1, flammability 3, and reactivity 1. The primary uses are as a solvent and in rubber cements. The structure and molecular formula for isopropyl ether are shown in Figure 5.54.

```
     H               H
     |               |
H — C — H       H — C — H
     |               |
H — C ——— O ——— C — H
     |               |
H — C — H       H — C — H
     |               |
     H               H
```

Isopropyl ether
$C_3H_7OC_3H_7$

Figure 5.54

Butyl ether (dibutyl ether) is a colorless, stable liquid, with a mild ether-like odor. It is immiscible in water, with a specific gravity of 0.8, which is lighter than water. Butyl ether is a moderate fire risk, and will form explosive peroxides on aging. The flammable range is 1.5 to 7.6% in air, with a boiling point of 286°F and a flash point of 77°F. The ignition temperature is 382°F, and the vapor density is 4.5, which is heavier than air. In addition to flammability, butyl ether is toxic on prolonged inhalation. The four-digit UN identification number is 1149. The NFPA 704 designation is health 2, flammability 3, and reactivity 1. The primary use is as a solvent. The structure and molecular formula for butyl ether are shown in Figure 5.55.

```
    H   H   H   H       H   H   H   H
    |   |   |   |       |   |   |   |
H — C — C — C — C — O — C — C — C — C — H
    |   |   |   |       |   |   |   |
    H   H   H   H       H   H   H   H
```

Butyl ether
$C_4H_9OC_4H_9$

Figure 5.55

Methyl *tert*-butyl ether is an ether hydrocarbon derivative and is highly flammable, with a wide flammable range. Methyl *tert*-butyl ether is not considered toxic; however, it is mildly irritating to the eyes and skin. If inhaled, it may cause suffocation. It is a gasoline additive, and there has been some controversy concerning potential health effects. Methyl *tert*-butyl ether is added to gasoline as an oxygenating compound that makes gasoline burn cleaner in the winter. The Centers for Disease Control conducted a study that was inconclusive: it did not vindicate the material; it just did not find enough evidence that the chemical is a health concern.

Methyl *tert*-butyl ether is a colorless, nonpolar liquid with an anesthetic-like odor. It has a boiling point of 131°F and a flash point of −14°F. Extinguishing agents for ethers and other nonpolar, nonmiscible, or slightly miscible liquids should be selected carefully. Small fires can be extinguished with dry-chemical foam with some difficulty; remember that ether has an oxygen atom in the compound, so excluding atmospheric oxygen may not be effective. Water may also be ineffective. Alcohol-type foams may be effective against materials that are slightly miscible. The higher-molecular weight liquids will attack the alcohol-type foam, and hydrocarbon foam will be needed. The vapor density is 3, which makes it heavier than air. The specific gravity is 0.74, which is less than the weight of water; therefore, it will float on top of water. Methyl *tert*-butyl ether has a four-digit UN identification number of 2398. It is used primarily as an octane booster for unleaded gasoline. There is no NFPA 704 data available for methyl *tert*-butyl ether, and the reference information on the material is sketchy. No ignition temperature or flammable range information is available in the common reference sources, including CAMEO (Computer-Aided Management of Emergency Operations). The structure and molecular formula for methyl *tert*-butyl ether are shown in Figure 5.56.

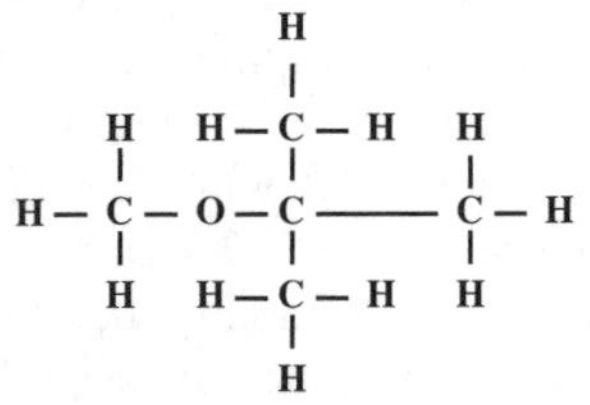

```
                 H
                 |
     H    H — C — H    H
     |        |        |
H — C — O — C ——— C — H
     |        |        |
     H    H — C — H    H
                 |
                 H
```

Methyl tert-butyl ether
$CH_3OtC_4H_9$

Figure 5.56

Propylene is a cyclic ether hydrocarbon derivative, although it does not follow the trivial naming system for ethers. It has a cyclic structure between the oxygen atom and the two carbon atoms, and is a colorless liquid with an ether-like odor. Propylene oxide is nonpolar, partially soluble in water, and is highly flammable, with a wide flammable range of 2 to 22% in air. The boiling point is 94°F, the flash point is 35°F, and the ignition temperature is 840°F. Small fires can be extinguished with dry-chemical foam with some difficulty. Remember that ether has an oxygen atom in the compound, so excluding atmospheric oxygen may not be effective. Water may also be ineffective. Alcohol-type foams may be effective against materials that are slightly miscible. The higher-molecular weight liquids will attack the alcohol-type foam, and hydrocarbon foam will need to be used. In addition to flammability, propylene oxide is an irritant, with a TLV of 20 ppm in air, and it is also corrosive. The vapor density is 2, so it is heavier than air. The specific gravity is 0.83, which is lighter than water, and it will float on the surface. The four-digit UN identification number is 1280. The NFPA 704 designation for propylene oxide is health 3, flammability 4, and reactivity 2. It is shipped in steel cylinders, tank trucks, railcars, and barges under the cover of nitrogen, which is an inert material. The structure and molecular formula for propylene oxide are shown below:

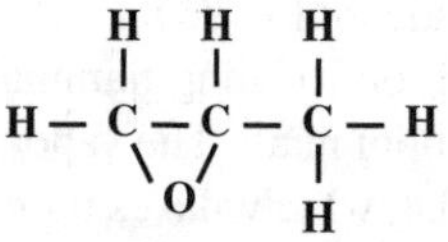

```
     H   H   H
     |   |   |
 H — C — C — C — H
      \ /    |
       O     H
```

Propylene oxide
CH_2OCHCH_3

Figure 5.57

Alcohol

The next flammable liquid hydrocarbon-derivative family is alcohol. In addition to being flammable, alcohols have wide flammable ranges from 1 to 36% in air, and are toxic to some degree. They have high boiling points, moderate flash points, and high ignition temperatures. Small fires involving alcohols should be fought with dry-chemical fire extinguishers. Large fires should be fought with alcohol-type foam; water may be ineffective. Alcohols are miscible with water. Water, as it mixes with alcohol, will at some point raise the boiling and flash points of the alcohol until the mixture of the water and the alcohol are no longer flammable. The problem with this is that the container must be large enough to hold the mixture, or this method of extinguishment will not work. Therefore, care must be taken in choosing the method of extinguishing alcohol fires. Methyl alcohol, or methanol as it is sometimes called, is toxic by ingestion and can cause blindness or death, with a TLV of 200 ppm. Ethyl alcohol or ethanol, also referred to as grain alcohol, is consumed in alcoholic beverages. It is classified as a depressant drug; too much of it can produce toxic effects and can lead to liver damage.

The alcohol functional group is identified by the general formula **R-O-H**. There is one radical attached to the oxygen atom. Alcohols are polar liquids because they have hydrogen bonding. Alcohol is the second most-polar functional group; the most polar are the organic acids. Because of polarity, alcohols are miscible with water and require the use of polar solvent or alcohol-type foam to extinguish fires. Because of polarity, alcohols as a family have high boiling and flash points.

An alcohol is named by identifying the radical attached to the oxygen. The radical is named first, and the compound ends in the word "alcohol." There are often various ways of naming the same chemical compounds. With alcohol, an "ol" ending may be added to the radical, indicating that it is an alcohol. For example, the radical "methyl" is attached to the oxygen in the alcohol functional group. The name of the compound is methyl alcohol or may be called methanol.

Methyl alcohol also known as methanol and wood alcohol, is an alcohol hydrocarbon derivative. It is a clear, colorless liquid that is highly polar and miscible in water. Alcohol has hydrogen bonding and is the second most-polar material of the hydrocarbon derivatives after organic acids. It is a dangerous fire risk, with a wide flammable range from 6 to 36.5% in air. Fighting fires will require the use of alcohol-type foam. The boiling point is 147°F, and the flash point is 52°F. Notice in the figure that the molecular weight of methyl alcohol is 32. It has only one carbon atom and one oxygen atom, yet the boiling point and flash point are high. The ignition temperature of methyl alcohol is 867°F. The effects of polarity can readily be seen in this example. In addition to being flammable, methyl alcohol is toxic by ingestion and has a TLV of 200 ppm in air. The vapor density is 1.1, which is heavier than air. The specific gravity is 0.8, which makes it lighter than water. Methyl alcohol is miscible in water, so it will mix rather than form defined layers. The four-digit UN identification number is 1230. The NFPA 704 designation is health 1, flammability 3, and reactivity 0. Methyl alcohol is shipped in glass bottles, 55-gal drums, tank trucks, and railcars. The structure and molecular formula for methyl alcohol are shown in Figure 5.58.

```
     H
     |
H— C — O— H
     |
     H
```

Methyl alcohol
CH_3OH

Figure 5.58

Ethyl alcohol, or ethanol, is a colorless, clear, volatile liquid. It is polar and soluble in water, with a specific gravity of 0.8, which is lighter than water. It is highly flammable with a flammable range of 3.3 to 19% in air. The boiling point is 173°F, and the flash point is 55° F. The vapor density is 1.6, which is heavier than air. The ignition temperature is 685°F, and ethanol is classified as a depressant drug with a TLV of 1000 ppm in air. The four-digit UN identification number is 1170. The NFPA 704 designation is health 0, flammability 3, and reactivity 0. The primary uses are as a solvent, in beverages, antifreeze, gasohol, pharmaceuticals, and explosives. The structure and molecular formula for ethyl alcohol is shown below.

```
     H    H
     |    |
H— C — C — O— H
     |    |
     H    H
```

Ethyl alcohol
C_2H_5OH

Figure 5.59

Propyl alcohol, 1-propanol, is a colorless liquid with an odor similar to ethanol. It is polar, soluble in water, and has a specific gravity of 0.8, which is lighter than water. It is a dangerous fire risk with a flammable range of 2 to 13% in air. The vapor density is 2.1, which is heavier than air. The boiling point is 207°F, the flash point is 74°F, and the ignition temperature is 775°F. It is toxic by skin absorption with a TLV of 200 ppm in air. The four-digit UN identification number is 1274. The NFPA 704 designation is health 1, flammability 3, and reactivity 0. The primary uses are in brake fluid, as a solvent, and as an antiseptic. The structure and molecular formula for propyl alcohol is shown below.

```
    H   H   H
    |   |   |
H — C — C — C — O — H
    |   |   |
    H   H   H
```

Propyl alcohol
C_3H_7OH

Figure 5.60

n-Butyl alcohol is an alcohol hydrocarbon derivative, and is a colorless liquid with a wine-like odor. Alcohols are highly polar and are miscible in water. Butyl alcohol is a moderate fire risk, with a flammable range of 1.4 to 11.2% in air. The boiling point is 243°F, the flash point is 98°F, and the ignition temperature is 650°F. Fires should be fought with alcohol-type foams. In addition to being flammable, butyl alcohol is toxic when inhaled for long periods, is irritating to the eyes, and is absorbed through the skin. The TLV is 50 ppm in air. The vapor density of butyl

Cone-roof tank for flammable liquids.

alcohol is 2.6, so it is heavier than air. The specific gravity is 0.8, which is lighter than water; however, it is miscible in water and will mix rather than form layers. The four-digit UN identification number is 1120. The NFPA 704 designation for butyl alcohol is health 1, flammability 3, and reactivity 0. It is shipped in glass bottles, pails, 55-gal drums, tank trucks, railcars, and barges. The structure and molecular formula for n-butyl alcohol is shown in Figure 5.61.

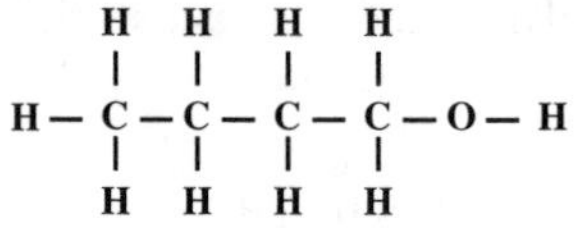

```
     H   H   H   H
     |   |   |   |
H— C — C — C — C — O — H
     |   |   |   |
     H   H   H   H
```

Butyl alcohol
C_4H_9OH

Figure 5.61

Isopropyl alcohol is an alcohol hydrocarbon derivative. The liquid is colorless, with a pleasant odor, highly polar, and miscible in water. Firefighting will require polar-solvent foam for extinguishment. Isopropyl alcohol is highly flammable, with a flammable range of 2 to 12% in air. The boiling point is 181°F, the flash point is 53°F, and the ignition temperature is 750°F. In addition to flammability, isopropyl alcohol is toxic by ingestion and inhalation, with a TLV of 400 ppm in air. The vapor density is 2.1, which is heavier than air. The specific gravity is 0.8, which is lighter than water; however, it is miscible in water and will mix rather than form layers. The four-digit UN identification number is 1219. The NFPA 704 designation is health 1, flammability 3, and reactivity 0. It is shipped in glass bottles, pails, 55-gal drums, tank trucks, railcars, and barges. The structure and molecular formula for isopropyl alcohol are shown in Figure 5.62.

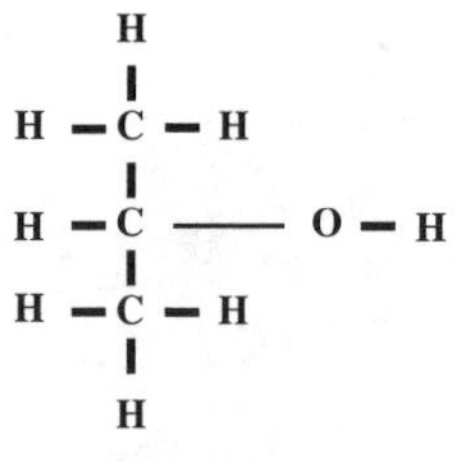

```
      H
      |
H — C — H
      |
H — C ——— O — H
      |
H — C — H
      |
      H
```

Isopropyl alcohol
C_3H_7OH

Figure 5.62

ISOMERS

Isomers were introduced in Chapter 4. An isomer is a compound with the same formula as the "normal" compound, but a different structure. Isomers are sometimes referred to as branched compounds. Branching of a compound has the effect of lowering the boiling point. In the Hydrocarbon section of Chapter 4, only the "iso" branch was discussed. With the hydrocarbon-derivative functional groups, there will also be an "iso" branch and, in addition, there will be secondary and tertiary branches.

When dealing with the hydrocarbon compounds, the branch is a carbon atom attached to one of the center carbon atoms of the chain. In the derivatives, the functional group is a part of that carbon chain and is considered when determining branching of the compound. The types of branches will be shown in this section because they occur commonly with the alcohol compounds. However, branching can occur in any of the hydrocarbon-derivative groups. Shown in the following illustration (see Figure 5.63), are examples of branches of a four-carbon alcohol. The branch is determined by the location of the functional group on the carbon chain. The first structure is the straight-chained compound butyl alcohol with a molecular formula of C_4H_9OH. Straight-chained compounds are sometimes referred to as the "normal" form. Normal butyl alcohol is represented by a small "n" in front of the compound name and molecular formula. The next structure is isobutyl alcohol. The "iso" branch is determined by locating the -O-H of the alcohol functional group and using it as an entry point into the structure. Then go to the first carbon that is attached to the -O-H. See how many carbon atoms are attached to the first carbon. In the case of the "iso" branch, only one carbon atom is attached. The third structure is secondary butyl alcohol. The -O-H is attached to a carbon that is attached to two other carbons. In the final structure, the compound is called tertiary butyl alcohol. The functional group is attached to a carbon atom, which is attached to three other carbon atoms. Notice that all of the compounds have the same molecular formula.

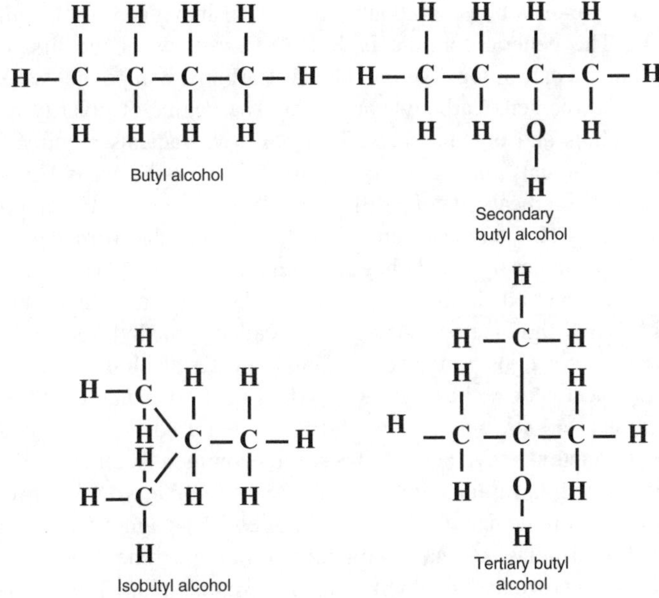

Figure 5.63

To distinguish between them, it is necessary to include a small letter indicating which branch is in the structure. A small "i" is used for "iso" branches, a small "s" for secondary, and a small "t" for tertiary branches.

There is one exception to the branching rules of hydrocarbon derivatives. Propane cannot be branched as a straight hydrocarbon because there are not enough carbon atoms to create a branch. However, in the derivatives, the functional group becomes a part of the carbon chain for the purpose of determining branching. It is possible to put a functional group on the center carbon atom of propane. The structure formed appears to be secondary, according to the examples shown. However, there is only one way to branch propane, so it is called the "iso" branch. That is something that has to be committed to memory as an exception to the branching rules. Another way to remember the exception is that secondary and tertiary are not used until there are four carbons in a compound. Isopropyl alcohol is shown in Figure 5.62.

Denatured alcohol is ethyl alcohol, or ethanol, to which another liquid has been added to make it unfit to use as a beverage. The primary reason for denaturing is for tax purposes. There are approximately 50 formulations of denatured alcohol. The hazards are the same as for ethanol. The primary uses for denatured alcohol are in the manufacture of acetaldehyde, solvents, antifreeze, brake fluid, and fuels.

Ketone

The next hydrocarbon-derivative family is ketone. As a group, the ketones are flammable and narcotic. They have moderate boiling and flash points, narrow flammable ranges, and high ignition temperatures. Ketones are polar, and fires should be fought with alcohol or polar-solvent type foams because water may be ineffective. They are made up of a carbon atom double-bonded to an oxygen atom, with a radical on each side. The general formula is **R-C-O-R**. Ketone is the first of several compounds that are part of the carbonyl family. Carbonyl compounds have a carbon double-bonded to oxygen. Carbonyls are polar. The degree of polarity is less than that of the alcohols and organic acids. There are two radicals required in ketone compounds. The radicals may be the same, in which case the prefix "di" is used to indicate two, or the radicals may be different. When naming these compounds, the smallest is named first, then the second radical, ending in the word "ketone." Some ketones have trade names by which they are commonly known. There are no naming rules that can be used to determine trade names. There are some hints, such as with acetone, which is a three-carbon ketone, also called dimethyl ketone. The "one" ending indicates ketone, just as the "ol" ending indicates alcohol. The ending would be a tip-off that acetone may be a ketone. DMK is the trade name used for dimethyl ketone, which is also known as acetone. MEK is often used as a shortened name or trade name for methyl ethyl ketone. These are common ketones, and the more familiar you are with hazardous materials, the more familiar you will become with alternate names and trade names. Shown in Figures 5.64–Figure 5.67 are the structures, molecular formulas, and names for four common ketones.

Acetone, also known as dimethyl ketone, is a ketone hydrocarbon derivative. It is a colorless, volatile liquid with a sweetish odor. Acetone is a carbonyl, is polar, and is miscible with water, which means it will require polar-solvent foam for fire extinguishment. This compound is highly flammable, with a range of 2.6 to 12.8% in air. The boiling point is 133°F, the flash point is -4°F, and the ignition temperature is 869°F. Fighting fires will require the use of alcohol-type foam. In addition to flammability, acetone is a narcotic at high concentrations, and is moderately toxic by ingestion and inhalation, with a TLV of 750 ppm in air. The vapor density is 2, which is heavier than air. The specific gravity is 0.8, so it is lighter than water, but is miscible and will mix with the water rather than form layers. The four-digit UN identification number is 1090. The NFPA 704 designation is health 1, flammability 3, and reactivity 0. It is shipped in pails, drums, tank trucks, railcars, and barges. Acetone is the solvent used to dissolve acetylene in cylinders to keep it stable. The structure and molecular formula are shown in Figure 5.64.

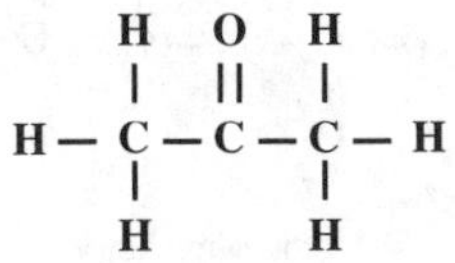

```
     H   O   H
     |   ||  |
H -  C - C - C - H
     |       |
     H       H
```

Dimethyl ketone
DMK
Acetone
CH_3COCH_3

Figure 5.64

Methyl ethyl ketone, **MEK**, is a colorless liquid with an acetone-like odor. MEK is polar, soluble in water, and has a specific gravity of 0.8, which is lighter than water. It is highly flammable, with a flammable range of 2 to 10% in air, a flash point of 1°F and a boiling point of 176°F. The ignition temperature of MEK is 759°F, and the vapor density is 2.5, which is heavier than air. It is toxic by inhalation, with a TLV of 200 ppm in air. The four-digit UN identification number is 1193. The NFPA 704 designation for MEK is health 1, flammability 3, and reactivity 0. The primary uses of MEK are as a solvent and in the manufacture of smokeless powder, cleaning fluids, in printing, and acrylic coatings. The structure and molecular formula for methyl ethyl ketone is shown in Figure 5.65.

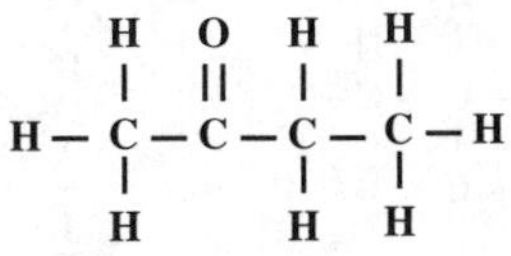

```
     H   O   H   H
     |   ||  |   |
H -  C - C - C - C - H
     |       |   |
     H       H   H
```

Methyl ethyl ketone
MEK
$CH_3COC_2H_5$

Figure 5.65

Methyl vinyl ketone, MVK, or vinyl methyl ketone, is a colorless liquid that is soluble in water. It is polar, with a specific gravity of 0.8636, which is lighter than water. It is flammable, with a flammable range of 2.1 to 15.6% in air, and the vapor density is 2.4, which is heavier than air. The boiling point is 177°F, the flash point is 20°F, and the ignition temperature is 915°F. MVK is a skin and eye irritant. The four-digit UN identification number is 1251. The NFPA 704 designation is health 4, flammability 3, and reactivity 2. The primary uses are as a monomer for vinyl resins and as an intermediate in steroid and vitamin A synthesis. The structure and molecular formula for methyl ethyl ketone are shown in Figure 5.66.

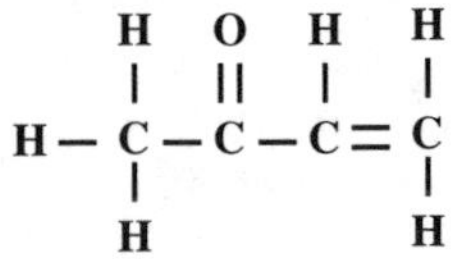

```
     H   O   H   H
     |   ||  |   |
H -  C - C - C = C
     |           |
     H           H
```

Methyl vinyl ketone
$CH_3COC_2H_3$

Figure 5.66

Methyl isobutyl ketone is a colorless, stable liquid with a pleasant odor. It is slightly soluble in water, with a specific gravity of 0.8, which is lighter than water. The vapor density is 3.5, which is heavier than air. It is highly flammable, with a flammable range of 1.2 to 8% in air. The boiling point is 244°F, the flash point is 64°F, and the ignition temperature is 840°F. It is toxic by inhalation, ingestion, and skin absorption, with a TLV of 50 ppm in air. The four-digit UN identification number is 1245. The NFPA 704 designation is health 2, flammability 3, and reactivity 1. The primary uses are as a solvent for paints, varnishes, and lacquers; in the extraction of uranium from fission products; and as a denaturant for alcohol. The structure and molecular formula for methyl isobutyl ketone are shown in Figure 5.67.

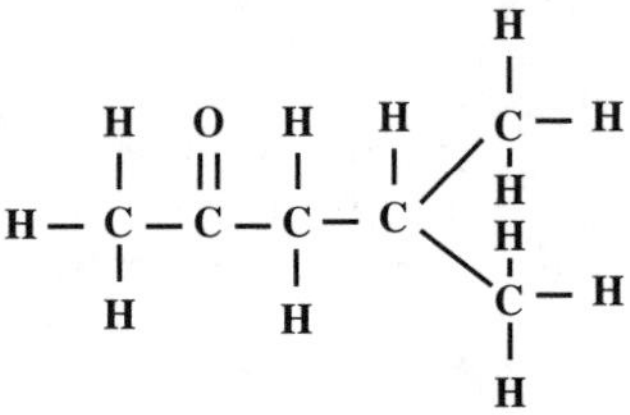

```
                           H
                           |
     H   O   H   H         C - H
     |   ||  |   |       / |
H -  C - C - C - C         H
     |       |       \     H
     H       H         \   |
                           C - H
                           |
                           H
```

Methyl isobutyl ketone
$CH_3COiC_4H_9$

Figure 5.67

Aldehyde

Aldehydes have a wide flammable range from 3 to 55% in air; they are toxic and may polymerize. They have moderate boiling and flash points and high ignition temperatures. Fires involving aldehydes should be fought with polar-solvent-type foams because water may be ineffective. Aldehydes may also form explosive peroxides as they age, much the same way ethers do. Aldehydes are composed of a carbon atom double-bonded to an oxygen atom with a hydrogen atom on the other carbon connection. Aldehydes are carbonyls and, therefore, polar. The degree of polarity is much the same as ketone and ester, and much less than alcohol and organic acid. They are miscible in water and require the use of polar-solvent foams to extinguish fires. Aldehydes have the general formula of **R-CHO**. There is one radical attached to the carbon atom of the aldehyde functional group. Aldehydes are one of the three derivatives in which the carbon atom in the functional group is counted when naming the compound. The alternate terms for one- and two-carbon radicals are also used with the aldehydes. A one-carbon aldehyde uses "form," and a two-carbon uses "acet." The aldehydes are named by identifying the radical, naming it, and ending with the word "aldehyde." Aldehydes may also be named in the same manner as the alternate names for alcohols; however, with the aldehydes, the ending is "al" instead of "ol." For example, a one-carbon aldehyde is called formaldehyde, with the alternate name of methanal. In the following examples, the structures, molecular formulas, and names are shown for one-, two-, and three-carbon aldehydes. The polarity of the carbonyls is somewhat less than that of the alcohols and the organic acids.

Acetaldehyde, also known as ethanal, is a colorless liquid with a pungent, fruity odor. The odor is detectable at 0.07 to 0.21 ppm in air. It is highly flammable and a dangerous fire and explosion risk, with a wide flammable range of 4 to 60% in air. The boiling point is 69°F, the flash point is –36°F, and the ignition temperature is 374°F. Acetaldehyde is miscible in water, and the specific gravity is 0.78, which is lighter than water. The vapor density is 1.52, which is heavier than air. In addition to flammability, it is toxic (narcotic) and has a TLV of 100 ppm in air. Eye irritation occurs at 25 to 50 ppm in air. The four-digit UN identification number is 1089. The NFPA 704 designation is health 3, flammability 4, and reactivity 2. Its primary use is in the manufacture of other chemicals and artificial flavorings. The structure and molecular formula are shown in Figure 5.68.

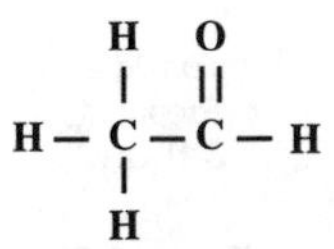

```
     H   O
     |   ||
 H — C — C — H
     |
     H
```

Acetaldehyde
Ethanal
CH_3CHO

Figure 5.68

Open floating-roof tank used for hydrocarbon fuels.

Propionaldehyde, also known as propanal and propylaldehyde, is a water-white liquid with a suffocating odor that is water-soluble. It is a dangerous fire and explosion risk, with a flammable range of 3 to 16% in air. The boiling point is 120°F, the flash point is 16°F, and the ignition temperature is 405°F. It is partially soluble in water and has a specific gravity of 0.81, which is lighter than water. The vapor density is 0.807, which is lighter than air. In addition to flammability, propionaldehyde is an irritant to the eyes, skin, and respiratory system. The four-digit UN identification number is 1275. The NFPA 704 designation is health 2, flammability 3, and reactivity 2. The primary uses are in the manufacture of other chemicals and plastics, as well as a preservative and disinfectant. The structure and molecular formula for propionaldehyde are shown in Figure 5.69.

```
     H   H   O
     |   |   ||
H  — C — C — C — H
     |   |
     H   H
```

Propionaldehyde
Propanal
C_2H_5CHO

Figure 5.69

Butyraldehyde, also known as butanal, is a water-white liquid with a pungent aldehyde odor. Butyraldehyde is a dangerous fire risk, with a flammable range of 2.5 to 12.5% in air. The boiling point is 168°F, the flash point is 10°F, and the ignition temperature is 446°F. It is slightly soluble in water, with a specific gravity of 0.8, which is lighter than water. The vapor density is 0.804, which is lighter than air. In addition to flammability, butyraldehyde is corrosive and causes severe eye and skin burns. It may be harmful if inhaled. The four-digit UN identification number is 1129. The NFPA 704 designation is health 3, flammability 3, and reactivity 2. The primary uses of butyraldehyde are in plastics and rubber and as a solvent. The structure and molecular formula for butyraldehyde are shown in Figure 5.70.

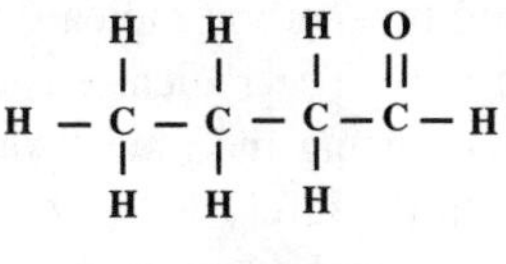

```
    H   H   H  O
    |   |   |  ||
H - C - C - C - C - H
    |   |   |
    H   H   H
```

Butyraldehyde
Butanal
C_3H_7CHO

Figure 5.70

Esters

The final flammable liquid hydrocarbon-derivative family is ester. In addition to being flammable, esters may polymerize. They have moderate boiling and flash points, narrow flammable ranges, and high ignition temperatures. Esters are made through a process referred to as esterification. An ester is formed when an alcohol is combined with an organic acid, with water as a by-product. This process is illustrated in Figure 5.71 by combining acrylic acid and methyl alcohol; the resulting ester compound is methyl acrylate.

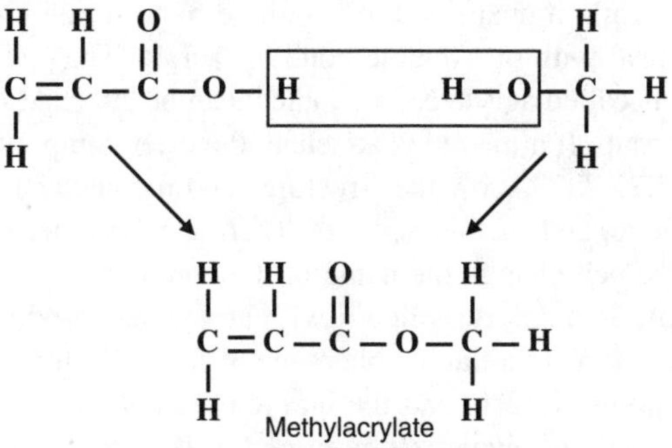

```
H   H   O                         H
|   |   ||                        |
C = C - C - O - H        H - O -  C - H
|                                 |
H                                 H

        H   H   O       H
        |   |   ||      |
        C = C - C - O - C - H
        |               |
        H               H
```

Methylacrylate
$C_2H_3COOCH_3$

Figure 5.71

Esters have a carbon atom double-bonded to one oxygen atom and a single bond with another oxygen atom. Esters are carbonyls and are polar. The degree of polarity is much less than organic acids and alcohols, and is similar to ketones and aldehydes. Esters are miscible in water and require polar-solvent foams when fighting fires. In this book, only three esters will be discussed because of their common commercial use. The general formula for ester is **R-C-O-O-R** or **R-C-O_2-R**. There are two radicals in the ester compounds.

Because nothing is ever called ester, the word "ester" will not appear in the name of the compound. The radical that is attached to the carbon atom in the functional group determines which ester compound it will be. Esters are one of the functional groups in which all the carbon atoms are counted, including the carbon atom in the functional group, when naming the type of ester compound. In addition, esters use the alternate name for one- and two-carbon radicals that are attached to the carbon atom in the functional group only. The radical on the other side is named in the normal way. The ester is named using the radical on the right first and ending in the name of the type of ester on the left.

Certain radicals attached to the carbon atom of the functional group will produce certain esters. For example, in a one-carbon ester, the carbon in the functional group is used as the one carbon. A single hydrogen atom is attached to the carbon atom in the functional group to complete the bonding requirements. A one-carbon ester uses the alternate name for one carbon, which is "form"; esters end in "ate," so the name of a one-carbon ester is formate.

Any radical can be attached to the oxygen atom in the functional group. If a methyl is added to the oxygen atom, the name of the ester compound is methyl formate. When a methyl radical is attached to the carbon atom in the functional group, it forms a two-carbon chain. The alternate name for two carbons, "acet," is used; the ending "ate" is used to indicate an ester, so a two-carbon ester is an acetate. The second radical added to the oxygen in the functional group determines what type of acetate the compound is. Theoretically, any radical can be used. If a vinyl radical were used, the compound would be vinyl acetate.

The last ester has a vinyl radical attached to the carbon in the functional group. This forms a three-carbon chain with one double bond in the chain. The name for a three-carbon radical with one double bond is "acryl." The ending for the ester is "ate"; so the ester is called acrylate. Any radical can be attached to the oxygen atom in the functional group. If a methyl is attached, the ester compound is called methyl acrylate. Figures 5.71–5.73 show the structures and molecular formulas for methyl formate, vinyl acetate, and methyl acrylate. There can be other radicals attached to the oxygen atom, which change the name of the compound.

Methyl formate is a colorless liquid with an agreeable odor. It is a dangerous fire and explosion risk, with a flammable range of 5 to 23% in air. The boiling point is 89°F, the flash point is –2°F, and the ignition temperature is 853°F. In addition to being flammable and a polymerization hazard, it is also an irritant, with a TLV of 100 ppm in air. It is water-soluble and has a specific gravity of 0.98, which is slightly lighter than water. The vapor density is 2.07, which is heavier than air. The four-digit UN identification number is 1243. The NFPA 704 designation is health 2, flammability 4, and reactivity 0. The primary uses of methyl formate are as a solvent, a fumigant, and a larvicide. The structure and molecular formula are shown in Figure 5.72.

Vinyl acetate is an ester hydrocarbon-derivative compound. It is a colorless liquid that has been stabilized with an inhibitor. Although it is a polar compound because of the carbonyl structure, it is only slightly miscible in water. Vinyl acetate is a highly flammable liquid, with a flammable range of 2.6 to 13.4% in air, and it

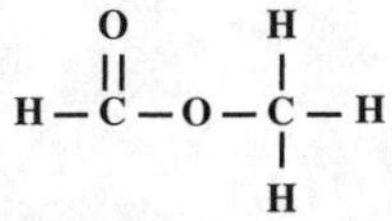

```
      O       H
      ||      |
H  —  C — O — C — H
              |
              H
```

Methyl formate
$HCOOCH_3$

Figure 5.72

may polymerize without the inhibitor or when exposed to heat or an oxidizer during an accident. The boiling point is 161°F, the flash point is 18°F, and the ignition temperature is 756°F. Fighting fires will require the use of alcohol-type foam. In addition to flammability, vinyl acetate is toxic by inhalation and ingestion, with a TLV of 10 ppm in air. The vapor density is 3, so it is heavier than air. The specific gravity is 0.9, which means it will float on water. The four-digit UN identification number is 1301. The NFPA 704 designation is health 2, flammability 3, and reactivity 2. It is shipped in 55-gal drums, tank trucks, railcars, and barges. It should be stored separately from oxidizing materials. The structure and molecular formula for vinyl acetate are shown in Figure 5.73.

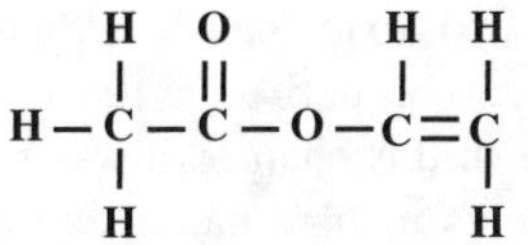

```
    H   O       H   H
    |   ||      |   |
H — C — C — O — C = C
    |               |
    H               H
```

Vinyl acetate
$CH_3COOC_2H_3$

Figure 5.73

Methyl acrylate (inhibited) is a colorless, volatile liquid. It is a dangerous fire and explosion risk, with a flash point of 2.8 to 25% in air. The boiling point is 177° F, the flash point is 27°F, and the ignition temperature is 875°F. It is immiscible in water and has a specific gravity of 0.96, which is lighter than water. The vapor density is 0.957, which is slightly lighter than air. In addition to flammability and polymerization hazards, methyl acrylate is toxic by inhalation, ingestion, and skin absorption. It is an irritant to skin and eyes, with a TLV of 10 ppm in air. The four-digit UN identification number is 1919. The NFPA 704 designation is health 3, flammability 3, and reactivity 2. The primary uses of methyl acrylate are in polymers, in vitamin B_1, and as a chemical intermediate. The structure and molecular formula are shown in Figure 5.71.

All of the chemicals mentioned in this section have a primary hazard of flammability; however, because they are different in chemical makeup, some require different firefighting tactics. Note the differences in physical and chemical characteristics from compound to compound and family to family. The secondary or hidden hazards of flammable liquids vary widely from one chemical to another. Among the

Geodesic dome retrofitted as closed floating-roof tank.

flammable liquids, many are also toxic, anesthetic, narcotic, and undergo polymerization. Many different hydrocarbon-derivative functional groups are represented in the flammable liquids. Earlier in this chapter, it was mentioned that almost 52% of all hazardous materials incidents involve flammable liquids. While many of those spills are hydrocarbon fuels, like gasoline and diesel fuel, they also involve industrial flammable liquids.

OTHER FLAMMABLE LIQUIDS

Acrylonitrile is sometimes referred to as vinyl cyanide. It is a colorless liquid with a mild odor, is nonpolar, and is partially miscible in water. Acrylonitrile is a dangerous fire risk, with a flammable range of 3 to 17% in air. Small fires can be extinguished with dry-chemical foam with some difficulty. Water may also be ineffective. Alcohol-type foams may be effective. The boiling point is 171°F, the flash point is 32°F, and the ignition temperature is 898°F. In addition to flammability, acrylonitrile is toxic by inhalation and skin absorption. Acrylonitrile may polymerize because of the double bond in the vinyl and the triple bond between the carbon and nitrogen; it is always shipped with an inhibitor. The TLV is 2 ppm in air and it is considered a human carcinogen. The vapor density is 1.8, which is heavier than air. The specific gravity is 0.8, which is lighter than water, so it will float on the surface. The four-digit UN identification number is 1093. The NFPA 704 designation is health 4, flammability 3, and reactivity 2. It is shipped in 55-gal drums, tank trucks, railcars, and barges. It should not be stored or shipped uninhibited. The structure and molecular formula are shown in Figure 5.74.

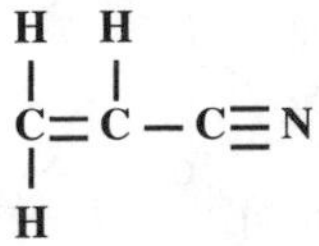

```
H   H
|   |
C = C — C ≡ N
|
H
```

Acrylonitrile
CH_2CHCN

Figure 5.74

Carbon disulfide is a nonmetal compound that is a clear, colorless, or faintly yellow liquid, and almost odorless. It is highly flammable, with a flammable range of 1.3 to 50% in air, and can be ignited by friction. The boiling point is 115°F, the flash point is –22°F, and the ignition temperature is 194°F. Contact with a steam pipe or a lightbulb could ignite carbon disulfide. It is slightly water-soluble and has a specific gravity of 1.26, which is heavier than water. In addition to being highly flammable, carbon disulfide is also a poison and is toxic by skin absorption, with a TLV of 10 ppm in air. The four-digit identification number is 1131. The NFPA 704 designation is health 3, flammability 4, and reactivity 0. The primary uses of carbon disulfide are as a solvent and in the manufacture of rayon, cellophane, and carbon tetrachloride. The structure and molecular formula for carbon disulfide are shown below:

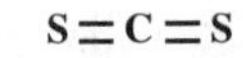

S = C = S

Carbon disulfide
CS_2

Figure 5.75

Cumene, also known as isopropylbenzene, is a colorless liquid that is insoluble in water. The specific gravity is 0.9, which is lighter than water. It is a moderate fire risk, with a flammable range of 0.9 to 6.5%. The boiling point is 306°F, the flash point is 96°F, and the ignition temperature is 795°F. Small fires may be extinguished with dry-chemical foam, and large fires with hydrocarbon-type foam. Water may be ineffective and should be applied gently to the surface of the liquid if used. The vapor density is 4.1, which is heavier than air. In addition to flammability, cumene is toxic by ingestion, inhalation, and skin absorption; it is also a narcotic. The TLV is 50 ppm in air. The four-digit UN identification number is 1918. The NFPA 704 designation is health 2, flammability 3, and reactivity 0. The primary uses are in the production of phenol, acetone, and methylstryene solvents. The structure and molecular formula for cumene are shown in Figure 5.76.

Cyclohexane is a cyclic alkane hydrocarbon with single bonds between the carbons. The liquid is colorless, nonpolar, and immiscible with water. Cyclohexane is highly flammable, with a flammable range of 1.3 to 8% in air. The boiling point is 179°F, the flash point is –4°F, and the ignition temperature is 473°F. Small fires may be extinguished with dry-chemical foam, and large fires with hydrocarbon-type foam. Water may be ineffective and should be applied gently to the surface of the

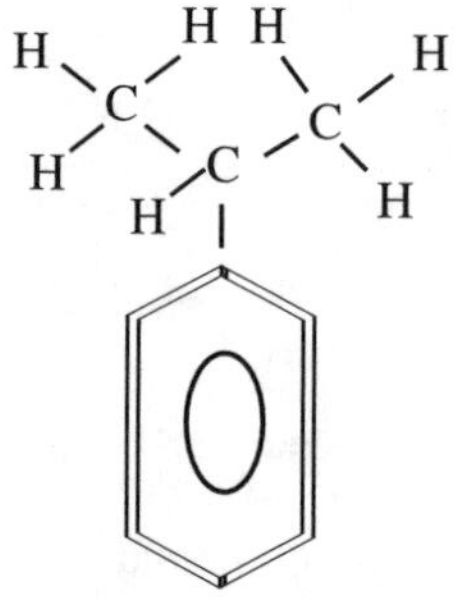

Cumene
$C_6H_5CH(CH_3)_2$

Figure 5.76

liquid if used. In addition to being flammable, cyclohexane is toxic by inhalation, with a TLV of 300 ppm in air. The vapor density is 29, so it is significantly heavier than air. The specific gravity is 0.8, which is lighter than water so it will float on the surface. The four-digit UN identification number is 1145. The NFPA 704 designation for cyclohexane is health 1, flammability 3, and reactivity 0. It is shipped in 55-gal drums, tank trucks, railcars, and barges. The structure and molecular formula of cyclohexane are shown in Figure 5.13.

Ethyl benzene is a colorless, aromatic hydrocarbon with a characteristic odor. It is a dangerous fire risk, with a flammable range of 0.8 to 6.7% in air. The boiling point is 277°F, the flash point is 70°F, and the ignition temperature is 810°F. Small fires may be extinguished with dry-chemical foam, and large fires with hydrocarbon-type foam. Water may be ineffective and should be applied gently to the surface of the liquid if used. In addition to flammability, it is toxic by ingestion, inhalation, and skin absorption, with a TLV of 100 ppm in air. The vapor density is 3.7, which makes it heavier than air, and the vapors will tend to stay close to the ground. It is nonpolar, with a specific gravity of 0.9, which means it will float on water. Ethyl benzene is immiscible in water. The four-digit UN identification number is 1175. The NFPA 704 designation is health 2, flammability 3, and reactivity 0. It is shipped in cans, bottles, 55-gal drums, tank trucks, railcars, and barges. It should not be stored near oxidizing materials. The primary uses of ethylbenzene are as a solvent and an intermediate in the production of styrene. The structure and molecular formula are shown in Figure 5.77.

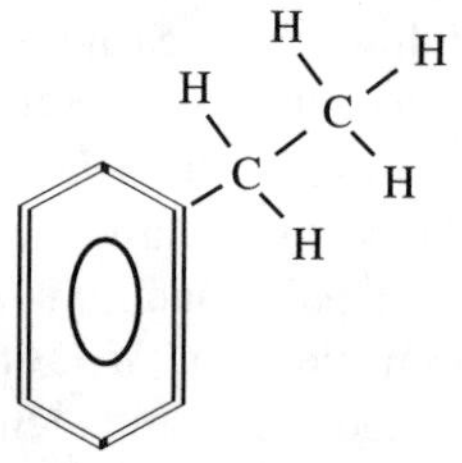

Ethyl benzene
$C_6H_5CH_2CH_3$

Figure 5.77

MC/DOT 306/406 tanker used to transport flammable fuels, such as gasoline and diesel fuel.

INCIDENTS

Most of the more commonly encountered flammable liquids are fuels: gasoline, diesel fuel, heating oil, and jet fuel. Other spills include materials such as alcohols, ketones, aldehydes, paint thinners, pesticides, benzene, toluene, and xylene, along with other industrial solvents. These flammable liquids can present responders with special problems as well as hidden hazards. Because they are liquids, they can flow away from the scene, following the terrain into storm drains, sanitary sewers, waterways, and other low-lying areas. In addition to the flammability hazard, responders will need to stop the flow of the product in some situations, if properly trained and equipped to do so. Incidents range from leaks in vehicle fuel tanks, to transportation accidents involving tank trucks resulting in leaks and fires, to large bulk-storage tank fires.

In Kansas City, Kansas, a fire occurred at a gasoline station that was also associated with bulk-fuel storage and loading areas. This incident occurred prior to the use of underground storage tanks for gasoline stations, and was one of the primary forces that led to the requirements for the use of underground tanks. The fire started at a loading rack on top of a tanker being loaded with fuel. Eventually, the fire spread to horizontal storage tanks used to supply the fuel dispensers at the gasoline station. The elevated horizontal tanks had unprotected steel supports. These supports quickly collapsed from the heat of the fire. When one of the tanks hit the ground, it opened up, sending burning fuel into the street where firefighters were directing large streams of water onto the fire. Six firefighters lost their lives when they were overrun by the burning fuel. This incident resulted in a number of code changes involving the storage of flammable liquids.

A fire in Norfolk, Virginia, occurred involving an MC 306 gasoline tanker truck. When firefighters arrived, they found the tanker fully involved, with leaking, burning fuel flowing down the street into storm sewers leading to a retention pond. They had fires on multiple fronts, including the tanker, parked cars in an adjacent parking lot, on the street, in the sewer, and on the retention pond. Exposures included a senior-citizen apartment building next to the fire in the retention pond. The fire was extinguished without incident. During mop-up operations, a firefighter was injured when cut by a jagged edge on the burned-out shell of the tanker as he fell into a portion of the tank.

REVIEW QUESTIONS CHAPTER 5

1. The boiling point of a liquid is defined as the point at which one of the following occurs:
   A. Where the proper mixture of fuel and air occur
   B. Vapor pressure equals atmospheric pressure
   C. Vapor produced will be lighter than air
   D. Vapor produced will easily ignite
2. Flash point has to do with which of the following temperatures?
   A. Temperature of the air around the liquid
   B. Temperature of the flammable liquid
   C. Temperature required for ignition to occur
   D. None of the above
3. Liquids that have low boiling points will also have low:
   A. Flash points
   B. Ignition temperatures
   C. Heat output
   D. Melting points
4. High flash point liquids have what type of vapor pressure?
   A. High
   B. Equal
   C. Low
   D. None
5. List the three factors that affect the boiling point of a liquid.
6. Match the following compounds with their degree of polarity.

Benzene	A. Polar
Propyl alcohol	B. Nonpolar
Ethyl ether	C. Super polar
Methyl ethyl ketone	D. Super-duper polar
Formic acid	
Pentane	

7. Polymers are long-chained molecules made up of individual building blocks called:
   A. Links
   B. Monomers
   C. Elastomers
   D. Inhibitors

8. Identify the following hydrocarbon formulae as to whether they are alkanes, alkenes, alkynes, or aromatics.

C_6H_{14} C_7H_8 C_7H_{14} C_6H_6 C_8H_{16}

9. Provide structures, names, and hazards for the following hydrocarbon-derivative compounds.

HCHO $C_2H_3COOCH_3$ HCOOH CH_3CHO CH_3COOH

10. Provide the names, formulas, and hazards for the following structures:

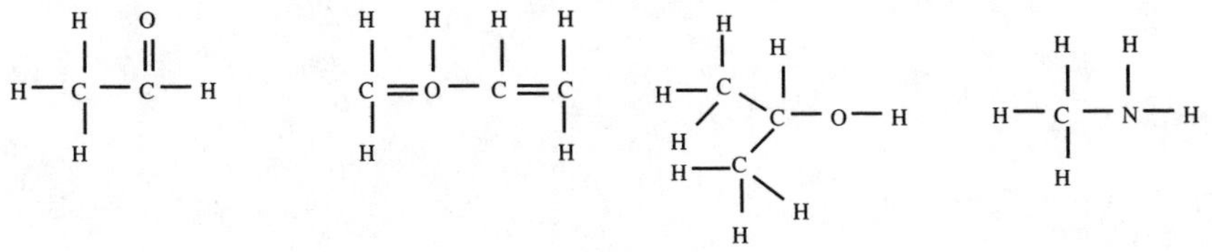

Figure 5.78

Chapter 6

Flammable Solids

Hazard Class 4 is composed of flammable solids. Additionally, some pyrophoric and water-reactive liquids are included under the 4.2 Spontaneously Combustible and 4.3 Dangerous When Wet divisions. Flammable solids can be categorized into five groups according to the hazards of the materials: flammable metals, spontaneous combustibles, intensely burning or difficult-to-extinguish flash point solids, and water reactives. The solid material may take on many physical forms, including fine powder, filings, chips, and various-sized solid chunks. The smaller the solid, the more dangerous it becomes in terms of flammability and potential explosiveness.

Some solid materials, such as white phosphorus, may be pyrophoric; that is, they spontaneously ignite on exposure to air. Calcium carbide must be wet before it becomes a hazard by releasing the flammable gas acetylene. When wet, some solids release flammable hydrogen gas. Many solids, when they contact water, release poisonous gases, such as phosphine, nitrous oxide, or chlorine. Some solid materials, such as sodium peroxide, when they are wet, release oxygen. This can present an added danger to firefighters because the primary extinguishing agent they use is water. Water can cause the release of oxygen from water-reactive materials. If fire is present, the oxygen will accelerate the combustion process and make the fire more difficult to extinguish. Solid materials may spontaneously combust when exposed to water; the reaction is exothermic, or heat-producing. If combustible materials are present, the heat of the reaction can cause combustion to occur. Some materials, like phosphorus, are shipped under water, whereas others, such as picric acid, are shipped with 10 to 50% water in the container.

Solids, such as picric acid, are considered wetted explosives. If allowed to dry out, they are classified as Class 1 Explosives. However, they present only a limited hazard as long as the water is present and, therefore, they are considered flammable solid materials.

The DOT divides the flammable solid hazard class into three divisions: 4.1 Flammable Solids, 4.2 Spontaneous Combustibles, and 4.3 Dangerous When Wet

Dry-bulk truck for transporting solid materials.

materials. Each division presents its own particular hazards when encountered in a hazardous materials incident. It is important that emergency responders have a thorough understanding of each of the divisions, the types of materials that make them up, and the hazards posed in a release.

CLASS 4.1 FLAMMABLE SOLIDS

Class 4.1 materials are flammable solids. The DOT defines 4.1 materials as: "1. Wetted explosives that, when dry, are Class 1 Explosives. 2. Self-reactive materials that are liable to undergo a strongly exothermic decomposition at normal or elevated temperatures caused by excessively high transport temperatures or by contamination. 3. Readily combustible solids that may cause a fire through friction, such as matches, show a burning rate faster than 0.087 inches per second, or any metal powders that can be ignited and react over the whole length of a sample in 10 minutes or less." Flammable solids can be elements, metals in various physical consistencies, or salts. The alkali metals in column 1 on the Periodic Table are considered flammable solids, particularly lithium, sodium, and potassium. With both families, the degree and intensity of combustion depends on the physical form of the solid material. For example, solid magnesium is difficult to ignite; however, if the magnesium is in the form of filings, shavings, or powder, it will burn explosively. Other metals, such as aluminum and titanium, can also be dangerously flammable and explosive as filings, shavings, or powders.

FLASH-POINT SOLIDS/SUBLIMATION

There is a small group of flammable solid materials that go through a process called sublimation at normal temperatures. Sublimation is a process by which solid materials go directly from a solid to a vapor without becoming a liquid. Even though these materials do not become liquids, they still have a flash point; hence, the term flash-point solid is given to the group. The flash points are generally above 100ºF. Once ignition occurs, the solid material melts and flows like a flammable liquid. In addition to being flammable, flash-point solids may also be narcotic and toxic.

Two common flash-point solids are camphor and paradichlorobenzene, also known as mothballs or flakes. Mothballs are placed in areas where clothing is stored to prevent moths from doing damage to the clothing. The fact that the mothballs are flash-point solids allows them to pass from a solid to a vapor without becoming a liquid. Because of this feature, the vapor from the mothballs repels the moths without harming the clothing.

Camphor, also known as gum camphor and 2-camphanone, is a naturally occurring ketone that comes from the wood of the camphor tree. Camphor is composed of colorless or white crystals, granules, or easily broken masses. It has a penetrating aromatic odor that sublimes slowly at room temperature. The flash point is 150ºF, and the autoignition temperature is 871ºF. Flammable and explosive vapors are evolved when heated. While searching old newspapers for articles on fires, the author discovered an article from the late 1800s that told of a bottle of camphor placed on a wood stove. The bottle exploded from the increased vapor pressure from the heat and severely burned the resident. Camphor is slightly water-soluble, undergoes sublimation, and is a flash-point solid. The molecular formula and structure of camphor are shown in Figure 6.1.

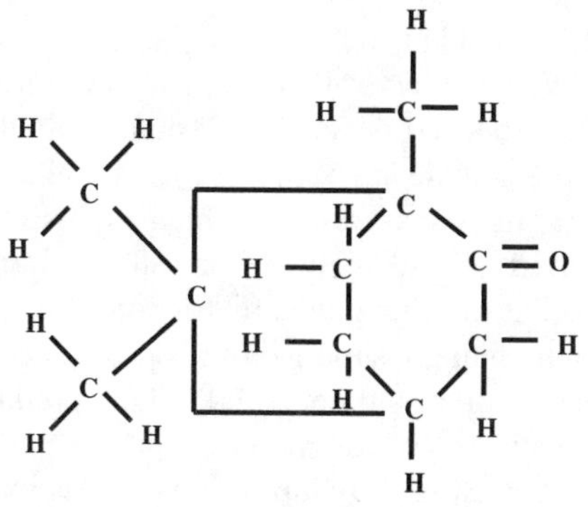

Camphor
$C_{10}H_{16}O$

Figure 6.1

Paradichlorobenzene, or **PDB**, also commonly known as mothballs, are white, volatile crystals with a penetrating odor. The boiling point is 345°F, the flash point is 150°F, and the melting point is 127°F. Paradichlorobenzene is insoluble in water and has a specific gravity of 1.5, which is heavier than water. The vapor density is 5.1, which is heavier than air. In addition to being flammable, it is also toxic by ingestion,and an irritant to the eyes, with a TLV of 75 ppm in air. The four-digit UN identification number is 1592. The NFPA 704 designation for paradichlorobenzene is health 2, flammability 2, and reactivity 0. The primary uses are as a moth repellent, a general insecticide, a soil fumigant, and in dyes. The structure and molecular formula are shown in Figure 6.2.

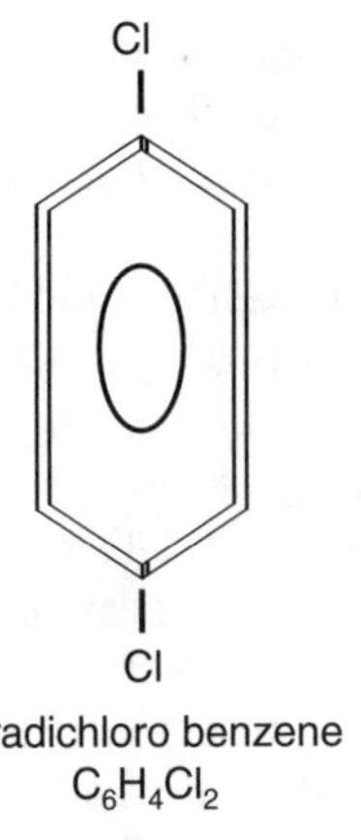

Paradichloro benzene
$C_6H_4Cl_2$

Figure 6.2

The following are some examples of 4.1 flammable solid materials, including wetted explosives and flammable metals.

Barium azide, $Ba(N_3)_2$, is a crystalline solid with not less than 50% water by mass that explodes when shocked or heated. Barium azide decomposes and gives off nitrogen at 240°F and is soluble in water. The four-digit UN identification number for barium azide is 1571. Its primary use is in high explosives.

Fusees are flares used on the highway or rail as warning devices and are considered flammable solids that are ignited by friction. They produce a temperature of 1200°F and a 70-candle flame visible for over one-quarter of a mile. Fusees are primarily composed of strontium nitrate (72%), which produces the characteristic red color and provides oxygen; potassium perchlorate (8%), which is an oxidizing agent; and sulfur (10%), which is an oxidizer and combustion controller. Oil, wax, and sawdust act as binding agents that aid in the control of burning. Flares absorb water and should be stored in a dry place. They are said to have an indefinite shelf life.

Magnesium, Mg, is a metallic element of the alkaline-earth metal family. It is found as a silvery, soft metal, as a powder, or as pellets, turnings, or ribbons. Magnesium is flammable and a dangerous fire hazard, and has an ignition temperature of about 1200°F. It is insoluble in water; however, when burning, it reacts violently with water. In the form of a powder, a pellet, or as turnings, it can explode

Flares used by police officers are shipped as flammable solid materials.

in contact with water. Dry sand or talc is used to extinguish small fires involving magnesium. The four-digit UN identification number for magnesium powder or magnesium alloys with more than 50% magnesium powder is 1418. The number for magnesium pellets, turnings, ribbons, or magnesium alloys with more than 50% magnesium powder is 1869. The NFPA 704 designation is health 0, flammability 1, and reactivity 1. The primary uses of magnesium are in die-cast auto parts, missiles, space vehicles, powder for pyrotechnics, flash photography, and dry and wet batteries.

Trinitrophenol, also known as picric acid, is composed of yellow crystals and is a nitro hydrocarbon derivative. It is shipped with not less than 10% water as a wetted explosive. There is a severe explosion risk when shocked or heated to 572°F, and it reacts with metals or metallic salts. In addition to being flammable and explosive, it is toxic by skin absorption. Picric acid has caused disposal problems in school and other chemistry laboratories where the moisture has evaporated from the container as the material ages. When the picric acid dries out, it becomes a high explosive closely related to TNT. Picric acid has been found in various amounts in school labs across the country. In a dry condition, picric acid is dangerous and should be handled by the bomb squad. The structure and molecular formula for picric acid are shown in Figure 6.3.

Ammonium picrate is a nitro hydrocarbon derivative. It is composed of yellow crystals with not less than 10% water by mass. Ammonium picrate is highly explosive when dry and a flammable solid when wet, and is slightly soluble in water. The four-digit UN identification number for ammonium picrate with not less than 10% water is 1310. The primary uses are in pyrotechnics and explosives. The structure and molecular formula are shown in Figure 6.4.

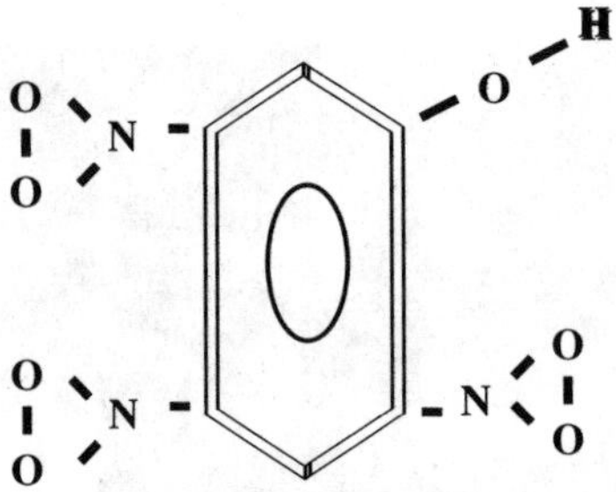

TriNitroPhenol(Picric Acid)
$C_6H_2OH(NO_2)_3$

Figure 6.3

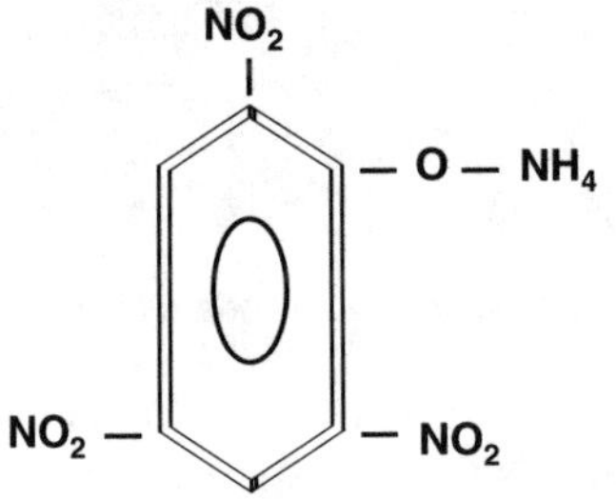

Ammonium picrate
$C_6H_2(NO_2)_3ONH_4$

Figure 6.4

Matches are shipped as flammable solids and will ignite by friction. The "safety" match was invented in 1855. The primary composition is mostly potassium chlorate (an oxysalt), which is a strong oxidizer, antimony III sulfide (a binary salt), and glue on the match head. The striking surface is composed of a mixture of red phosphorus, antimony III sulfide, a little iron III oxide (a binary oxide), and powdered glass held in place by glue. The match functions when a strike produces heat that converts a tiny trace of red phosphorus to white phosphorus, which instantly ignites. The heat ignites the chemicals in the match head, and their short blaze ignites the wood or paper of the matchstick. The "strike-anywhere" match was invented in 1898. It is primarily composed of potassium chlorate, tetraphosphorus trisulfide, ground glass, and the oxides of zinc and iron. This match functions when the strike gives enough heat to initiate a violent reaction between the $KClO_3$ and the P_4S_3; the heat from the reaction ignites the matchstick. The four-digit UN identification numbers for matches are as follows: fusee matches 2254, safety matches 1944, strike-anywhere matches 1331, and wax (Vesta) matches 1945.

Sulfur, S, is a nonmetallic element composed of pale yellow crystals. It is a dangerous fire and explosion risk in a finely divided form. In the molten form, it has a flash point of 405ºF, a boiling point of 832ºF, and an ignition temperature of 450°F. Sulfur is insoluble in water, and its specific gravity is 1.8, which is heavier than water.

The four-digit UN identification number as a dry solid is 1350. The NFPA 704 designation is health 2, flammability 1, and reactivity 0. The primary uses of sulfur are in the manufacture of sulfuric acid, carbon disulfide, and petroleum refining.

Titanium, Ti, is a metallic element that is a silvery solid or dark-gray amorphous powder. Titanium is shipped and stored in a number of physical forms. These include powder, sheets, bars, tubes, wire, rods, sponges, and single crystals. Titanium is a dangerous fire and explosion risk, and will burn in a nitrogen atmosphere. It has an ignition temperature of 2192°F when suspended in air. Water and carbon dioxide are ineffective extinguishing agents for fires involving titanium. The four-digit UN identification number for dry titanium powder is 2546. The number for titanium powder with not less than 20% water is 1352. Titanium sponges, granules, or powder have a number of 2878. The primary uses of titanium are in the manufacture of structural material in aircraft, jet engines, missiles, marine equipment, surgical instruments, and orthopedic appliances.

Urea nitrate, $CO(NH_2)_2HNO_3$, is a colorless crystal shipped with not less than 20% water by mass. It is a dangerous fire and explosion risk, and is slightly soluble in water. It decomposes at 152°C. The four-digit UN identification number is 1357. It is used in the manufacture of explosives and urethane.

Zirconium, Zr, is a metallic element with a grayish, crystalline scale or gray amorphous powder form. It is flammable or explosive in the form of a powder or dust and as borings and shavings. The powder should be kept wet in storage. Zirconium is a suspected carcinogen, with a TLV of 5 mg/m^3 of air and it is insoluble in water. The four-digit UN identification numbers depend on the form and amount of water present. The number for zirconium, dry, as wire, sheeting, or in the form of strips is 2009. The number for zirconium, dry, as a wire, sheeting, or as strips that are thinner than 254 μm, but not thinner than 18 μm, is 2858. Dry zirconium metal powder is 2008. Wet zirconium powder is 1358. Zirconium metal in a liquid suspension is 1308. The primary uses are as a coating on nuclear fuel rods, photo flashbulbs, pyrotechnics, explosive primers, and laboratory crucibles.

COMBUSTIBLE DUSTS

There are solid materials that are not listed in a DOT hazard class, but may become flammable solids because of their physical state. These materials are combustible dusts, which are finely divided particles of some other ordinary Class A combustible material. This group may include such ordinary materials as sawdust, grain dusts, flour, and coal dust. Although the DOT does not consider these materials a hazard class, they may, however, be shipped in dry-bulk transportation containers. The combustible dusts become a problem when they are suspended in air in the presence of an ignition source. If this happens in a fixed facility, such as a grain elevator or flourmill, an explosion may occur. These materials may also be suspended in air in a transportation accident, and in the presence of an ignition source, may create an explosion. Even though combustible dusts are not considered in any of the DOT hazard classes, they present a significant fire and explosion hazard under certain conditions, and responders should be aware of this hazard.

For several years, Nebraska had the unwelcome distinction of having more deaths as a result of grain-elevator explosions than any other state. The state fire marshal implemented an intensive grain-elevator inspection program. The inspections focused on housekeeping, maintenance, and compliance with state and NFPA codes and regulations. The initial inspection of all grain elevators throughout the state took approximately two years to complete. Since that time, it is the author's understanding that there have not been any deaths from grain-elevator explosions in the state. These types of explosions are also discussed in Chapter 3.

CLASS 4.2 SPONTANEOUS COMBUSTIBLES

Class 4.2 materials are spontaneously combustible. The DOT defines them as pyrophoric materials. Even though this hazard class is flammable solids, these materials may be found as solids or liquids. They can ignite without an external ignition source within 5 min after coming into contact with air. There are other 4.2 materials that may be self-heating; i.e., in contact with air and without an energy supply (ignition source), they are liable to self-heat, which can result in a fire involving the material or other combustible materials nearby.

Spontaneous Ignition

Activated carbon is flammable, and the dust is toxic by inhalation. When some carbon-based materials, such as activated carbon or charcoal briquettes, are in contact with water, an oxidation reaction occurs between the carbon material, the water, and pockets of trapped air. The reaction is exothermic, which means heat is produced

Containers of spontaneously combustible materials in a warehouse.

Examples of Materials Subject to Spontaneous Heating

Alfalfa meal	Hides	Peanut oil
Used burlap	Castor oil	Charcoal
Coal	Powdered eggs	Lanolin
Coconut oil	Lard oil	Linseed oil
Cottonseed oil	Manure	Soybean oil
Fertilizers	Metal powders	Fish meal
Fish oil	Olive oil	Whale oil

Figure 6.5

in the reaction, and slowly builds up until ignition occurs spontaneously. Materials subject to spontaneous heating are listed in Figure 6.5.

Carbon-based animal or **vegetable oils**, such as linseed oil, cooking oil, and cottonseed oil, can also undergo spontaneous combustion when in rags or other combustible materials. This type of spontaneous heating *cannot* occur in the case of petroleum oils. Petroleum oils do not have double bonds in the compounds. The oxidation reaction that occurs with animal and vegetable oils is different from the reaction with the carbon-based materials. The oxygen from the air trapped in the mass reacts with the double bonds present in the animal and vegetable oils. The breaking of the double bonds creates heat, which ignites the materials. Petroleum oils do not contain these double bonds and, therefore, cannot undergo this type of spontaneous heating to cause a fire. Fires started by this spontaneous heating process can be difficult to extinguish because they usually involve deep-seated fires. In order for enough heat to be sustained to cause combustion, there must be insulation. This insulation can be the material itself or may be in the form of some other combustible material, such as rags.

Charcoal briquettes are a dangerous fire risk. They may undergo spontaneous ignition when they become wet. However, this is a slow process, and the heat generated must be confined as it builds up and ignites.

Pyrophoric Solids and Liquids

Diethyl zinc is an organo-metal compound and is a dangerous fire hazard. It spontaneously ignites in air and reacts violently with water, releasing flammable vapors and heat. It is a colorless, pyrophoric liquid with a specific gravity of 1.2, which is heavier than water, so it will sink to the bottom. It decomposes explosively at 248°F. It has a boiling point of 243°F, a flash point of –20°F, and a melting point of –18°F. The four-digit UN identification number is 1366. The NFPA 704 designation is health 3, flammability 4, and reactivity 3. The white space at the bottom of the diamond has a W with a slash through it to indicate water reactivity. The primary uses of diethyl zinc are in the polymerization of olefins, high-energy aircraft, and missile fuel and in the production of ethyl mercuric chloride. The molecular formula and structure are shown in Figure 6.6.

Pentaborane is a nonmetallic, colorless liquid with a pungent odor. It decomposes at 300°F, if it has not already ignited, and will ignite spontaneously in air if

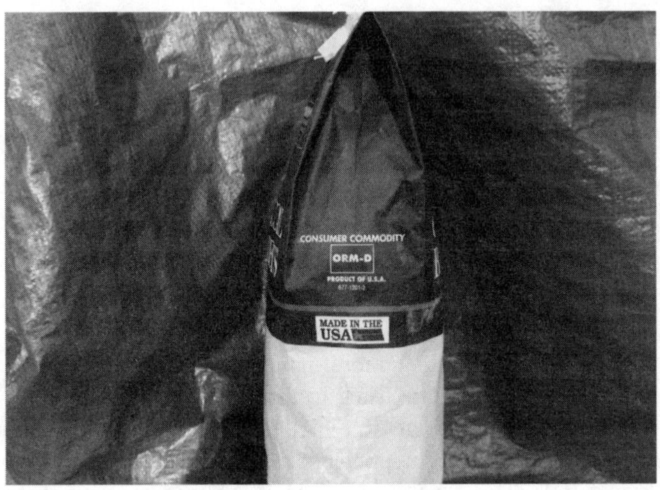

A bag of charcoal, an ORM-D material, that may undergo spontaneous heating upon contact with water. If the heat is confined, combustion may occur.

```
     H   H        H   H
     |   |        |   |
H — C — C — Zn — C — C — H
     |   |        |   |
     H   H        H   H
```

Diethyl zinc
$Zn(C_2H_5)_2$

Figure 6.6

impure. It is a dangerous fire and explosion risk, with a flammable range of 0.46 to 98% in air. The boiling point is 145°F, the flash point is 86°F, and the ignition temperature is 95°F, which is extremely low. Any object that is 95°F or above can be an ignition source. Ignition sources can be ordinary objects on a hot day in the summer, such as the pavement, metal on vehicles, and even the air. In addition to extreme flammability, it is also toxic by ingestion or inhalation and is a strong irritant. The TLV is 0.005 ppm in air, and it is immiscible in water. The four-digit UN identification number is 1380. The NFPA 704 designation for pentaborane is health 4, flammability 4, and reactivity 2. The primary uses are as fuel for air-breathing engines and as a propellant. The structure and molecular formula are shown below:

```
     H   H   H   H   H
     |   |   |   |   |
H — B = B = B = B = B — H
     |               |
     H               H
```

Pentaborane
B_5H_9

Figure 6.7

Aluminum alkyls are colorless liquids or solids. They are pyrophoric and may ignite spontaneously in air. They are often in solution with hydrocarbon solvents. Aluminum alkyls are pyrophoric materials in a flammable solvent. The vapors are heavier than air, water-reactive, and corrosive. Decomposition begins at 350°F. The four-digit UN identification number is 3051. The NFPA 704 designation is health 3, flammability 4, and reactivity 3. The white space at the bottom of the diamond has a W with a slash through it, indicating water reactivity. They are used as catalysts in polymerization reactions.

Aluminum phosphide, AlP, is a binary salt, one of the NCHP acronym (see Chapter 2). These salts have the specific hazard of giving off poisonous and pyrophoric phosphine gas when in contact with moist air, water, or steam. They will also ignite spontaneously upon contact with air. This compound is composed of gray or dark yellow crystals and is a dangerous fire risk. Aluminum phosphide decomposes upon contact with water and has a specific gravity of 2.85, which is heavier than water. The four-digit UN identification number is 1397. The NFPA 704 designation is health 4, flammability 4, and reactivity 2. The white section at the bottom of the diamond has a W with a slash through it, indicating water reactivity. Aluminum phosphide is used in insecticides, fumigants, and semiconductor technology.

Potassium sulfide, K_2S, is a binary salt. It is a red or yellow-red crystalline mass or fused solid. It is deliquescent in air, which means it absorbs water from the air, and it is also soluble in water. Potassium sulfide is a dangerous fire risk and may ignite spontaneously. It is explosive in the form of dust and powder. It decomposes at 1562°F and melts at 1674°F. The specific gravity is 1.74, which is heavier than air. The four-digit UN identification number is 1382. The NFPA 704 designation is health 3, flammability 1, and reactivity 0. Potassium sulfide is used primarily in analytical chemistry and medicine.

Sodium hydride, NaH, is a binary salt that has a specific hazard of releasing hydrogen upon contact with water. It is an odorless powder that is violently water-reactive. The four-digit UN identification number is 1427. The NFPA 704 designation is health 3, flammability 3, and reactivity 2. The white space at the bottom of the diamond has a W with a slash through it, indicating water reactivity.

White phosphorus, P, also known as yellow phosphorus, is a nonmetallic element that is found in the form of crystals or a wax-like transparent solid. It ignites spontaneously in air at 86°F, which is also its ignition temperature. White phosphorus should be stored under water and away from heat. It is a dangerous fire risk, with a boiling point of 536°F and a melting point of 111°F. White phosphorus is toxic by inhalation and ingestion, and contact with skin produces burns. The TLV is 0.1 mg/m^3 of air, and it is insoluble in water, with a specific gravity of 1.82, which is heavier than air. White phosphorus is shipped and stored under water to keep it from contacting air. The four-digit UN identification number is 2447. The NFPA 704 designation is health 4, flammability 4, and reactivity 2. The primary uses are in rodenticides, smoke screens, and analytical chemistry.

INCIDENTS

A series of fires have occurred in laundries around the country since 1989. One in six commercial, industrial, or institutional laundries reports a fire each year, which results in over 3000 fires. The primary cause is thought to be spontaneous combustion. Chemicals, including animal and vegetable oils, may be left behind in fabrics after laundering. The heat from drying may cause the initiation of the chemical reaction that causes spontaneous ignition. Spontaneous combustion, according to the *Handbook of Fire Prevention Engineering,* "is a runaway temperature rise in a body of combustible material, that results from heat being generated by some process taking place within the body." On June 16, 1992, a fire in a nursing home laundry in Litchfield, Illinois, caused $1.5 million in damage. The cause was determined to be spontaneous ignition of residual chemicals in the laundered fabric reacting to heat from the dryer. In Findlay, Ohio, on July 2, 1994, a fire destroyed a commercial laundry, causing over $5 million in damage. Traces of linseed oil were found in a pile of clean, warm garments piled in a cart waiting to be folded.

Fires have also occurred involving residual animal or vegetable oils in cleaning rags in restaurants. The oils are never completely removed by laundering. When placed in the dryer, the rags are heated. When put away on the storage shelf, this heat can become trapped, along with the oil remaining on the rags when confined. The spontaneous combustion process begins slowly, and the heat of the reaction increases until combustion occurs.

An incident occurred in Gettysburg, Pennsylvania, involving phosphorus shipped under water in 55-gal drums (phosphorus is air-reactive). One of the drums developed a leak, and the water drained off. This allowed the phosphorus to be exposed to air, which caused it to spontaneously ignite. The fire spread to the other containers and eventually consumed the entire truck. The ensuing fire was fought with large volumes of water and in the final stages covered with wet sand. Cleanup created problems because as the phosphorus and sand mixture was shoveled into over-packed drums, the phosphorus was again exposed to air and reignited small fires. When phosphorus burns, it also gives off toxic vapors.

A train derailment in Brownson, Nebraska, resulted in a tank car of phosphorus overturning and the phosphorus igniting upon contact with air. Phosphorus is shipped under water, so there was water inside the tank car. CHEMTREC (Chemical Transportation Emergency Center) was called, and responders were told correctly that the phosphorous would not explode. However, the water inside the tank car turned to steam from the heat of the phosphorus fire. The pressure from the steam caused a boiler-type of explosion that had nothing to do with the phosphorus. This is just another example of the hidden hazards that emergency responders must be aware of when dealing with hazardous materials. Not only do the hazardous materials have to be considered, but also the container and any "inert" materials that may be involved with the product.

CLASS 4.3 DANGEROUS WHEN WET

Class 4.3 materials are dangerous when wet. The DOT definition is "a material that, by contact with water, is liable to become flammable or to give off flammable or toxic gas at a rate greater than 1 liter per kilogram of the material per hour." Examples of water-reactive materials include zinc powder, trichlorosilane, sodium phosphide,

Rail tank car of air-reactive phosphorus shipped under water.

sodium aluminum hydride, and the metallic elements potassium, sodium, and lithium. Potassium, lithium, and sodium come from family 1 on the Periodic Table, known as the alkali metals. They are in the first column on the table and, as with other families on the chart, they have similar chemical characteristics. They are silvery, soft metals that are reactive with air and violently reactive with water. Contact with water causes spattering, the release of free hydrogen gas, and the production of heat. The heat can be so great that it ignites the hydrogen gas.

Metallic elements, such as **magnesium** and **calcium**, are from family 2 on the Periodic Table. These materials are known as the alkaline-earth metals. Unlike the alkali metals, magnesium must be burning before it reacts with water or it must be in a finely divided form, such as filings and powder. The filings, flakes, dusts, and powders can ignite explosively upon contact with water to evolve flammable hydrogen gas and heat. The heat may be great enough to ignite the hydrogen gas. Magnesium is insoluble in water. The ignition temperature of magnesium is about 1200°F, as is the melting point. When magnesium ignites, the temperatures can reach 7200°F. In contact with burning magnesium, water produces a violent explosion. Water in contact with magnesium fillings or powder can produce a spontaneous explosion. Talc, dry sand, Met-L-X, foundry flux, and G-1 powder should be used to extinguish small magnesium fires. Large fires should be fought with flooding volumes of water from unmanned monitors and aerial devices. Calcium can be expected to behave in a similar fashion.

Calcium carbide, **CaC_2**, is a binary salt. It is a grayish-black, hard solid that reacts with water to produce acetylene gas, a solid corrosive that is calcium hydroxide, and release heat. Acetylene gas is manufactured by reacting calcium carbide with water. Because acetylene is so unstable, it is not shipped in bulk quantities.

Closed containers of water-reactive phosphorus pentasulfide, which produces toxic hydrogen-sulfide gas upon contact with water.

Calcium carbide is shipped to acetylene-generating plants where it is reacted with water in a controlled reaction. After the reaction process, the acetylene gas is placed into specially designed containers for shipment and use. Calcium carbide has a specific gravity of 2.22, which is heavier than water. The four-digit UN identification number for calcium carbide is 1402. The NFPA 704 designation is health 3, flammability 3, and reactivity 2. The white section at the bottom of the diamond contains a W with a slash through it, indicating water reactivity. It is shipped in metal cans, drums, and specially designed covered bins on railcars and trucks. When shipped and stored, it should be kept in a cool, dry place. The primary uses are in the generation of acetylene gas for welding, vinyl acetate monomer, and as a reducing agent.

Phosphorus pentasulfide, P_4S_{10}, is a nonmetallic, inorganic compound. It is a yellow to greenish-yellow crystalline mass with an odor similar to hydrogen sulfide. It is a dangerous fire risk and ignites by friction or in contact with water. The boiling point is 995°F, and the ignition temperature is 287°F. It decomposes upon contact with water or moist air, liberating toxic and flammable hydrogen-sulfide gas. The specific gravity is 2.09, so it is heavier than water. It is toxic by inhalation, with a TLV of 1 mg/m^3 of air. The four-digit UN identification number is 1340. The NFPA 704 designation is health 2, flammability 1, and reactivity 2. The primary uses are in insecticides, safety matches, ignition compounds, and sulfonation.

Methyl dichlorosilane, CH_3SiHCl_2, is a colorless liquid with a sharp, irritating odor. It is a dangerous fire risk, corrosive, and water-reactive. The flammable range is wide, from 6% on the lower end to 55% on the upper end. The boiling point is 107°F, the flash point is 15°F, and the ignition temperature is more than 600°F. The specific gravity is 1.11, which is heavier than water. Vapors are heavier than air and will travel to ignition sources. It is immiscible in water and decomposes on contact to release hydrogen chloride gas. Methyl dichlorosilane is toxic by inhalation and skin absorption; it is irritating to the skin, eyes, and respiratory system. Contact with the material may cause burns to the eyes and skin. The four-digit UN identification number is 1242. The NFPA 704 designation is health 3, flammability 3, and reactivity 2. The white space at the bottom of the diamond has a W with a slash through it, indicating water reactivity. The primary use is in the manufacture of siloxanes, which are straight-chained compounds similar to paraffin hydrocarbons.

Potassium, **K**, is a metallic element, also known as kalium. It is an alkali metal that is soft, silvery, and rapidly oxidizes in moist air. Potassium is a combustible solid that may ignite spontaneously on contact with moist air. It is a dangerous fire risk and it reacts violently with water and moisture in the air to release hydrogen gas and form potassium hydroxide, which is a corrosive liquid. The boiling point is 1410°F, and the melting point is 146°F. The specific gravity is 0.86, which is lighter than water. The reaction with water is also exothermic, and the heat produced is enough to ignite the hydrogen gas that is released. Potassium metal is usually stored under kerosene to keep it from reaching the air. As it ages, it can form explosive peroxides, much like ethers do. When these peroxides are present, it may explode violently if handled or cut. Potassium that is coated with peroxides should be destroyed by burning. The four-digit UN identification number is 2257. The NFPA 704 designation is health 3, flammability 3, and reactivity 2. The white section at the bottom of the diamond has a W with a slash through it, indicating water reactivity.

Carbon black is a finely divided form of carbon. It may ignite explosively if suspended in air in the presence of an ignition source or slowly undergo spontaneous combustion upon contact with water. In addition, it is toxic by inhalation, with a TLV of 3.5 mg/m^3 in air. The primary uses are in the manufacture of tires, belt covers, plastics, carbon paper, colorant for printing inks, and as a solar-energy absorber.

Open- and closed-leg grain elevators are a primary source of dust explosions.

INCIDENTS

A fire in a warehouse in Chicago involved barrels of oil-soaked magnesium shavings and filings. Fighting the fire with water produced violent explosions, resulting in whiplash injuries to firefighters who were on aerial apparatus over the burning magnesium. The facility was in a residential neighborhood, so firefighters had to try to control the fire with water even though the material is water-reactive when burning.

In New York City, three firefighters died and several others were injured in a fire involving sodium metal. The firefighters were extinguishing a fire in a 55-gal drum of molten sodium, when a small amount of water on a shovel came in contact with the sodium. This triggered a chemical reaction and explosion producing temperatures in excess of 2000°F, splattering the molten sodium on the firefighters. The sodium burned through their turnouts, station uniforms, and underwear. Contact with moisture on the skin caused burning of the tissue below. Water in contact with water-reactive and molten metals can produce violent reactions. Had the firefighters tried to extinguish the fire in the drum with a hose line, many more might have died.

FIRE-EXTINGUISHING AGENTS

Class 4.3 materials are water-reactive. When large amounts of these materials are involved in fire, water is the only extinguishing agent available in quantities large enough to extinguish the fires; just understand that when water is used, there may be violent reactions and explosions. Preparations need to be made for the safety of personnel based upon the hazards of the materials. Small fires of water-reactive materials, especially metallic-based materials, can be extinguished with a dry-powder extinguishing agent. For other flammable solid materials, water is also the agent of choice in most cases. It is important, however, to make sure there is a positive identification of the product, as with all hazardous materials. Once the product is identified, the proper extinguishing agent for any individual material can be identified through reference materials such as the *North American Emergency Response Guide Book,* the CAMEO (Computer-Aided Management of Emergency Operations) computer database, CHEMTREC, or some other reference source.

REVIEW QUESTIONS FOR CHAPTER 6

1. List the three subclasses of Class 4 hazardous materials.
2. Sublimation is the process in which a solid goes directly to which physical state?
   A. Liquid
   B. Slurry
   C. Particles
   D. Gas

3. Phosphorus is a Class 4.1 Flammable Solid and will do what when removed from its container?
   A. Dry out
   B. Spontaneously combust
   C. Oxidize
   D. Nothing
4. Which of the following may provide combustible dusts?
   A. Flour
   B. Coal
   C. Grain
   D. All of the above
5. Which of the following compounds or mixtures may undergo spontaneous heating?
   A. Fish oils
   B. Motor oils
   C. Linseed oil
   D. Cottonseed oil
   E. Gasoline
   F. Methyl ethyl ketone
   G. Isopropyl ether
6. Which of the following extinguishing agents would be appropriate for large fires involving water-reactive materials?
   A. Carbon dioxide
   B. Dry-chemical foam
   C. Dry powder
   D. Water
7. Which families from the Periodic Table are water-reactive materials in their elemental state?
   A. 1 and 7
   B. 1 and 2
   C. 2 and 8
   D. Only 1
8. When carbide salts contact water, what is produced?
   A. Phosgene
   B. Chlorine
   C. Carbon
   D. Acetylene
9. Class 4.2 Spontaneously Combustible contains which types of materials?
   A. Solids
   B. Liquids
   C. Gases
   D. Both a and b
10. Class 4.1 Flammable Solids contains which of the following materials?
    A. Wetted explosives
    B. Road flares
    C. Phosphorus
    D. All of the above

CHAPTER 7

Oxidizers

Hazard Class 5 Oxidizers are separated into two divisions: 5.1 and 5.2. Class 5.1 materials are solids and liquids that, according to the DOT, "by yielding oxygen, can cause or enhance the combustion of other materials." Although that is not a technical definition from a chemistry standpoint, it gets right to the point for emergency response purposes. While oxidizers themselves do not burn, if present in a fire situation, they will make the fire burn faster and become more difficult to extinguish.

The NFPA classifies oxidizers into four groups. These groups are identified in Figure 7.1. Common groups of oxidizers include oxysalts, inorganic peroxides (salt peroxides), certain acids, elements, and organic peroxides. Examples of oxidizers from each of the NFPA classes are illustrated in Figure 7.2. Oxygen is probably the most recognized oxidizer. Even though oxygen is essential for life to exist, it can be a dangerous material. Oxygen is found in transportation and storage as a compressed gas and as a cryogenic liquid with a temperature of –183°F. Oxygen does not burn, but in contact with organic materials, it can become explosive. Oxygen-enriched atmospheres can be deadly to emergency responders if there is a fire or heat source nearby. Many times, the enriched atmosphere is not visible or detectable to responders without the use of monitoring instruments. While the oxidizers in this chapter are solids and liquids, through physical and chemical reactions, they can release oxygen gas, which can cause some of the same problems as compressed oxygen.

Class 5.2 materials are organic peroxides. Unlike peroxide salts, which contain metals, these are organic compounds that contain carbon in their formula. These materials contain oxygen in the bivalent -O-O- structure and may be considered a derivative of hydrogen peroxide. The structure and molecular formula for hydrogen peroxide are shown in Figure 7.3.

In organic-peroxide compounds, both of the hydrogen atoms in hydrogen peroxide have been replaced by organic radicals. One of the major hazards of 5.2 organic peroxides is the instability of the compounds. The oxygen-to-oxygen single bond is an unstable bond. It is this same bond that is responsible for the explosiveness of

NFPA Classes of Oxidizers

Class 1	Solid or liquid that readily yields oxygen or oxidizing gas or that readily reacts to oxidizer combustible materials.
Class 2	Oxidizing material that can cause spontaneous ignition when in contact with combustible materials.
Class 3	Oxidizing material that can undergo vigorous self-sustained decomposition when catalyzed or exposed to heat.
Class 4	Oxidizing material that can undergo an explosive reaction when catalyzed or exposed to heat, shock, or friction.

Figure 7.1

Examples of Oxidizers in the NFPA Classes

Class 1
- Aluminum nitrate
- Calcium peroxide
- Potassium persulfate
- Sodium nitrite

Class 2
- Calcium hypochlorite
- Nitric acid (above 70%)
- Sodium peroxide
- Potassium permanganate

Class 3
- Ammonium dichromate
- Calcium hypochlorite (over 50%)
- Hydrogen peroxide (52–91%)

Class 4
- Perchloric acid (60–72%)
- Hydrogen peroxide (>90%)
- Potassium superoxide

Figure 7.2

$$H-O-O-H$$

Hydrogen Peroxide
H_2O_2

Figure 7.3

the nitro compounds discussed in Chapter 3. Oxidizers, especially the organic peroxides, should be treated with a great deal of respect. They can be just as dangerous and explosive as Class 1 compounds.

CLASS 5.1 OXIDIZERS

Oxidizers may be elements, acids, or salts classified into families, with specific hazards associated with each family. There are elements found on the Periodic Table that are oxidizers in their elemental state. These include oxygen, chlorine, fluorine, bromine, and iodine.

Oxygen, O_2, can be encountered as a gas, cryogenic liquid, and liquid or solid in compound with other materials. Although nontoxic, it is reactive with hydrocarbon-based materials. Oxygen is a strong oxidizer. Liquid oxygen in contact with an asphalt surface, such as a parking lot or highway, can create a contact explosive. Dropping an object on the area, driving over, or even walking on the area can cause an explosion to occur. Oxygen is a nonmetallic gaseous element. Oxygen makes up approximately 21% of the atmosphere. The boiling point of oxygen is −297°F. It is nonflammable, but supports combustion. Liquid oxygen can explode when exposed to heat or organic materials.

Chlorine, **fluorine**, and **bromine** in their elemental forms are all strong oxidizers even though they are placarded and labeled as poisons. Two terms commonly associated with oxidizers are oxidation (or oxidation reaction) and reduction. **Oxidation** is the loss of electrons by one reactant, and **reduction** is the gaining of electrons by another. Metals usually lose electrons, and nonmetals usually gain electrons. The elements in the upper-right corner of the Periodic Table are electronegative, or electron-drawing. Fluorine is the most electron-drawing element known. Chlorine and oxygen are also electron-drawing. Oxidation and reduction always occur together. No substance is ever oxidized unless something else is reduced (Figure 7.4).

For example, when sodium and chlorine combine with an ionic bond, the electron of sodium is given to chlorine; sodium has been oxidized. Chlorine receives the electron of sodium; chlorine has been reduced. The substance that accepts the electrons is known as the oxidizing agent. Therefore, in the reaction between sodium and chlorine, chlorine is the oxidizing agent and sodium is the reducing agent. Chlorine is reduced to the chloride ion in the reaction with sodium. In summary, the substance that is oxidized is the reducing agent (gives up its electrons). The substance that is reduced is the oxidizing agent (receives electrons).

Chlorine is a dense, greenish-yellow gas. Although it may be a gas or a liquefied gas, it can also be released from solid compounds that are oxidizers. Chlorine is not combustible; however, it will support combustion just like oxygen. Chlorine does not occur freely in nature. It is found in compounds within the minerals halite (rock salt), sylvite, and carnallite, and as the chloride ion in seawater.

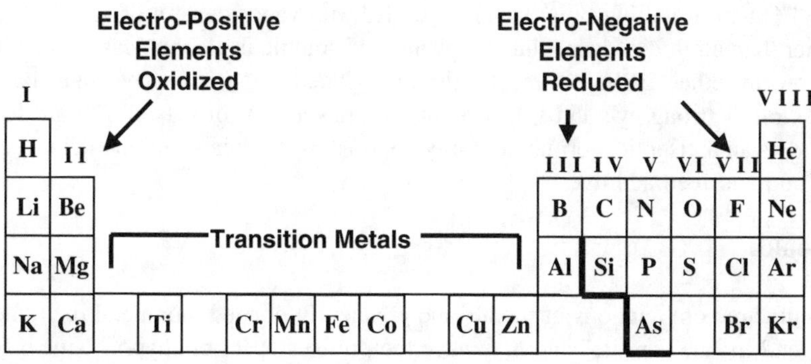

Figure 7.4

Chlorine, although placarded as a poison, is also a strong oxidizer and will support combustion just like oxygen.

Fluorine, F, is the most powerful oxidizing agent known. Like its relative, chlorine, it is classified as a 2.3 poison gas by the DOT and is toxic by inhalation, with a TLV of 1 ppm. Liquid fluorine is such a strong oxidizer that it can cause concrete to burn. It is a pale-yellow gas or cryogenic liquid with a pungent odor. It reacts violently with a wide range of organic and inorganic compounds, and is a dangerous fire and explosion risk when in contact with these materials.

Bromine is a dark, reddish-brown liquid with irritating fumes. It attacks most metals and reacts vigorously with aluminum and explosively with potassium. It is a strong oxidizing agent, and may ignite combustibles on contact.

Iodine is the least reactive of the halogens. It consists of heavy, grayish-black granules or solid plates, readily sublimed to a violet vapor. Iodine has a boiling point of 184ºC and a melting point of 113.5ºC. It has a vapor density of 4.98, which is heavier than air. Iodine is insoluble in water and soluble in alcohol, carbon disulfide, chloroform, ether, carbon tetrachloride, and glycerol. It is toxic by ingestion and inhalation, a strong eye and skin irritant, and has a TLV of 0.1 ppm in air. Iodine is used as an antiseptic, germicide, x-ray contrast, water treatment, medicinal soaps, and food and feed additive.

Oxysalts

Oxysalts are combinations of metals and covalently bonded, nonmetal oxyradicals. They end in "ate" or "ite" and may have the prefixes "per" or "hypo." Nine oxysalt radicals will be presented with this group (see Figure 7.5). The first six all have –1 charges: FO_3 (fluorate), ClO_3 (chlorate), BrO_3 (bromate), IO_3 (iodate), NO_3 (nitrate), and MnO_3 (manganate). The next two have –2 charges: CO_3 (carbonate) and SO_4

Oxysalts

Fluorate	FO_3
Chlorate	ClO_3
Bromate	BrO_3
Iodate	IO_3
Nitrate	NO_3
Manganate	MnO_3
Carbonate	CO_3
Sulfate	SO_4
Phosphate	PO_4

Figure 7.5

(sulfate). The last oxyradical is PO_4 (phosphate), which has a –3 charge (see Figure 2.13, Chapter 2). All of these radicals are considered to be in their base state, that is, containing the "normal" number of oxygen atoms present in that oxyradical.

When a metal is added to any oxyradical in the base state, the compound ends in "ate," such as sodium phosphate. Oxyradicals may be found with varying numbers of oxygen atoms. There may be more or less oxygen atoms in a compound than the base state. Regardless of the number of oxygen atoms on the oxyradical, the charge of the radical *does not* change. When naming compounds with one additional oxygen atom, the prefix "per" is used; to indicate excess oxygen over the base state, the ending is still "ate." When the number of oxygen atoms is one less than the base state of an oxyradical, the ending of the oxyradical name is "ite." An example is magnesium sulf*ite*. Furthermore, an oxyradical can have two less oxygen atoms than the base state; the oxyradical name has a prefix "hypo" and ends in "ite." An example is aluminum *hypo*phosph*ite*.

All oxysalt compounds are salts and have the hazard of being oxidizers. They contain, among other things, fluorine, chlorine, and oxygen, which are all strong oxidizers. Most oxysalts do not react with water, but are soluble in water. In the process of mixing with water, they may liberate oxygen, fluorine, or chlorine. Oxysalts have varying numbers of oxygens in their compounds. The common oxysalt compound has three or four oxygen atoms, which is known as the base state. Some oxysalts are loaded with oxygen, such as the "per-ate" compounds, which will have four or five oxygen atoms. The compounds that have two oxygen atoms will end in "ite." The "hypo-ite" compounds will have one oxygen atom. The fact that the "hypo-ites" have only one oxygen atom does not mean that they are not dangerous oxidizers. Some of the "hypo-ite" compounds, in addition to oxygen, also have chlorine atoms, which are oxidizers. Perchlorates and other oxyradicals in the "per-state" contain one more oxygen atom than the base state chlorates. They are loaded with oxygen and want to give it up readily. Perchlorates can form explosive mixtures with organic, combustible, or oxidizable materials. Contact with acids, such as sulfuric acid, can form explosive mixtures.

Lithium perchlorate, $LiClO_4$, is an oxysalt that is a colorless, deliquescent crystal. Oxysalt "per-ate" compounds are loaded with excess oxygen and will readily give it up in a reaction. Lithium perchlorate is a powerful oxidizing agent. It has more available oxygen than does liquid oxygen on a volume basis. Lithium perchlorate has a specific gravity of 2.429, which is heavier than water, and is water-soluble. It is a dangerous fire and explosion risk in contact with organic materials, and is an irritant to skin and mucous membranes. The primary use of lithium perchlorate is as a solid rocket propellant.

Chlorates are strong oxidizing agents. When heated, they give up oxygen readily. Contact with organic or other combustible materials may cause spontaneous combustion or explosion. They are incompatible with ammonium salts, acids, metal powders, sulfur, and finely divided organic or combustible substances.

Potassium chlorate, $KClO_3$, is a transparent, colorless crystal or white powder. It is soluble in boiling water and decomposes at approximately 750ºF, giving off oxygen gas. Potassium chlorate is a strong oxidizer and forms explosive mixtures with combustible materials, such as sugar, sulfur, and others. Potassium chlorate is incompatible with sulfuric acid, other acids, and organic material. The four-digit UN identification number is 1485. Its primary uses are as an oxidizing agent in the manufacture of explosives and matches; in pyrotechnics; and as a source of oxygen. Sodium and potassium chlorates have similar properties. Chlorites are powerful oxidizing agents. They have one less oxygen than the base-state oxysalts. They form explosive mixtures with combustible materials and, in contact with strong acids, they can release explosive chlorine dioxide gas.

Calcium chlorite is an oxysalt with a molecular formula of **$Ca(ClO_2)_2$**. It is a white, crystalline material that is soluble in water. It is a strong oxidizer and a fire risk in contact with organic materials. The four-digit UN identification number is 1453. Hypochlorites have two less oxygen atoms than the base-state compounds. They can cause combustion in high concentrations when in contact with organic materials. When heated or in contact with water, they can give off oxygen gas. At ordinary temperatures, they can give off chlorine and oxygen when in contact with moisture and acids. They are commonly used as bleaches and swimming pool disinfectants.

Calcium hypochlorite, $Ca(ClO)_2$, is an oxysalt; it is a crystalline solid and an oxidizer that decomposes at 212ºF. Calcium hypochlorite is a dangerous fire risk in contact with organic materials. It is also a common swimming pool chlorinator and decomposes in contact with water, releasing chlorine into the water. If a container of calcium hypochlorite becomes wet in storage, the result can be an exothermic reaction. If combustible materials are present, a fire may occur. The chlorine in the compound will be released by contact with the water and will then accelerate the combustion process. The four-digit UN identification number for dry mixtures with not less than 39% available chlorine (8.8% oxygen) is 1748; hydrated with not less than 5.5% and not more than 10% water, the number is 2880; mixtures that are dry, with not less than 10% but not more than 39% available chlorine, are numbered as 2208. The NFPA 704 designation for calcium hypochlorite is health 3, flammability 0, and reactivity 1. The white section at the bottom of the diamond has the prefix "oxy," indicating an oxidizer. The primary uses are as a bleaching agent, a swimming pool disinfectant, a fungicide, in potable-water purification, and as a deodorant.

Metal nitrates are oxysalts, and as a group have a wide range of hazards. Common to many of them, however, is the fact they are oxidizers and are heat- and shock-sensitive. When heated, they will melt, releasing oxygen, which will increase the combustion process. Molten nitrates react violently with organic materials. When solid streams of water are used for fire suppression, steam explosions may occur upon contact with the molten materials. Nitrates can be dangerous oxidizers and will explode if contaminated, heated, or shocked. Most nitrates have similar properties.

Aluminum nitrate, $\mathbf{Al(NO_3)_3}$, is a white, crystalline material that is soluble in cold water. It is a powerful oxidizing agent that decomposes at approximately 300ºF. Aluminum nitrate should not be stored near combustible materials. The four-digit UN identification number is 1438. The primary uses are in textiles, leather tanning, as an anticorrosion agent, and as an antiperspirant.

Sodium nitrate, also known as Chile saltpeter and soda niter, has a molecular formula of $\mathbf{NaNO_3}$. Sodium nitrate is a colorless, odorless, transparent crystal. It oxidizes when exposed to air and is soluble in water. This material explodes at 1000ºF, much lower than temperatures encountered in many fires. Sodium nitrate is toxic by ingestion, and has caused cancer in test animals. When used in the curing of fish and meat products, it is restricted to 100 ppm. Sodium nitrate is incompatible with ammonium nitrate and other ammonium salts. The four-digit UN identification number is 1498. Sodium nitrate is used as an antidote for cyanide poisoning and in the curing of fish and meat.

Potassium nitrate (saltpeter) has a molecular formula of $\mathbf{KNO_3}$. It is found as a transparent to white crystalline powder and as crystals. Potassium nitrate is water-soluble and is a dangerous fire and explosion risk when heated or shocked or in contact with organic materials. It is a strong oxidizing agent, with a four-digit UN identification number of 1486. Potassium nitrate is used in the manufacture of pyrotechnics, explosives, and matches. It is often used in the illegal manufacture of homemade pyrotechnics and explosives.

Persulfates are strong oxidizers and may cause explosions during fires. Oxygen may be released by the heat of the fire and cause explosive rupture of the containers. Explosions may also occur when persulfates are in contact with organic materials.

Potassium persulfate, $\mathbf{K_2S_2O_8}$, is composed of white crystals that are soluble in water, and it decomposes below 212ºF. Potassium persulfate is a dangerous fire risk in contact with organic materials. It is a strong oxidizing agent and an irritant, with a four-digit UN identification number of 1492. The primary uses are in bleaching, as an oxidizing agent, as an antiseptic, as a polymerization promoter, and in the manufacture of pharmaceuticals.

Permanganates mixed with combustible materials may ignite from friction or spontaneously in the presence of inorganic acids. Explosions may occur with either solutions or dry mixtures of permanganates.

Potassium permanganate, $\mathbf{KMnO_4}$, is composed of dark purple, odorless crystals with a blue metallic sheen. It is soluble in water, decomposes at 465ºF, and is a powerful oxidizing material. Potassium permanganate is a dangerous fire and explosion risk in contact with organic materials. Potassium permanganate is incompatible with sulfuric acid, glycerin, and ethylene glycol. The four-digit UN identification number is 1490. The primary uses of potassium permanganate are as an oxidizer, bleach, or dye; during radioactive decontamination of the skin; and in the manufacture of organic chemicals.

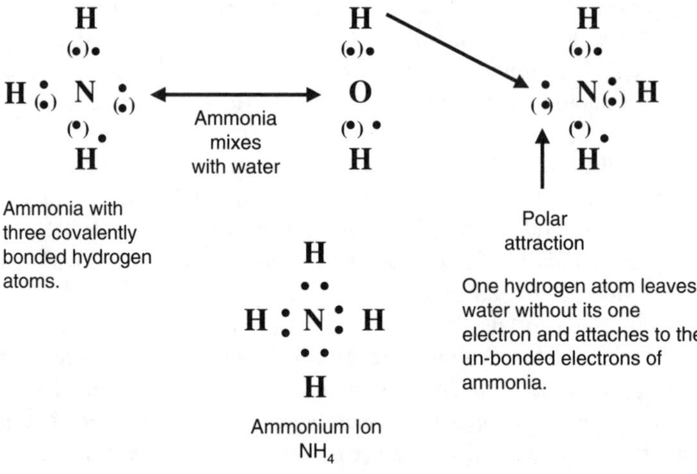

Figure 7.6

Some ammonium compounds are oxysalts. Although ammonia is not a metal, in the case of the ammonium ion, it acts like a metal when attached to the oxyradicals. When ammonia gas is added to water, it readily dissolves and remains as NH_3. One of the hydrogen atoms leaves water, but leaves its electrons behind. The protons of the hydrogen then attach to the unbonded electrons on nitrogen to complete its duet. This hydrogen is loosely held to the nitrogen and comes off easily. The hydrogen ions in the water are attracted to the negative side of the ammonia molecule where that unbonded pair of electrons is located (you can think of the ammonia as a slightly polar molecule). The ammonium ion is positive because the hydrogen ion contributes no electrons. It is not important to understand why this happens, but rather that the hazards of the compounds will be similar to the rest of the oxysalts, i.e., they are oxidizers. This is a complex covalent-sharing arrangement and is one of those chemistry concepts that should be accepted rather than explained for the purposes of emergency response. The ammonium ion is shown in Figure 7.6.

Ammonium iodate, $\mathbf{NH_4IO_3}$, is an oxidizing agent and a dangerous fire and explosion risk in contact with organic materials. It is a white, odorless powder.

Ammonium chlorate is an ammonium compound with a molecular formula of $\mathbf{NH_4ClO_3}$. It is a colorless or white crystal that is soluble in water. Ammonium chlorate is a strong oxidizer, and when contaminated with combustible materials, can spontaneously ignite. It is shock-sensitive and can detonate when exposed to heat or vibration. One of its primary uses is in the manufacture of explosives.

Peroxide Salts

Peroxide salts should not be confused with organic peroxides. Peroxide salts contain a metal and a peroxide radical. Organic peroxides are made up of nonmetal hydrocarbon radicals and the peroxide functional group. Peroxide salts are water-reactive and give off oxygen, evolve heat, and produce a corrosive liquid in contact with water.

The corrosive liquid is the hydroxide of the metal in the compound. The heat produced may be sufficient to ignite nearby combustible materials. Metal peroxides may also decompose when exposed to heat, with results similar to their reaction with water. The hazards and physical and chemical characteristics of metal peroxides are similar. Sodium peroxide and barium peroxide are common metal peroxides and are detailed in the following paragraphs. Keep in mind that they are not the only metal peroxides.

Sodium peroxide has a molecular formula of Na_2O_2 and is an inorganic peroxide salt. It is a yellowish-white powder that turns yellow when heated. Sodium peroxide absorbs water and carbon dioxide from the air, and is soluble in cold water. It is a strong oxidizing agent. It is corrosive and can cause burns to the eyes and skin, and is also toxic by ingestion and inhalation. It is water-reactive and a dangerous fire and explosion risk in contact with water, alcohol, or acids. Sodium peroxide forms self-igniting mixtures with powdered metals and organic materials. It is incompatible with ethyl or methyl alcohol, glacial acetic acid, carbon disulfide, glycerin, ethylene glycol, and ethyl acetate. The four-digit UN identification number is 1504. The NFPA 704 designation is health 3, flammability 0, and reactivity 1. The 704 diamond has the prefix "oxy" in the white space at the bottom. It is used as bleach and as an oxygen-generating material for diving bells and submarines.

Barium peroxide, BaO_2, is a grayish-white powder that is slightly soluble in water. Barium peroxide is a dangerous fire and explosion risk in contact with organic materials, and decomposes around 1450°F. It is also toxic by ingestion, is a skin irritant, and should be kept cool and dry in storage. The four-digit UN identification number is 1449. The primary uses of barium peroxide are in bleaching, in thermal welding of aluminum, as an oxidizing agent, and in dyeing of textiles.

Inorganic Acid Oxidizers

At higher concentrations, some acids can be strong oxidizers and cause combustion in contact with organic materials. Nitric acid is a dangerous oxidizing acid; it is also corrosive, and its vapors are toxic.

Nitric acid, HNO_3, is considered an oxidizer above 40% concentration. It is shipped in bulk quantities in MC/DOT 312 tanker trucks. The MC 312 tanker is a small-diameter tank with reinforcing rings around the outside. The corrosive materials carried in these tanks are heavy. By recognizing the shape of the MC 312, you will realize that there is a corrosive material in the tank even though it may be placarded with the oxidizer placard. Nitric acid will be discussed further in Chapter 10.

Chromic acid, CrO_3, is composed of dark, purplish-red, odorless crystals that are soluble in water. The specific gravity is 2.7, which is heavier than water. It is a powerful oxidizing agent and may explode on contact with organic materials. Chromic acid is a poison, corrosive to the skin, and has a TLV of 0.05 mg/m^3 of air. Chromic acid is a known human carcinogen. The four-digit UN identification number is 1463. The NFPA 704 designation is health 3, flammability 0, and reactivity 1. The white section at the bottom of the 704 diamond has an "oxy" prefix, indicating it is an oxidizer.

Perchloric acid has a molecular formula of **$HClO_4$**. At concentrations of more than 50% but less than 72%, by volume, it is placarded as an oxidizer. Concentrations above 72% are forbidden in transportation. It is a colorless, fuming liquid, and is unstable in its concentrated form. It is a strong oxidizing agent, is corrosive, and is placarded as a Class 8 corrosive in concentrations less than 50%. Perchloric acid will ignite vigorously upon contact with organic materials or detonate by shock or heat. It is toxic by inhalation and ingestion, and is a strong irritant. Perchloric acid is water-soluble and has a specific gravity of 1.77, which means it is heavier than water. However, it is water-soluble and will mix rather than form layers. Upon contact with water, heat is produced. The boiling point is 66°F, and the vapor density is 3.46, which is heavier than air. Perchloric acid is incompatible with acetic anhydride, bismuth and its alloys, alcohols, paper, wood, and other organic materials. The four-digit UN identification number for concentrations greater than 50% is 1873. Concentrations less than 50% have a four-digit identification number of 1802. The NFPA 704 designation is health 3, flammability 0, and reactivity 3. There is an "oxy" prefix in the white section of the 704 diamond, indicating an oxidizer.

Oxidizers, like many of the other hazard classes, can have more than one hazard. They can be corrosive, poisonous, and explosive under certain conditions. Many are water-reactive, and some may react violently with other chemicals, particularly organic materials.

Oxygen is one of the components of the original fire triangle and the more recent fire tetrahedron. The combustion process involves heat, oxygen, fuel, and a chemical chain reaction. Combustion is simply a rapid oxidation reaction that is accompanied by the emission of energy in the form of heat and light. When we think about combustion, we usually think about atmospheric oxygen allowing combustion to occur. In the combustion process, the fuel source is heated; the molecules start to vibrate rapidly and, if the vibrations are strong enough, the molecules break into small fragments with incomplete bonds.

These "free radical" fragments are unstable and cannot remain as fragments, so they want to bond with some other element to complete their electron requirements. As the heat increases, the radical fragments rise and encounter oxygen from the air. The oxygen is electronegative (electron-drawing) and quickly attracts the electrons from the radical fragments. Bonding occurs between the radical fragments and the oxygen, forming a chemical bond. This bonding process is exothermic, and the heat energy is fed back into the combustion process, allowing the combustion to continue. Without oxygen or chemical oxidizers, most combustion cannot occur.

Materials that contain oxygen in their compounds can support combustion even in the absence of atmospheric oxygen. When extra oxygen is added, over and above atmospheric oxygen, the process of combustion is accelerated. The more oxygen that is present, the more bonds that can take place between the radical fragments and the oxygen. The higher the oxygen concentration, the more accelerated the combustion process.

Oxidizers can undergo spontaneous combustion by three different means. First, there is slow oxidization. This occurs when an oxidizer comes in contact with a material that has double bonds (also known as Pi bonds), such as animal or vegetable oils and alkenes. Animal and vegetable oils are actually large esters, with the exception of turpentine, which is a pure hydrocarbon compound. The problem is the same, however; the double bonds in turpentine can be attacked by oxygen from the air and can undergo spontaneous ignition. The structure and molecular formula for turpentine are shown in Figure 7.7; notice the double bonds.

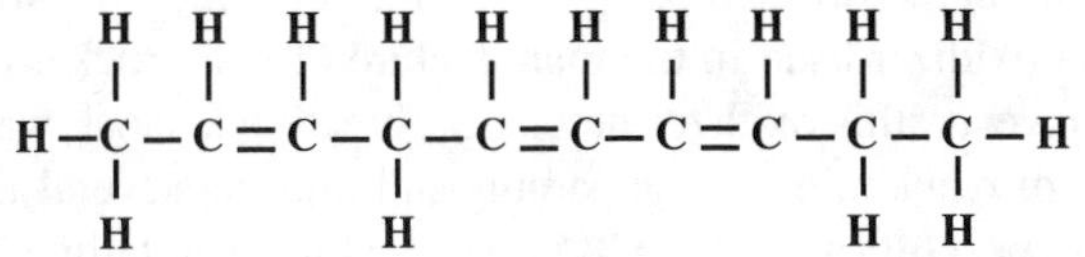

```
    H   H   H   H   H   H   H   H   H   H
    |   |   |   |   |   |   |   |   |   |
H — C — C = C — C — C = C — C = C — C — C — H
    |           |                   |   |
    H           H                   H   H
```

Turpentine
$C_{10}H_{16}$

Figure 7.7

As the oxidizer (which can be oxygen in the air) breaks the double bonds, heat is produced. If there is enough insulating material present, the heat can build up to the point of spontaneous ignition. Second, the reaction of an inorganic oxidizer with water can be exothermic, i.e., it gives off heat. Oxidizers also generate oxygen gas in contact with water. The heat and oxygen represent two sides of the fire triangle or tetrahedron. All that is needed is fuel. If combustible materials are present, combustion can occur spontaneously. When that occurs, there is also excess oxygen present that accelerates the rate of combustion. Third, the corrosive action of a strong acid, such as nitric acid, can generate heat. The heat can be sufficient to ignite combustible materials to which the corrosive is exposed.

Some oxidizers mix with water rather than react. The liquid mixture can impregnate combustible and noncombustible materials. When the water evaporates from the material, the oxidizer is left behind. The materials will now burn with great intensity if exposed to heat or fire because of the oxidizers present. Firefighter turnouts can become impregnated with oxidizer materials during firefighting operations. While firefighter turnouts are resistant to combustion, the oxidizer in the fabric can cause the turnouts to burn with great intensity if exposed to heat or fire. The turnouts must be decontaminated prior to being used again, or the firefighters will be in unnecessary danger when exposed to heat or fire.

Some compounds can undergo spontaneous combustion when they contact water. This is a slow process, as in the case of charcoal. Wetted charcoal is oxidized by oxygen in the air. If the heat produced by the water reaction is accumulated in the material, combustion may occur. This air oxidation can also occur with animal and vegetable oils when they are present in combustible materials, such as rags. The double bond, also referred to as a Pi bond, in the animal and vegetable oils is oxidized by the oxygen in the air. This action on the double bond creates heat. If the heat is confined, spontaneous combustion may occur. The corrosive action of strong acids can generate heat. The heat may be enough to ignite combustible materials. When the corrosive material is also an oxidizer, the oxidizer will contribute to the acceleration of the combustion process.

An explosion is really nothing more than a rapid combustion. A chemical explosion, as discussed in Chapter 3, requires a chemical oxidizer to be present. Without the chemical oxidizer, the combustion does not accelerate fast enough to allow an explosion to occur. Oxidizers that allow for explosions can themselves explode when heated or shocked.

In Henderson, Nevada, a fire occurred in the Pepcon chemical plant. The fire heated ammonium perchlorate to the point that a detonation occurred. Ammonium perchlorate is an oxidizer used in the manufacture of solid rocket fuel, but it is not an explosive. However, this oxidizer detonated, creating a shock wave that blew the windshields out of responding fire apparatus and injuring several firefighters.

Ammonium perchlorate, NH_4ClO_4, is a white, crystalline material that is soluble in water. It is a strong oxidizing agent and a skin, eye, and respiratory irritant. Ammonium perchlorate is shock-sensitive and may explode or detonate when exposed to heat-reducing agents or by spontaneous chemical reaction. (A reducing agent is a material that removes the oxygen from the compound.) Oxidizers can be dangerously explosive materials even though they are not classified as explosives. Closed containers can rupture violently when heated. Ammonium perchlorate decomposes at 464°F and produces oxides of nitrogen, hydrogen chloride, and ammonia. It is incompatible with acids, alkalis, powdered metals, and organic materials. It has a four-digit UN identification number of 1442. The NFPA 704 classification is health 1, flammability 0, and reactivity 4. The prefix "oxy" is placed in the white section at the bottom of the 704 diamond. Ammonium perchlorate is usually shipped in fiber drums, bags, steel drums, and tote bins. It is primarily used in the manufacture of explosives.

Other Oxidizer Compounds

Sodium carbonate, Na_2CO_3, also known as soda ash, is an oxysalt. It can be found naturally or can be synthetic. It is a grayish-white powder or lumps containing up to 99% sodium carbonate. Sodium carbonate is soluble in water. It is not a particularly hazardous material and is not regulated in transportation by the DOT. The primary uses are in the manufacture of other chemicals and products, including glass, paper, soaps, cleaning compounds, petroleum refining, and as a catalyst in coal liquefaction.

Ammonium nitrate, NH_4NO_3, is a colorless crystal that is soluble in water. It is a strong oxidizer. The specific gravity is 1.72, which is heavier than water. Ammonium nitrate is soluble in water and decomposes at 410°F, evolving nitrous oxide gas. It may explode under confinement and at high temperatures. Large amounts of water should be applied using unmanned appliances to fight fires, with all personnel evacuated to a safe distance. Ammonium nitrate is incompatible with acids, flammable liquids, metal powders, sulfur, chlorates, and any finely divided organic or combustible substance. The four-digit UN identification numbers are listed in Figure 7.8. The NFPA 704 designation for ammonium nitrate is health 0, flammability 0, and reactivity 3. In the white space at the bottom of the diamond, the prefix "oxy" indicates it is an oxidizer. The primary uses of ammonium nitrate are

Ammonium Nitrate 4-Digit Identification Numbers

More than 0.2% combustible material—0222
Not more than 0.2% combustible material—1942
With organic coating—1942
Ammonium Nitrate Fertilizer—2067
Fertilizer that is liable to explode—0223
Fertilizer with ammonium sulfate—2069
Fertilizer with calcium carbonate—2068
Fertilizer with not more than 0.4% combustible material—2071
Fertilizer with phosphate or potash—2070
Fertilizers—2071, 2072
Fertilizers N.O.S.—2072
Ammonium nitrate fuel oil mixture—0331
Mixed fertilizers—2069

Figure 7.8

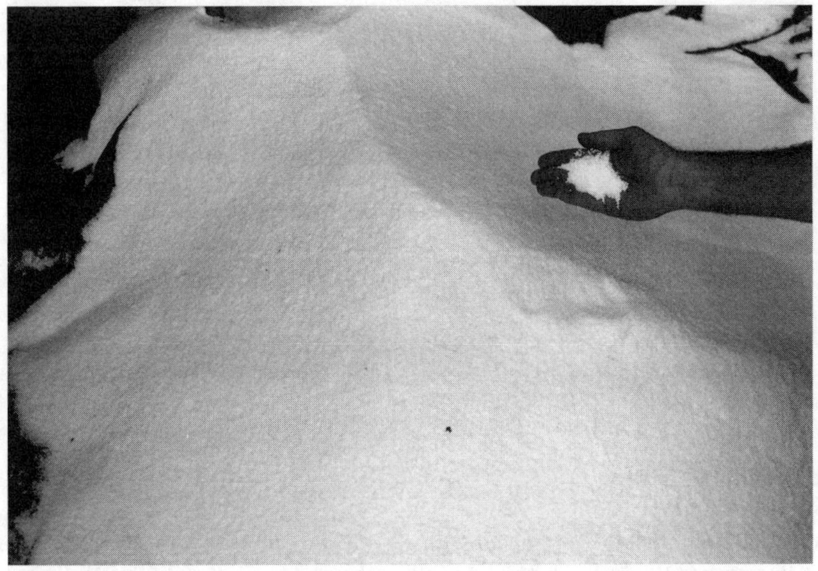

Commercial-grade ammonium nitrate, an oxidizer used to make blasting agents.

in fertilizers, explosives, pyrotechnics, herbicides, and insecticides. It is also used as an oxidizer in solid rocket fuel.

Ammonium sulfate, $(NH_4)_2SO_4$, an ammonium compound, is a brownish-gray to white crystal. Ammonium sulfate is soluble in water and is nonflammable. This compound is an oxidizer with a specific gravity of 1.77, which is heavier than water. The primary uses are in fertilizers, water treatment, fermentation, fireproofing compositions, and as a food additive.

Relative Strengths of Some Oxidizing Agents

1 - Fluorine
2 - Ozone
3 - Hydrogen Peroxide
4 - Metallic Chlorates
5 - Nitric Acid
6 - Chlorine
7 - Sulfuric Acid (concentrated)
8 - Oxygen
9 - Bromine
10 - Iron III compounds
11 - Iodine
12 - Sulfur

Figure 7.9

Hopper truck hauling ammonium nitrate.

Incidents

Ammonium nitrate fertilizer, mixed with a hydrocarbon fuel, was used in the explosion that rocked the Federal Building in Oklahoma City in April 1995, killing 166 people, including 19 children, and injuring 450 others. The damage to the building was so extensive that it had to be demolished. In addition, several other buildings in the downtown area were damaged by the explosion. Ammonium nitrate was also used in the bombing of the World Trade Center in New York City in 1994.

Commercial-grade ammonium nitrate was involved in the explosion in Kansas City, Missouri, in 1988 that killed six firefighters. The firefighters were responding to a construction site where explosives were in a box trailer used for storage. The storage trailer was on fire; the firefighters may have been unaware the explosives were stored there and fought the fire. The resulting explosion totally destroyed one fire engine and damaged another beyond repair and, in addition, killed six firefighters.

As a result of that tragic explosion in Kansas City, OSHA issued a new regulation involving the use of DOT placards and labels in fixed storage. All hazardous materials that require DOT placarding and labeling in transportation must continue to be placarded and labeled in fixed storage. The placards and labels must remain on the containers until the materials are used up and the containers have been purged or properly discarded.

Ammonium nitrate fertilizer can be made resistant to flame and detonation by an exclusive process involving the addition of 5 to 10% ammonium phosphate.

CLASS 5.2 ORGANIC PEROXIDES

The DOT assigns Class 5.2 organic peroxides to seven generic types (see Figure 7.10), and they are classified by the extent to which they will detonate or deflagrate. Organic peroxides are a hydrocarbon derivative, and the major hazard is explosion. They are highly dangerous materials used as initiators and catalysts for polymerization reactions. Organic peroxides are highly reactive because of the presence of the oxidizer and the fuel within the formula. They can start their own decomposition process when contaminated, heated, or shocked. Organic peroxides are nonpolar and immiscible in water. Some examples of organic peroxide compounds and compounds subject to peroxide formation are listed in Figure 7.11.

Genetic Types of Organic Peroxides

TYPE A Organic peroxide that can detonate or deflagrate rapidly as packaged for transport. Transportation of type A organic peroxides is forbidden.

TYPE B Organic peroxide as packaged for transport, neither detonates nor deflagrates rapidly, but can undergo a thermal explosion.

TYPE C Organic peroxide as packaged for transport, neither detonates nor deflagrates rapidly and cannot undergo a thermal explosion.

TYPE D Organic peroxide which:

(I) Detonates only partially, but does not deflagrate rapidly and is not affected by heat when confined;

(II) Does not detonate, deflagrates slowly, and shows no violent effect if heated when confined; or

(III) Does not detonate or deflagrate, and shows a medium effect when heated under confinement.

TYPE E Organic peroxide which neither detonates nor deflagrates and shows low, or no, effect when heated under confinement.

TYPE F Organic peroxide which will not detonate in a cavitated state, does not deflagrate, shows only a low, or no, effect if heated when confined, and has low, or no, explosive power.

TYPE G Organic peroxide that will not detonate in a cavitated state, will not deflagrate, shows no effect when heated under confinement, has no explosive power, and is thermally stable.

Figure 7.10

Organic Peroxides and Peroxidizable Compounds

Methyl Ethyl Ketone Peroxide	Dioxane
Benzoyl Peroxide	Furan
Ether Peroxides	Butadiene
Peracetic Acid	Vinyl Chloride
Potassium Metal	Styrene
Vinylidene	Cyclohexane
Methyl Acetylene	Tetrahydrofuran
Cyclopentane	Cumene

Figure 7.11

The organic peroxide functional group is composed of two oxygens single-bonded to each other. There is a hydrocarbon radical on each side of the single-bonded oxygens. Their general formula is **R-O-O-R**. Peroxides are named much the same way as ethers and ketones. There must be two radicals, which are named smallest to largest with the word "peroxide" at the end of the name. If the radicals are the same, the prefix "di" is used to indicate two of the same radicals. For example, if the radicals "methyl" and "ethyl" are attached to the peroxide, the compound is named methyl ethyl peroxide. Shown in the following illustrations are the names, molecular formulas, and structures of two common organic peroxides. Note that benzoyl peroxide does not follow the trivial naming system for peroxides; however, the "peroxide" in the name provides the family information that will help in determining the hazard, which is explosive.

Organic peroxides are widely used in the plastics industry as polymerization reaction initiators. All organic peroxides are combustible. Many can be decomposed by heat, shock, or friction; some, such as methyl ethyl ketone peroxide, can detonate. Organic peroxides can be liquids or solids, and are usually dissolved in a flammable or combustible solvent. Organic peroxides can be dangerously explosive materials. Organic peroxides have a self-accelerating decomposition temperature (SADT), and they are shipped and stored under refrigeration to keep them cool. The SADT temperatures range from 0° to 50°F and higher. SADT is the temperature at which the compounds will start to decompose. This decomposition reaction may result in a violent detonation that cannot be stopped by anything responders might try to do. The best bet is to make sure that the materials *do not* reach their SADT. Some organic peroxides are so unstable that they are forbidden in transportation. Organic peroxides exhibit the following hazards to emergency responders: they are unstable, flammable, and highly reactive; may explode in a fire; are corrosive; may be toxic; and are oxidizers. They all have SADTs, and once this reaction starts, there is little responders can do to stop it. Responders should withdraw and treat the material as Class 1 explosives.

Methyl ethyl ketone peroxide is an organic peroxide, even though there is also a ketone functional group in the compound. It is a colorless liquid and a strong oxidizing agent. It is a fire risk in contact with organic materials. Methyl ethyl ketone peroxide is a strong irritant to the skin and tissues. The TLV ceiling is 0.2 ppm in air. The four-digit UN identification number is 2550. The primary uses are in the

Organic peroxides have self-accelerating decomposition temperatures and may decompose explosively upon exposure to heat.

production of acrylic resins and as a hardening agent for fiberglass-reinforced plastics. The molecular formula and structure are shown in Figure 7.12.

```
                 H           H
                 |           |
  H   H      H — C — H   H — C — H   H   H
  |   |          |  /O—O\    |       |   |
H—C — C ———— C           C ———— C — C — H
  |   |          \O—O/               |   |
  H   H                              H   H
```

Methyl ethyl ketone peroxide
MEK peroxide
$C_8H_{16}O_4$

Figure 7.12

Ditertiary butyl peroxide is a clear, water-white liquid. It has a specific gravity of 0.79, which is lighter than water, and it will float on the surface. It is nonpolar and insoluble in water. Ditertiary butyl peroxide is a strong oxidizer and may ignite organic materials or explode if shocked or in contact with reducing agents. In addition to being an oxidizer, ditertiary butyl peroxide is highly flammable. It has a boiling point of 231°F and a flash point of 65°F. The NFPA 704 designation is health 3, flammability 2, and reactivity 4. The prefix "oxy" for oxidizer is placed in the white section at the bottom of the 704 diamond. The molecular formula and structure are shown in Figure 7.13.

Intermodal tank of hydrogen peroxide, a strong oxidizer.

```
              H                            H
              |                            |
     H   H — C — H                 H — C — H   H
     |        |                            |        |
H — C ——— C ——— O — O ——— C ——— C — H
     |        |                            |        |
     H   H — C — H                 H — C — H   H
              |                            |
              H                            H
```

Di-*tert*-butyl peroxide

$C_4H_9O_2+C_4H_9$

Figure 7.13

Hydrogen peroxide is a colorless, liquid organic peroxide that is a powerful oxidizer. It is also a dangerous fire and explosion risk and is toxic in high concentrations, with a TLV of 1 ppm in air. It is soluble in water, with a boiling point of 25°F. Hydrogen peroxide is also corrosive and a strong irritant. It is shipped and stored in 40 to 60% and in greater than 60% concentrations. Common commercial strengths are 27.5, 35, 50, and 70%. Hydrogen peroxide is incompatible with most metals and their salts, alcohols, organic substances, and any flammable substances. The four-digit UN identification number for the 40 to 60% concentrations is 2014. The NFPA 704 designation is health 2, flammability 0, and reactivity 1. The white section at the bottom of the 704 diamond has an "oxy" for oxidizer. The four-digit UN identification number for concentrations greater than 60% is 2015. The NFPA 704 designation is health 2, flammability 0, and reactivity 3. The white section at the bottom of the 704 diamond has an "oxy" for oxidizer. The structure and molecular formula for hydrogen peroxide are shown in Figure 7.1.

In its pure form, **benzoyl peroxide** is a white, granular, crystalline solid that ignites easily, burns with great intensity, and may explode. It has a faint odor of benzaldehyde and is tasteless. Benzoyl peroxide is slightly water-soluble. It may explode spontaneously when dry with less than 1% water. It decomposes explosively above 105°C. Its autoignition temperature is 176°F. It should not be mixed unless at least 33% water is present. Benzoyl peroxide is highly toxic by inhalation, with a TLV of 5 mg/m³ of air. The burning characteristics are similar to black powder. Benzoyl peroxide decomposes rapidly when heated, and if the material is confined, detonation will occur. The molecular formula and structure for benzoyl peroxide are shown in Figure 7.14.

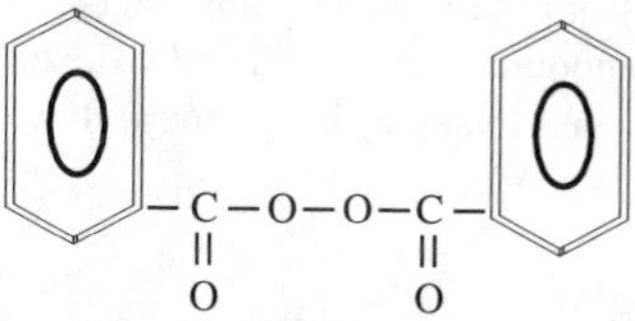

Benzoyl Peroxide
$(C_6H_5CO)_2O_2$

Figure 7.14

Most ethers, when stored for more than six months, will form explosive **ether peroxides** in their containers. The primary ethers to be concerned about are ethyl ether, ethyl tertiary butyl ether, ethyl tertiary amyl ether, and isopropyl ether. Isopropyl ether is considered the worst hazard in storage.

Ether peroxides that form inside containers are organic peroxides and are sensitive to shock and heat. When these peroxides are concentrated or heated, they may detonate. Peroxide formation can be detected as early as one month in storage. Ethers are usually stored in amber glass bottles. Light and heat are important contributors to peroxide formation, although light seems to have more effect than heat. There is no effective means of inhibiting peroxide formation in ether containers. If aging containers of ethers are encountered, they should be treated like bombs. The bomb squad should be called for handling and disposal. There have been cases where employees and response personnel have been severely injured when an aging can of ether has exploded in their hands. Ethers may be found in high schools and college laboratories and throughout industry. Ether is also used in the processing of illegal drugs, and may be encountered in clandestine drug lab operations.

The dangers presented by all oxidizers are similar. They are dangerous in contact with organic materials. They accelerate combustion. They can be a serious fire and explosion hazard. Even though some are water-reactive, water is the extinguishing agent of choice for fires involving oxidizers. Extinguishing agents that work by excluding atmospheric oxygen will not always work with oxidizers. Oxidizers have their own oxygen supply within the compound and do not need atmospheric oxygen to support combustion. Emergency responders should treat oxidizers with the same respect as incidents involving explosives, because they may be just as dangerous.

Incident

Two workers were killed and 13 people injured, including three firefighters, in an explosion and fire at a chemical plant. The facility manufactured organic peroxides, including methyl ethyl ketone (MEK) peroxide, which was involved in the fire. Also involved in the fire were bunkers for storing the MEK peroxide and benzoyl peroxide. Arriving companies were ordered to stage until the explosions subsided and an aggressive attack could be mounted safely.

REVIEW QUESTIONS FOR CHAPTER 7

1. Name four elements from the Periodic Table that are oxidizers.
2. Which families of compounds are considered oxidizers?
3. Oxidizers may be shipped under which of the following placards?
   A. Dangerous
   B. Oxidizer
   C. Poison or poison gas
   D. All of the above
4. Name the following oxysalt compounds and balance the formulas if needed. (Any compounds with transition metals are balanced.)

   $AlSO_5$ $LiClO_2$ $NaFO_3$ $MgPO_2$ $CuClO_3$

5. Provide formulas and structures for the following organic peroxide hydrocarbon-derivative compounds.

   Methyl ethyl peroxide Di-vinyl peroxide Isopropyl butyl peroxide

6. When in contact with water, some peroxide salts may produce which of the following hazards?
   A. Release heat
   B. Release oxygen
   C. Impregnate other materials
   D. All of the above
7. Some organic peroxides may undergo polymerization if inhibitors are released during an accident. Which of the following best describes polymerization?
   A. Violent reaction
   B. Releases heat
   C. Self-reaction
   D. All of the above
8. Oxidizers, when present during a fire, will do which of the following?
   A. Help extinguish the fire
   B. Accelerate combustion
   C. Purify runoff water
   D. None of the above

CHAPTER 8

Poisons

Hazard Class 6 materials are poisons that are solids and liquids. Some of the liquids are volatile and produce vapors, which are an inhalation hazard. Volatile poisons that are an inhalation hazard require the transport vehicle to be placarded regardless of the quantity. Class 6 is divided into two subclasses: 6.1 and 6.2. The DOT defines a Class 6.1 poison as a material, other than a gas, known to be so toxic to humans as to afford a health hazard during transportation or which, in the absence of adequate data, is presumed to be toxic to humans because it falls within any one of the following categories when tested on laboratory animals.

Oral toxicity. A liquid with a lethal dose (LD) of 50 or not more than 500 milligrams per kilogram, or a solid with an LD_{50} of not more than 200 mg/kg of the body weight of the animal. (LD_{50} is the single dose that will cause the death of 50% of a group of test animals exposed to it by any route other than inhalation.)

Dermal toxicity. A material with an LD_{50} for acute dermal toxicity of not more than 1000 mg/kg of the body weight of the research animal.

Inhalation toxicity. A dust or mist with a lethal concentration (LC) of 50 for acute toxicity upon inhalation of not more than 10 mg/kg of body weight of the laboratory animal. (LC_{50} is the concentration of a material in air that, on the basis of laboratory tests through inhalation, is expected to kill 50% of a group of test animals when administered in a specific period.)

Simply stated, the DOT definition really says that a small amount of a poison is dangerous to life and that the material should be considered dangerous. A medical definition for a poison is "the ability of a small amount of a material to produce injury by a chemical action." A chemical definition is "the ability of a chemical to produce injury when it comes in contact with a susceptible tissue." Perhaps a better definition of a poison would be that "a poison is a chemical that, in relatively small amounts, has the ability to produce injury by chemical action when it comes in contact with a susceptible tissue."

Corrosives are not usually thought of as poisonous materials. However, if the combined chemical definition is applied to corrosives, then, in fact, they are poisonous to the tissues they contact. Allyl alcohol (DOT/UN identification number 1098) is toxic by absorption, inhalation, and ingestion. However, like many other hazardous materials, it also has multiple hazards. Allyl alcohol is placarded as a 6.1 poison primary hazard,

OSHA Permissible Exposure Limits (PELs)

Chemical	8 hr. PEL	Odor Threshold
Ozone	0.1 ppm	0.1 ppm
Chlorine	0.5 ppm	0.3 ppm
Benzene	1 ppm	5–12 ppm
Hydrogen Sulfide	10 ppm	Fatigues nose
Carbon Monoxide	35 ppm	Odorless
Trichloroethylene	50 ppm	20 ppm
Toluene	100 ppm	2.0 ppm
Gasoline (TLV)	300 ppm	10 ppm
Freon 113	1000 ppm	350 ppm
Methane	Simple asphyxiant	Odorless

Figure 8.1

but it is also a flammable liquid. Materials in subclass 6.2 are infectious substances, meaning "they are viable microorganisms or their toxins, which may cause diseases in humans or animals." This section also includes regulated medical waste.

Poisons are among the most dangerous materials for emergency responders. Many of the effects of poisons do not present themselves right away; in fact, the toxic effect may not appear for days, months, or years. Because the effects may not present themselves right away, responders may be led to believe there is no danger. One of the main reasons decontamination is done for hazardous materials incidents is to prevent the spread of toxic materials away from the "hot zone."

Toxicology is the science of the study of poisons and their effects on the human body. It is also the study of detection in the body systems and of antidotes to counteract poisonous effects. The science of toxicology is relatively new. It evolved out of the concern for worker health and safety that became a concern in the early 20th century.

Some of the first concerns for worker health and safety came from the trade unions. Eventually, the Occupational Health and Safety Act was passed by Congress in the 1970s, creating the Occupational Safety and Health Administration (OSHA). These events led to the eventual formation of the science of industrial hygiene, which involved the protection of workers in the workplace. Industrial hygiene is the application of industrial hygiene concepts to the work environment. Many of the toxicological measuring terms found in reference sources come from the recommendations of safe workplace exposures (see Figure 8.1). When responding to poison incidents, it is important to remember that, first of all, the emergency scene is the workplace for emergency responders. Second, the concentrations that emergency responders will encounter at a spill will often be much higher than the "normal" or acceptable workplace measurements.

TYPES OF EXPOSURE

Types of exposure include:

Acute: A one-time, short-duration exposure. Depending on the concentration and duration, there may or may not be toxic effects. A one-time exposure can cause illness or death; however, it cannot be cumulative. In order for cumulative effects to occur, there must be multiple exposures. If multiple exposures occur, it is considered chronic rather than acute.

Subacute: Involves multiple exposures with a period of time between exposures. The effect is actually less than an acute exposure. The theory is that as long as there are periods of time between exposures, there will be no ill effects. There are no cumulative effects of subacute exposure because of the time between exposures. This concept is similar to the time factor when dealing with radioactive materials. Personnel can be exposed to certain levels of radioactivity for short periods of time without any ill effects.

Chronic: As with subacute, chronic exposures are multiple exposures; however, with chronic exposure, there can be cumulative effects. Cumulative effects are simply a buildup of poison in the body. After the first exposure, some or all of the toxic material stays in the body. The first exposure may not cause any illness or damage. As additional exposures occur and the poison builds up in the body, it can reach toxic levels where illness, damage, or death can occur.

All of these exposures are usually considered workplace events. The emergency responders' workplace is the incident scene. An emergency responder can have an acute exposure to a toxic material at the scene of a HazMat incident or other type of emergency response. A single exposure may not produce symptoms or illness, but multiple, or chronic, exposures may cause damage. It is important to monitor exposures of personnel to determine whether they experience any illnesses following a response to an incident. Sometimes, it takes several days for a toxic material to reach the susceptible target organ. The tendency is to assume that when a person is exposed to a poison at an incident scene, the ill effects, if there are going to be any, will occur right away. This is not always the case with many poisons. When an illness occurs following a hazardous materials incident, the personnel should be checked out just in case.

Once a poison has entered the body, it can behave in a number of ways. First, the effect on the body may be localized, i.e., it only affects the tissue that it has directly contacted. Second, the effect may be systemic, or a whole-body effect. In this instance, the effect on the contact tissue is little, if any. The poison enters the bloodstream and travels throughout the body until a target organ is reached, or the material is secreted through the body's waste removal system. Last, the effects may be a combination of localized and systemic.

ROUTES OF EXPOSURE

In order for a person to be affected by a poison, the poison must directly contact the body or enter into the body. There are four routes by which toxic materials can enter the body and cause damage (see Figure 8.2): inhalation, absorption, ingestion, and injection.

Inhalation requires the poison gas or vapor to enter the body through the respiratory system, where most of the damage usually occurs. Once in the lungs,

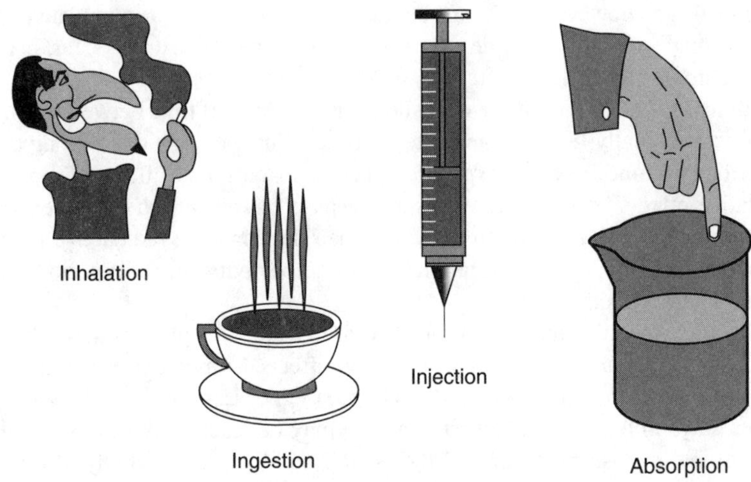

Figure 8.2

Toxicants that Produce Disease of the Respiratory Tract

Toxicant	Acute Effect	Chronic Effect
Ammonia	Irritation, edema	Bronchitis
Arsenic	Bronchitis, irritation, pharyngitis	Cancer
Chlorine	Cough, irritation, asphyxiation	
Phosgene	Edema	Bronchitis, fibrosis, pneumonia
Toluene	Bronchitis, edema, bronchospasm	

Figure 8.3

the poison can injure the respiratory tissues, enter the bloodstream, or both. Examples of toxicants that produce disease of the respiratory tract are shown in Figure 8.3.

Absorption occurs when solid, liquid, or gaseous poisons enter the body through the skin, eyes, or other tissues (see Figure 8.4). Damage may occur at the point of contact, or the material may travel to the susceptible target organ and cause harm there. Effects may be local, such as irritation and death of body tissues through direct contact. For example, when naphthalene contacts the eyes, it can cause cataracts and retina damage. Phenothiazine (insecticide) damages the retina. Thallium causes cataracts and optic nerve damage. Methanol causes optic nerve damage.

Examples of Chemicals Toxic by Skin Absorption

Acetaldehyde	Acetone
Acrolein	Ammonia
Aniline	Arsenic
Benzene	Barium
Camphor	Carbon Disulfide
Carbon Tetrachloride	Chlordane
Butyric Acid	Chlorine
Cumene	Bromine

Figure 8.4

Ingestion occurs when a solid, liquid, or gaseous poison enters the body through the mouth and is swallowed. Damage may occur to the tissues contacted, or the poison may enter the bloodstream. Absorption can also occur after ingestion. A poisonous material may be absorbed through the tissues in the mouth, stomach, intestines, or other tissues it contacts after ingestion occurs.

Injection involves a jagged or sharp object that has been contaminated with a toxic material creating or entering an open wound in the skin. The poison enters the bloodstream once injected into the skin. The exposure method may greatly impact the severity of the damage produced. A chemical that is extremely poisonous by one route of exposure may have little, if any, effect by other routes. For example, carbon monoxide is toxic by inhalation. Just a 1% concentration in air, if inhaled, is fatal in one minute. However, you could stay in a 100% concentration indefinitely provided there was an outside air supply, such as self-contained breathing apparatus (SCBA). Carbon monoxide is not absorbed through the skin. Rattlesnake venom is poisonous if it gets in the bloodstream; it damages the cells. It must be injected to cause damage. If ingested, it may cause nausea, but it will not enter the bloodstream.

EFFECTS OF EXPOSURE

Toxicology relates to the physiological effect, source, symptoms, and corrective measures for toxic materials. Poisons can be divided into several general categories: asphyxiants, corrosives, sensitizers, carcinogens, mutagens, teratogens, and irritants.

Short-Term Effects

When a poison contacts a tissue and a chemical action occurs, the poison will produce an injury to that tissue. The effects of an exposure to a poison, however, will differ based on the poison involved, the type of exposure, and the method of exposure. There are three types of effects that can occur as the result of an exposure to any given poison: immediate, long-term, and etiologic. Immediate effects depend on the dosage received by the person exposed to the poison. Dosages can be large or small, short-term or long-term, and can be acute, subacute, or chronic. Effects from the type of exposure can vary from none, to slight discomfort, to illness, and even death. Effects may also depend on the person exposed. Not all people are affected in the same way by the same poison, the same dosage, and the same exposure. Differences are based upon individual body chemistry, age, health, sex, and size. Immediate effects can include asphyxiation, corrosive damage, sensitizing, and irritation.

Asphyxiants act upon the body by displacing oxygen. Asphyxiants can be simple or chemical. Gases, such as nitrogen, hydrogen, helium, and methane, are known as simple asphyxiants, because they dilute the oxygen in the air below the level required for life to exist. Victims die because there is not enough oxygen in the air they breathe. In the case of Class 6 poisons, however, asphyxiants act by interfering with the blood's ability to convert or carry oxygen in the bloodstream. These are known as chemical asphyxiants. Death may also result from a lack of oxygen, but it is not a result of a lack of oxygen in the air. Examples of chemical asphyxiants include hydrogen cyanide, benzene, toluene, and aniline.

Corrosives are acids or bases that, in small amounts, can cause damage to tissue. Tissue is damaged in much the same way as a thermal burn; however, the burn is much more damaging. The type of damage is the same whether exposed to acids or bases. Examples of corrosive materials include nitric acid, sulfuric acid, phosphoric acid, sodium hydroxide, and potassium hydroxide. They are covered in detail in Chapter 10.

Sensitizers, on first exposure, cause little or no harm in humans or test animals, but on repeated exposure, they may cause a marked response not necessarily limited to the contact site. This response is similar to the process that occurs in allergies that humans develop. It is a physiological reaction to a sensitizing material. For example, a person who moves into an area that has high pollen counts and other airborne allergens may not experience any effects at first, but the longer the exposure occurs, the more symptoms that develop. Examples of sensitizers are isocyanates and epoxy resins.

Irritants are materials that cause irritation to the respiratory system, body organs, or the surface of the skin. The irritation may be a corrosive action, localized irritation, inflammation, pulmonary edema, or a combination. Effects may include minor discoloration of the skin, rashes, or tissue damage. Examples of irritants are tear gas and some disinfectants.

Several factors influence toxic effects, one of which is the concentration of a material (see Figure 8.5). Concentrations may be expressed in terms of percentage or parts per million or billion per milligram or kilogram; the higher the concentration, the more serious the effect. Figure 8.6 illustrates some parts-per-million approximations. Some materials can also have an anesthetic effect. This may range from loss

Expressions of Concentration

- Percentages
- Parts per million (ppm) or billion (ppb)
- Milligrams per cubic meter, foot, kilogram, and cubic liter

Figure 8.5

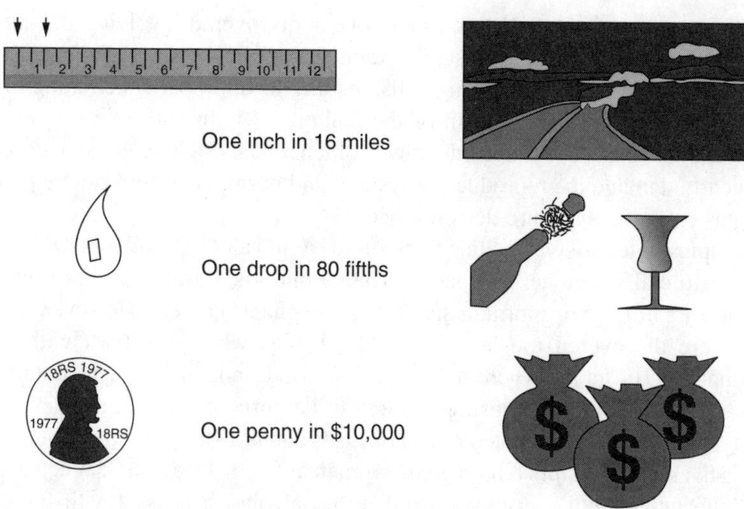

Figure 8.6

of feeling and sensation to unconsciousness. Anesthetics include nitrous oxide, ethers, and hydrocarbons.

Long-Term Effects

Long-term effects can be divided into three types: carcinogenic, teratogenic, and mutagenic. All three types of effects take, from the time of exposure, up to 15 years or longer to develop symptoms. Effects can be aggravated by exposures to other materials or unrelated health problems.

Carcinogens are materials that cause cancer. This is a chronic toxic effect. Information about cancer-causing agents has been obtained by studying populations exposed to chemicals for long periods, usually in the workplace. Data has also been obtained through tests using laboratory animals. This does not mean that a material will actually cause cancer in humans, but it is a good indication. Examples of known carcinogens include benzene, asbestos, arsenic, arsenic compounds, vinyl chloride, and mustard gas.

Mutagens cause mutations or alterations in genetic material, thereby altering human DNA. Changes may be chromosomal breaks, rearrangement of chromosome pieces, gain or loss of entire chromosomes, or damage within a gene. Effects may involve the current generation that was exposed or future generations. Examples of mutagens include arsenic, chromium, dioxin, mercury, ionizing radiation from x-rays or other radioactive material, caffeine, LSD, marijuana, and nitrous oxide. Additionally, ethylene oxide, ethyleneimine, hydrogen peroxide, benzene, and hydrazine are also mutagens.

Teratogens cause one-time birth defects in offspring resulting from maternal or paternal exposure to toxic materials. The word "teratology" is derived from the Latin meaning "the study of monsters." It is actually the study of congenital malformations, which started with the study of the correlation of German measles to birth defects.

Later, there was an industrial link to teratogens discovered involving the chemical methyl mercury. Results of exposure to teratogenic chemicals on living organisms include the alteration of developing cells, leading to improper functioning of these cells. This may result in the death of the embryo or fetus. Specific types of birth defects can be caused by specific types of chemicals. These chemicals do not permanently damage the reproductive system, and normal children can be produced as long as repeated exposure does not occur.

Examples of teratogens include thalidomide, ethyl alcohol, and *o*-benzoic sulfimide (the artificial sweetener saccharin). Thalidomide is a drug that was used in the 1960s as a treatment for morning sickness in pregnant mothers. However, the toxic effects were discovered too late for 10,000 babies who were born with various malformations. In Japan, where fish is a staple food, mothers ate fish contaminated with mercury compounds, causing children to be born with cerebral palsy.

Ethyl alcohol is found in alcoholic beverages and is a known teratogenic material, which is the reason doctors tell pregnant women not to drink alcohol. When a mother drinks, the unborn child drinks as well. Ethyl alcohol causes growth failure and impaired brain development. Unborn children exposed to alcohol may suffer the effects of Fetal Alcohol Syndrome when they are born. Symptoms of Fetal Alcohol Syndrome include sleep disturbance, jitteriness, a higher incidence of impaired vision and hearing, lack of motor coordination, balance problems, abnormal thyroid function, and a decrease in immune system effectiveness. Additional teratogens include heavy metals, methyl mercury, mercury salts, lead, thallium, selenium, penicillin, tetracyclines, excess Vitamin A, and carbon dioxide.

Etiologic Effects

Class 6.2 infectious substances are the last type of toxic effect that will be presented here. Etiological toxins are among the most poisonous materials known. For example, the bacterium *Clostridium botulinum*, the cause of botulism, is a single cell that can release a toxin so potent that four-hundred thousandths of an ounce is enough to kill 1,000,000 laboratory guinea pigs.

Variables of Toxic Effects

Not all people are affected by toxic materials in the same way. Variables include age, genetic makeup, sex (in terms of the genetic difference between males and females, most of which are related to reproduction and nurturing functions), weight, general health, body chemistry (the chemicals produced in each individual are unique, not only by form, but by quantity), and physical condition. These variables can all affect the way individuals respond to any toxic material. Infants and children are often more sensitive to some toxic materials than younger adults. Elderly persons have diminished physiological capabilities to deal with toxic materials. This age group may be more susceptible to toxic effects at relatively lower doses. Chemicals may be more toxic to one sex than the other. Males may be affected by chemicals that do not affect females; some chemicals can affect both. Chemicals may affect the reproductive system of either the male or female. Females who are pregnant

Infectious medical waste is a DOT Class 6.2 Infectious Substance.

may be affected by a toxic material that will cause damage to the fetus, whereas a male may not be affected at all. Damage to the male reproductive system may include sterility, infertility, abnormal sperm, low sperm count, or reduced hormonal activity. Chemicals that cause these types of effects include lead; mercury; PCBs; 2,4,D; paraquat; ethanol; vinyl chloride; and DDT. Females are affected by DDT, parathion, PCBs, cadmium, methyl mercury, and anesthetic gases.

DOSE/RESPONSE

Dose can be expressed in three ways: the amount of the substance actually in the body, the amount of material entering the body, and the concentration in the environment. Response is the health damage resulting from a specified dose. All chemicals, no matter what their makeup, are toxic if taken in a large enough dose. Even water can be toxic if too much is ingested at one time. Paracelsus (1493–1541) observed that "all substances are poisons; there is none which is not a poison. The right dose differentiates a poison and a remedy." There is a no-adverse-effects level for almost all materials. Dose is more than just the amount of a toxic material that has caused the exposure. More correctly, dose is related to the weight of the individual exposed. This is the basis for many of the toxicological terms that have been developed based upon tests on laboratory animals. For example, a 1-lb rat given 1 oz of a toxic material should have the same response as a 2000-lb elephant given 2000 oz of the same toxic material. In both cases, the amount per weight is the same: 1 oz. for each pound of animal weight. This amount per weight is known as the dose.

Response of an individual plant or animal species to a chemical is based on concentration; length, type, and route of exposure; and susceptible target organ.

Target Organs and Chemicals that Affect Them

Lungs	Halogens, hydrogen sulfide
Liver	Vinyl chloride, aromatics, chlorinated HC
Kidneys	Mercury, calcium, carbon tetrachloride
Blood	Carbon monoxide, chlorinated HC
Neurologic	Organophosphates, carbon monoxide
Skeletal	Fluorides, selenium
Skin	Arsenic, chromium, beryllium

Figure 8.7

Additionally, other health-related variables include age, sex, physical condition, and size (mass). This relationship is referred to as dose/response, or in other words, the amount of the exposure and the resulting biological effect. Each plant or animal species has its own individual response to a given chemical. A substance administered at a dose large enough to be lethal to rabbits may have a lesser effect on rats or dogs.

SUSCEPTIBLE TARGET ORGANS

When a poison enters the body through one of the four routes of exposure, it will cause damage to some bodily function. This is referred to as the susceptible target organ (see Figure 8.7). Depending on the dose, it may take a toxic material several hours or several days to reach a susceptible target organ and cause damage. The target organ should not be confused with routes of exposure, which are the manner in which the poison enters the body. The susceptible target organ is the organ or system to which the poison does its damage once it enters the body (see Figure 8.8). For example, a pesticide may enter the body through inhalation, but may have an effect on the central nervous system. In this instance, the route of exposure is inhalation through the lungs; the target organ is the central nervous system. Target organs in the body include the respiratory system, liver, kidneys, central nervous system, blood, bone marrow, skin, cardiovascular system, and other body tissues. The skin is the largest single organ of the body. When unbroken, it provides a barrier between the environment and other organs, except for the lungs and eyes, and is a defense against many chemicals. Studies show that 97% of chemicals to which the body is exposed are deposited on the skin.

EXPOSURE RATE

Exposure rate is the measurement of workplace exposure of hazardous materials based upon tests conducted on laboratory animals. Values are translated to humans based on the weight ratio between an animal and a human. These are only estimations and should be used with caution. Toxicology information is expressed in parts per million (ppm) or billion (ppb) and milligrams per cubic meter (mg/m^3), which are terms indicating the concentration of the toxic material. These terms are not related,

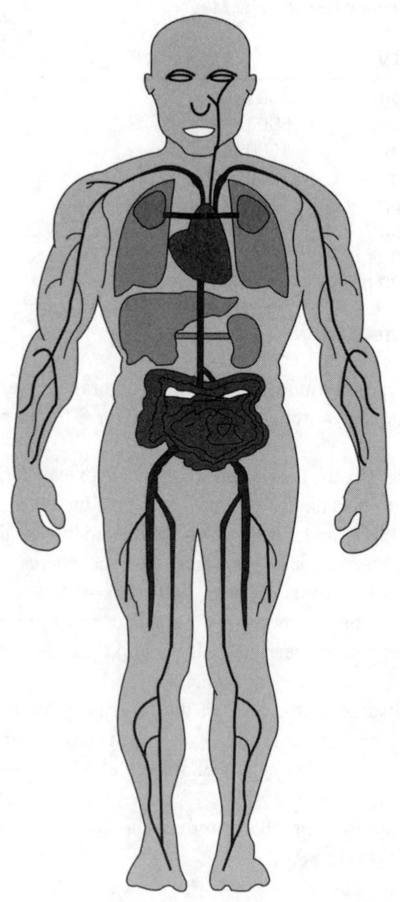

Figure 8.8

but they do have one thing in common: the smaller the numerical value, the more toxic the material measured. The emergency responder's workplace is the incident scene, and the concentrations of toxic materials here are usually much higher than the values indicated through the following terms:

TLV-TWA: Threshold limit value, time-weighted average is the average concentration for a normal 8-hour workday and a 40-hour workweek, to which nearly all workers can be exposed repeatedly, day after day, without adverse effect and without protective equipment. Some examples are shown in Figure 8.9.

TLV-STEL: Threshold limit value, short-term exposure limit is the maximum concentration averaged over a 15-min period to which healthy adults can be exposed safely. Exposures should not occur more than four times a day, and there should be at least 60 min between exposures.

TLV-C or TLV-ceiling: The concentration that should not be exceeded during any part of the workday. *This is the only reliable measurement that should be used by emergency responders on incident scenes.*

Toxicity Comparisons of Some HazMat

Common Name	TLV	LC_{50}
Hydrogen cyanide	10 ppm	300 ppm
Hydrogen sulfide	10 ppm	600 ppm
Sulfur dioxide	5 ppm	1000 ppm
Chlorine	1 ppm	1000 ppm
Carbon monoxide	50 ppm	1000 ppm
Ammonia	50 ppm	10,000 ppm
Carbon dioxide	5,000 ppm	10%
Methane	90,000 ppm	Simple asphyxiant

Figure 8.9

PEL: The permissible exposure limit is the maximum concentration averaged over 8 hours, to which 95% of healthy adults can be repeatedly exposed for 8 hours per day, 40 hours per week.

IDLH: Immediately dangerous to life and health, the maximum amount of a toxic material that a healthy adult can be exposed to for up to 15 min (National Institute for Occupational Safety and Health [NIOSH]) and escape without irreversible health effects. There are also IDLH values established for oxygen-deficient atmospheres and explosive or near-explosive atmospheres (above, at, or near the LELs). At the scene of a hazardous materials incident, it should be assumed that the concentrations present are above the TLV-ceiling, and only responders wearing SCBA and proper chemical protective clothing should be allowed near the incident scene.

LD_{50} is the lethal dose by ingestion or absorption for 50% of the laboratory animals exposed. (There is a wide variation among species. The LD_{50} for one type of animal could be thousands of times less than for another type.) Examples of LD_{50} values for common chemicals are shown in Figure 8.10.

LC_{50} is the lethal concentration by inhalation for 50% of the laboratory animals exposed. Some examples are shown in Figure 8.9.

NOAEL is no-observable-adverse-effect level.

GRAS is generally recognized as safe.

Ld_{50} Values for Common Chemicals

Sucrose (table sugar)	29,000 mg/kg
Ethyl alcohol	14,000
Sodium chloride	3,000
Vitamin A	2,000
Vanillin	1,580
Aspirin	1,000
Chloroform	800
Copper sulfate	300
Caffeine	192
Phenobarbital, sodium salt	162
DDT	113
Sodium nitrate	85
Nicotine	53
Sodium cyanide	6.4
Strychnine	2.5

Figure 8.10

Toxicity measures are based on dose and size or weight of the test animal. This is then projected to humans, based on the weight ratio of the dose, the weight of the animal, and the weight and health of the human. All of the values should be considered nothing more than an educated guess. Many factors can influence the way any given individual will react if exposed to a toxic material. Some factors may be allergies and previous illnesses or operations. Individuals that have had a splenectomy are much more susceptible to poisons than those who have not. Individuals with such a history should be aware that they are at additional risk whenever they are exposed to toxic chemicals. Consider all toxicity data as an estimate; do not stake your life on it.

DEFENSE MECHANISMS FOR TOXIC MATERIALS

There are three types of defenses against toxic materials: internal, antidotal, and external. Three things can happen once a chemical is taken into the body: metabolism, storage, and excretion. Internal defenses are the ability of the body to get rid of a toxic material, sometimes referred to as metabolism. The body normally excretes waste materials through the feces or urine. Additionally, women can also excrete through the ova and breast milk. In these instances, the excretions from the mother represent exposure to the offspring.

Two main organs in the body filter materials and remove them through excretion: the liver and the kidneys. All of the blood in the body passes through these two organs and is filtered. Filtered material is passed from the liver to the intestine through the gallbladder or to the kidney through the blood, after some degree of chemical breakdown (metabolizing). If the liver is unable to break down a poison, it may store the toxic material within its own tissue. Toxic materials, such as lead, can be stored in the bone tissue. Some may be bound by blood proteins and stored in the blood. It is this storage of the toxic material that causes most of the long-term damage. Kidneys also filter poisons from the blood and may incur damage in the process. Materials filtered are then passed on to the urine, which is held in the bladder. This can lead to bladder cancer through chronic exposure to the toxic materials.

Toxic materials that enter the body cannot always be excreted and may be stored in the fatty tissues. Examples include the organochlorines, DDT, PCBs, and chlordane. When a person loses weight, for whatever reason, toxic materials that have been stored in the fatty tissues are released back into the body as the levels of fat decrease. When this occurs, the body is reexposed to the chemicals and illness can occur. Chemicals that cannot be excreted from the body may also be bound by blood proteins and stored in the blood. Lead is an example of a material that cannot be excreted from the body and is stored in the bones.

Polarity has an effect on the body's ability to excrete toxic materials. If a poison is polar, it is usually soluble in water, because water is also polar and is more easily removed from the body. The body also has a system of converting nonpolar compounds so that they can also be removed; however, the body cannot convert all nonpolar compounds. Therefore, some poisons are difficult to convert and may stay

Antidotes for Poisons

Cyanide	Amyl nitrate, sodium nitrate, sodium thiosulfate
Organophosphate pesticides	Atropine, pralidoximine
Methanol or ethylene glycol	IV ethanol, hemodialysis
Nitrites	Oxygen, methylene blue
Hydrocarbons	Oxygen

Figure 8.11

in the body for long periods of time. For example, table salt (sodium chloride) is polar and is easily excreted. DDT, a now-banned pesticide, is nonpolar and is not easily excreted, so it stays in the body for long periods. This is one of the main reasons DDT is no longer allowed to be used on food crops. Characteristics of polarity or nonpolarity can have a crucial impact on the effect a toxic material has on the body.

Two other mechanisms by which the body defends against toxic materials is through breathing and sweating. The lungs are also able to remove materials from the blood. This can be noted by the odor of alcohol on the breath of a person who has been drinking. Odor detected from a person who is sweating is toxic material being removed from the body.

While all three methods contribute to the removal of toxic materials from the body, the last two are of only minor significance. Internal defenses of the body against toxic materials do a good job against many types of chemicals. However, not all toxic materials can be removed by the body's systems. It is best not to rely on the body to remove toxic materials, but rather to take precautions to ensure that the toxic materials do not enter the body in the first place.

Antidotes are administered to counteract the effects of some toxic materials. The definition of an antidote is "any substance that nulls the effects of a poison on the spot and prevents its absorption or blocks its destructive action once absorbed." The problem with antidotes is there are only a few in existence and they do not work on all types of poisons (see Figure 8.11). Another problem with antidotes is availability — an antidote must be given immediately after exposure to a poison or the victim may die anyway. Most EMS units do not carry antidotes for toxic exposures; by the time a person is taken to a medical center where the antidote may be available, it may be too late.

External defenses against toxic materials are by far more effective than the internal defenses of the body or antidotes. What it amounts to in simple terms is: do not let toxic materials into the body to begin with; the idea is to place a barrier between responders and the poison. These barriers include chemical protective clothing and SCBA. Chemical protective clothing provides protection against absorption and contact tissue damage. SCBA prevents inhalation and ingestion of toxic materials. Firefighter turnouts do not provide any type of chemical protection from toxic materials. Turnouts may prevent some types of injection; however, the toxic material will contaminate the turnouts. There are also some preventive measures that can be taken to prevent ingestion of toxic materials. Decontamination is also an external defense. The primary reason for conducting decontamination is the removal of toxic materials. Do not eat, drink, smoke, or place anything in your mouth until decontamination has been completed.

TOXIC ELEMENTS

A number of elements are naturally toxic. These include arsenic; mercury; heavy metals, such as lead and cadmium; and the halogens: fluorine, chlorine, bromine and iodine. Fluorine and chlorine are covered in Chapter 4 in the Poison Gas section. It is important to note the uses of elements, because they give an indication where these materials may be found in storage and manufacturing.

Arsenic, As, is a nonmetallic element that is a silver-gray, brittle, crystalline solid that darkens in moist air. Arsenic is acutely toxic, depending on the dose, and is a carcinogen and mutagen. It is insoluble in water and reacts with nitric acid. Primary routes of exposure are through inhalation absorption, skin or eye contact, and ingestion. The exposure limit is 0.002 mg/m^3. The IDLH is 100 mg/m^3 in air. The target organs are the liver, kidneys, skin, lungs, and lymphatic system. Arsenical compounds are incompatible with any reducing agents. The four-digit UN identification number is 1558. The primary uses are as alloying additives for metals, especially lead and copper, and in boiler tubes, high-purity semiconductors, special solders, and medicines.

Mercury, Hg, is a liquid metallic element that is silvery in color and heavy. It is insoluble in water and has a specific gravity of 13.59, which is heavier than water. Mercury is highly toxic by skin absorption and inhalation of fumes or vapor. The TLV is 0.05 mg/m^3 of air, and the IDLH is 10 mg/m^3. All inorganic compounds of mercury are toxic by ingestion, inhalation, and absorption. Most organic compounds of mercury are also highly toxic. The target organs affected are the central nervous system, kidneys, skin, and eyes. Mercury is incompatible with acetylene and ammonia. The four-digit UN identification number is 2809. The uses of mercury are in electrical appliances, instruments, mercury vapor lamps, mirror coating, and as a neutron absorber in nuclear power plants.

Bromine, Br_2, is a nonmetallic, fuming, liquid element of the halogen family on the Periodic Table. It is dark reddish-brown in color with irritating fumes. Bromine is slightly soluble in water and attacks most metals. The boiling point is 138°F, and the specific gravity is 3.12, which is heavier than water. Bromine is toxic by ingestion and inhalation, and is a severe skin irritant. The TLV is 0.1 ppm in air, and the IDLH is 10 ppm. The target organs are the respiratory system, the eyes, and the central nervous system. It is also a strong oxidizing agent and may ignite combustible materials on contact. The DOT lists it as a Class 8 corrosive; however, it carries the corrosive and poison label. The four-digit UN identification number for bromine is 1744. The NFPA 704 designation for bromine and bromine solutions is health 3, flammability 0, and reactivity 0. The white section at the bottom of the 704 diamond has the prefix "oxy," indicating an oxidizer. The primary uses are in anti-knock compounds for gasoline, bleaching, water purification, as a solvent, and in pharmaceuticals.

Iodine, (I), is a nonmetallic element of family seven, the halogens. It is heavy, grayish-black in color, has a characteristic odor, and is readily sublimed to a violet vapor. It has a vapor density of 4.98, which is heavier than air. It melts at 113.5ºC, has a boiling point of 184ºC, and is insoluble in water. Iodine is toxic by ingestion and inhalation, and is a strong irritant to eyes and akin. The TLV ceiling is 0.1 ppm in air. Iodine is used for antiseptics, germicides, x-ray contrast material, food and feed additives, water treatment, and medicinal soaps. The four-digit UN identification number for iodine is only for the compounds iodine monochloride and iodine pentafluoride, and they are 1792 and 2495, respectively. The DOT lists iodine monochloride as a Class 8 corrosive, and iodine pentafluoride carries an oxidizer and poison label. Iodine does not have an NFPA 704 designation.

TOXIC SALTS

Binary salts have varying hazards, one of which is toxicity. Some of the binary salts are highly toxic, such as sodium fluoride, calcium phosphide, and mercuric chloride. Cyanide salts are also highly toxic, such as sodium cyanide and potassium cyanide. The remaining salts, binary oxides, peroxide, hydroxides, and oxysalts are generally not considered toxic.

Calcium phosphide, Ca_3P_2, is a binary salt made up of reddish-brown crystals or gray, granular masses. It is water-reactive and evolves phosphine gas in contact with water, which is highly toxic and flammable. The four-digit UN identification number is 1360. The primary uses are as signal fires, torpedoes, pyrotechnics, and rodenticides.

Sodium fluoride, NaF, is a binary salt that is a clear, lustrous crystal or white powder. The insecticide grade is frequently dyed blue. It is soluble in water and has a specific gravity of 2.558, which is heavier than water. Sodium fluoride is highly toxic by ingestion and inhalation, and is also strongly irritating to tissue. The TLV is 2.5 mg/m^3 of air. The four-digit UN identification number is 1690. The primary uses are fluoridation of municipal water at 1 ppm, as an insecticide, rodenticide, and fungicide, and in toothpastes and disinfectants.

Mercuric chloride (mercury II chloride), $HgCl_2$, is a binary salt composed of white crystals or powder. It is odorless and soluble in water. It is highly toxic by ingestion, inhalation, and skin absorption. The TLV is 0.05 mg/m^3 of air. The four-digit UN identification number is 1624. The primary uses of mercuric chloride are in embalming fluids, insecticides, fungicides, wood preservatives, photography, textile printing, and dry batteries.

Sodium cyanide, NaCN, is a cyanide salt that is a white, deliquescent, crystalline powder and is soluble in water. The specific gravity is 1.6, which is heavier than water. Sodium cyanide is toxic by inhalation and ingestion, with a TLV of 4.7 ppm and 5 mg/m^3 of air. The target organs are the cardiovascular system, central nervous system, kidneys, liver, and skin. Reactions with acids can release flammable and toxic hydrogen cyanide gas. Cyanides are incompatible with all acids. The four-digit UN identification number is 1689. The NFPA 704 designation is health 3,

A 55-gallon drum of parathion, a highly toxic pesticide.

flammability 0, and reactivity 0. The primary uses are in gold and silver extraction from ores, electroplating, fumigation, and insecticides.

Potassium cyanide, **KCN**, is a cyanide salt that is found as a white, amorphous, deliquescent lump or crystalline mass with a faint odor of bitter almonds. It is soluble in water and has a specific gravity of 1.52. It is a poison that is absorbed through the skin. Target organs are the same as for sodium cyanide. Reaction with acids releases flammable and toxic hydrogen cyanide gas. The four-digit UN identification number is 1680. The NFPA 704 designation is health 3, flammability 0, and reactivity 0. The primary uses are in gold and silver ore extraction, insecticides, fumigants, and electroplating.

HYDROCARBONS

Most of the alkane, alkene, and alkyne hydrocarbon compounds are considered to be flammable as their major hazard, and the toxicity is considered as moderate to low. The vapors are more likely to be asphyxiant than toxic. TLVs range from 50 ppm for hexane to 300 ppm for octane. Decane is listed as having a narcotic effect. Many of these hydrocarbons are found in mixtures, and it will be necessary to look at the Material Safety Data Sheets (MSDS) to obtain toxicity information on particular mixtures. Benzene, toluene, and xylene are aromatic hydrocarbons. They are considered highly toxic and human carcinogens. Benzene has a TLV of 0.1 ppm in air, according to the *NIOSH Guide 1997 Addition*, and an STEL of 1 ppm. The OSHA STEL is 5 ppm and a PEL of 1 ppm. Toluene is toxic by ingestion, inhalation, and skin absorption. The TLV for toluene is 100 ppm in air. Xylenes are toxic by inhalation and ingestion, with a TLV of 100 ppm. The target organs are the blood, skin, bone marrow, eyes, central nervous system, and respiratory system.

HYDROCARBON DERIVATIVES

Several hydrocarbon derivatives are toxic as a primary hazard. While some compounds in each of the groups are toxic, not all compounds are toxic. It is important, however, to consider them all toxic within a group until the specific material can be researched and the exact hazards verified. Families with toxicity as a primary hazard are the alkyl halides and amines. Other groups that are toxic in certain concentrations and doses, although they may not be the primary hazard, include the alcohols, aldehydes, and organic acids. Ethers are considered anesthetic; however, there are some ethers that the DOT lists as Class 6.1 poisons. They are compounds that have chlorine added to the ether. The toxicity comes from the chlorine. For example, 2,2-dichlorodiethyl ether and dichlorodimethyl ether have NFPA 704 health designations of 3 and 4, respectively. Epichlorohydrin, also known as chloropropylene oxide, is a Class 6.1 poison and is an ether with chlorine added to the compound. Methychloromethyl ether is another compound with chlorine added. The NFPA 704 designation for health is 3.

Ketones as a group are considered narcotic. The primary hazard of the esters is polymerization. Many of them are flammable liquids and polymers. The DOT does list some ester compounds as Class 6.1 poisons and, again, these compounds have chlorine added, which accounts for their toxicity. For example, ethyl chloroformate is an ester that has an NFPA health designation of 4. Remember that too much of any chemical can be toxic, so too much of an anesthetic or a narcotic can be toxic.

Alkyl Halides

The alkyl halide functional group is composed of a hydrocarbon radical and some combination of halogens from family seven on the Periodic Table. The halogens are all toxic and, therefore, it is not difficult to see that the alkyl halides are also going to be toxic. The general formula for alkyl halide is **R-X**. The "R" represents one or more hydrocarbon radicals, and the "X" represents one or more halogens. The "X" can be replaced by fluorine (F), chlorine (Cl), bromine (Br), iodine (I), or combinations of two or more. It is important to remember that hydrocarbon derivatives started out as hydrocarbons before hydrogen was removed and other elements were added. Many of the hydrocarbon names are still used in the naming process with the alkyl halides.

There are three ways alkyl halides can be named. They are all correct naming conventions, and the compounds may be listed under any one of the possibilities. When researching the compounds in reference books, you may have to look under the alternate names to find information on the compound. The first naming convention is one in which the radical is named first, the "ine" is dropped from the halogen, and an "ide" ending is added. For example, if the compound has one carbon, the radical for one carbon is methyl. If there is chlorine attached to the methyl radical,

```
     H
     |
H —  C — Cl
     |
     H
```

Chloromethane
Methyl chloride
CH_3Cl

Figure 8.12

the alkyl halide compound is named methyl chloride. The structure, molecular formula, and names for it are shown in Figure 8.12.

If fluorine is attached to a one-carbon radical, the name is methyl fluoride, and so on.

The second convention is to name the halogen first and then the hydrocarbon radical. In this case, the "ine" ending is dropped from the halogen, and an "o" is added to the abbreviated name for the halogen. In the case of chlorine, it is "chloro." The radical is on the end of the name. When the radical is on the end, the name reverts back to the hydrocarbon that was used to form the radical. For example, methyl is a radical of the one-carbon alkane, methane. So if the halogen chlorine is added to methane, and the halogen is named first, the name is chloromethane. If bromine is the halogen, the name is bromomethane. If the radical is a two-carbon radical and the halogen is fluorine, the name of the compound is fluoroethane, and so on. Illustrated in Figure 8.13 are the names and structural formulas for some one-, two-, and three-carbon alkyl halides.

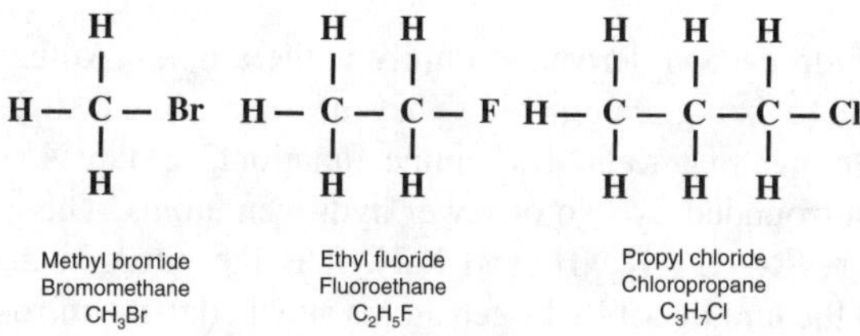

Figure 8.13

It is possible to use more than one halogen to form alkyl halide compounds. If the multiple halogens are the same type, the prefix "di" is used for two, "tri" for three, and "tetra" for four. Some chemicals, as has been previously mentioned, have trade names. There are also trade names for some of the alkyl halides. For example, a one-carbon radical with three chlorine atoms attached is called trichloro methane, or methyl trichloride; however, the trade name for the compound is chloroform. A methyl radical with four chlorines attached is named tetrachloro methane, or methyl tetrachloride. The trade name for the compound is carbon tetrachloride, a material that was used as a fire-extinguishing agent. It is no longer approved as an extinguishing agent because when it contacts a hot surface, it decomposes to phosgene

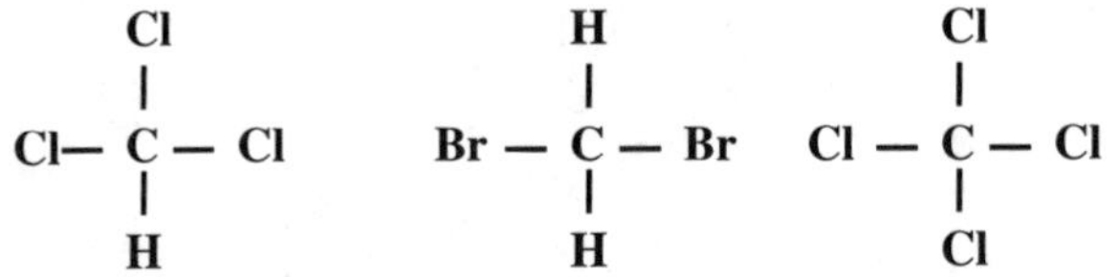

```
      Cl                 H                  Cl
      |                  |                  |
 Cl— C — Cl        Br — C — Br       Cl — C — Cl
      |                  |                  |
      H                  H                  Cl
```

Trichloro methane
Methyl trichloride
$CHCl_3$

Dibromo methane
Methyl dibromide
CH_2Br_2

Tetrachloro methane
Methyl tetrachloride
Carbon tetrachloride
CCl_4

Figure 8.14

gas. Shown in Figure 8.14 are the names, molecular formulas, and structures for some alkyl halides with multiple numbers and combinations of halogens in the compounds.

Some alkyl halide compounds also have double bonds: dichloroethylene, dichloropropene, dichlorobutene, and trichloroethylene.

Trichloroethylene, $CHClCCl_2$, is a stable, low-boiling, colorless liquid with a chloroform-like odor. It is not corrosive to the common metals even in the presence of moisture. It is slightly soluble in water and is nonflammable. It is toxic by inhalation, with a TLV of 50 ppm and an IDLH of 1000 ppm in air. The FDA has prohibited its use in foods, drugs, and cosmetics. The four-digit UN identification number is 1710. The NFPA 704 designation is health 2, flammability 1, and reactivity 0. Its primary uses are in metal degreasing, dry cleaning, as a refrigerant and fumigant, and for drying electronic parts.

Amines

The next toxic hydrocarbon-derivative family is the amines. Amines are toxic irritants, in addition to being flammable. They are considered slightly polar when compared to nonpolar materials. The amine functional group is represented by a single nitrogen surrounded by two or fewer hydrogen atoms. The general formulas for the amines are **$R\text{-}NH_2$, R_2NH,** and **R_3N.** It is the nitrogen that identifies the amine group, not the number of hydrogen atoms attached to the nitrogen. The amines are covered in detail in Chapter 5 under the Hydrocarbon Derivatives section. The degree of toxicity of amines varies from compound to compound. Many of them are strong irritants. TLV values range in the low double digits from 5 to 10 ppm. Diethylamine is toxic by ingestion and is a strong irritant. It has a TLV of 10 ppm in air. Butyl amine is a skin irritant with a TLV of 5 ppm in air. It is important to obtain further information when dealing with amines. Look the materials up in reference books and MSDS sheets to determine the toxic characteristics of a given amine compound.

Aniline, $C_6H_5NH_2$, also known as phenylamine, is a colorless, oily liquid with a characteristic amine odor and taste. It rapidly turns brown when exposed to air. It is soluble in water, with a specific gravity of 1.02, which is slightly heavier than water. Aniline is an allergen and is toxic if absorbed through the skin. The TLV is 2 ppm in air, and the IDLH is 100 ppm. The target organs are the blood, cardiovascular system, liver, and kidneys. Aniline is incompatible with nitric acid and hydrogen peroxide. The four-digit UN identification number is 1547. The NFPA 704 designation is health 3, flammability 0, and reactivity 1. The primary uses are in dyes, photographic chemicals, isocyanates for urethane foams, explosives, herbicides, and petroleum refining. The structure and molecular formula for aniline are shown in Figure 8.15.

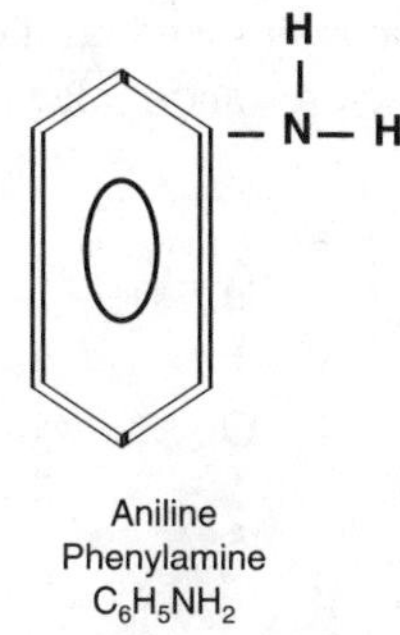

Figure 8.15

Alcohols

All alcohols are toxic to some degree. Ethyl alcohol is drinking alcohol and, when consumed in moderation, has limited toxic effects, as shown in Figure 8.16. However, ethyl alcohol, if taken in excess, can have toxic effects. In the short-term, the effects of ethyl alcohol are sedative and depressant; in the long run, cancer or liver damage may occur. Methyl alcohol is toxic by ingestion and can cause blindness. It has a TLV of 200 ppm. Propyl alcohol is toxic by skin absorption. It has a TLV of 200 ppm. However, isopropyl alcohol is used as rubbing alcohol and is applied to the skin. Isopropyl alcohol is toxic by ingestion and inhalation with a TLV of 400 ppm. Butyl alcohol is toxic by prolonged inhalation, is an eye irritant, and is absorbed through the skin. The TLV ceiling is 50 ppm in air. There are also some alcohols that have chlorine added to the compound, which increases their toxicity. For example, ethylene chlorohydrin is an alcohol that has an NFPA 704 designation for health of 4. As you can see, the hazards and routes of entry vary widely among the alcohols. It should be assumed the alcohol in a spill is the worst-case scenario until the exact hazard of the compound can be researched. The alcohols are covered in detail in Chapter 4 under Hydrocarbon Derivatives.

Cresols (*o-*,*m-*,*p-*,), $CH_3C_6H_4OH$, are alcohol hydrocarbon derivatives. They are colorless to yellowish or pinkish liquids. They are found in the "ortho," "meta," and

Toxic Effects of Ethyl Alcohol in the Blood

0.08%	2.0%	3.0%	4.0%	5.0%
Happy	Very Happy	Drunk	Falling Down Drunk	Death

Figure 8.16

"para" isomers, like the xylenes in Chapter 5. Cresol has a characteristic phenolic odor, and is soluble in water, with a specific gravity of 1.05, which is slightly heavier than water. It is an irritant, corrosive to the skin and mucous membranes, and is absorbed into the skin. The TLV is 5 ppm in air, and the IDLH is 250 ppm. The target organs are the central nervous system, respiratory system, liver, kidneys, skin, and eyes. The four-digit UN identification number is 2076. The NFPA 704 designation is health 3, flammability 2, and reactivity 0. The primary uses are as a textile scouring agent, herbicide, phenolic resins, and ore flotation; the para isomer is used in synthetic food flavors. The structures and molecular formulas for the isomers of cresol are shown in Figure 8.17.

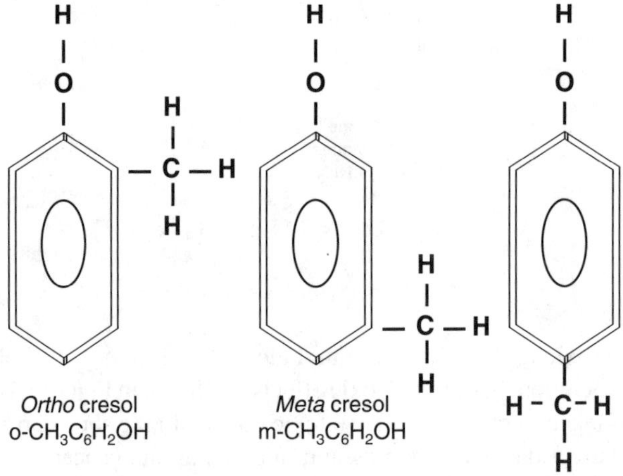

Ortho cresol
o-$CH_3C_6H_2OH$

Meta cresol
m-$CH_3C_6H_2OH$

Para cresol
p-$CH_3C_6H_2OH$

Figure 8.17

Aldehydes

Aldehydes are highly toxic compounds that are known to be human carcinogens. Formaldehyde is toxic through inhalation and is a strong irritant. The TLV is 1 ppm in air. Acetaldehyde is toxic, with narcotic effects. The TLV is 100 ppm in air, and the IDLH is 10,000 ppm. Propionaldehyde is an irritant. The toxicity hazards of the aldehydes also vary from one compound to another. Care should be taken to determine the exact hazards of any given compound.

Acrolein, CH_2CHCHO, also known as acrylaldehyde, is an aldehyde hydrocarbon derivative and it is a colorless or yellowish liquid with a disagreeable, suffocating odor. Acrolein is soluble in water, with a specific gravity of 0.84, which is lighter than water. It polymerizes readily, is reactive, and is not shipped or stored without an inhibitor. Acrolein is toxic by inhalation and ingestion, and is a strong irritant to the skin and eyes. The TLV is 0.1 ppm in air, and the IDLH is 5 ppm. The target organs are the heart, eyes, skin, and respiratory system. Acrolein is a dangerous fire and explosion risk, with a wide flammable range of 2.8 to 31% in air. The four-digit UN identification number for acrolein is 1092. The NFPA 704 designation is health 4, flammability 3, and reactivity 3. The primary uses of acrolein are in the manufacture of polyester resins, polyurethane resins, pharmaceuticals, and as a herbicide. The structure and molecular formula for acrolein are shown in Figure 8.18.

```
H   H    O
|   |    ||
C = C  — C — H
|
H
```

Acrolein
Acrylaldehyde
CH_2CHCHO

Figure 8.18

Agricultural feed store, a primary source of restricted-use agricultural pesticides.

ORGANIC ACIDS

Organic acids are toxic and corrosive. Corrosivity is a form of toxicity to the tissues that the acid contacts. However, the organic acids have other toxic effects. Formic acid is corrosive to skin and tissue. It has a TLV of 5 ppm in air and an IDLH of 30 ppm. Pure acetic acid is toxic by ingestion and inhalation. It is a strong irritant to skin and tissues. The TLV is 10 ppm in air, and the IDLH is 1000 ppm. Propionic acid is a strong irritant, with a TLV of 10 ppm in air. Butyric acid is a strong irritant to skin and tissues. The degree of toxicity varies with the different organic acid compounds. Review reference sources and MSDS sheets to determine the exact hazards of specific acids.

MISCELLANEOUS TOXIC MATERIALS

A wide variety of toxic liquids and solids do not fit neatly into any of the families discussed so far. Some of the most dangerous and more common ones will be listed here, but this will, by no means, be a comprehensive listing. The intent is to foster familiarity with the many different types of toxic chemicals that may be encountered in the real world, both in transportation and fixed facilities. The chemicals presented will be drawn from the DOT Hazardous Materials Tables and the NFPA Hazardous Chemicals Data listing.

Chloropicrin, CCl_3NO_2, is a slightly oily, colorless, refractive liquid. It is relatively stable and slightly water-soluble. The specific gravity is 1.65, which is heavier than air. It is toxic by ingestion and inhalation, and is a strong irritant. The TLV is 0.1 ppm in air. The four-digit UN identification number is 1580. The NFPA 704 designation is health 4, flammability 0, and reactivity 3. The primary uses of chloropicrin include dyestuffs, fumigants, fungicides, insecticides, rat exterminator, and tear gas. The structure and molecular formula for chloropicrin are shown in Figure 8.19.

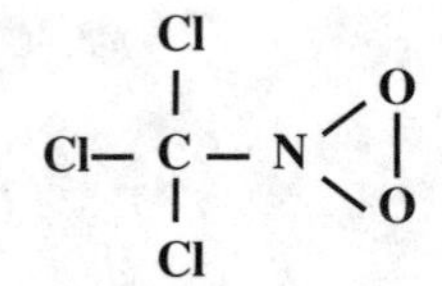

```
      Cl
      |        O
Cl— C — N ⟨  |
      |        O
      Cl
```

Chloropicrin
CCl_3NO_2

Figure 8.19

Epichlorohydrin, $ClCH_2CHOCH_2$, an epoxide, is a highly volatile, unstable liquid, with a chloroform-like odor. It is slightly water-soluble and has a specific gravity of 1.18, which is heavier than water. Epichlorohydrin is toxic by ingestion, inhalation, and skin absorption. It is a strong irritant and a known carcinogen. The TLV is 2 ppm in air, and the IDLH is 250 ppm. The target organs affected are the respiratory system, skin, and kidneys. The four-digit UN identification number is 2023. The NFPA 704 designation is health 3, flammability 3, and reactivity 2. The primary uses are as a raw material for epoxy and phenoxy resins, in the manufacture of glycerol, and as a solvent. The structure and molecular formula for epichlorohydrin are shown in Figure 8.20.

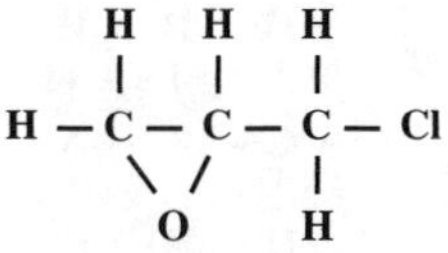

```
    H    H    H
    |    |    |
H — C —  C —  C — Cl
     \  /     |
      O       H
```

Epichlorohydrin
CH_2CHOCH_2Cl

Figure 8.20

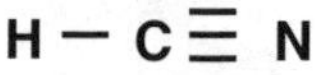

$H-C\equiv N$

Hydrogen cyanide
HCN

Figure 8.21

Hydrogen cyanide, **HCN**, also known as hydrocyanic acid, is a water-white liquid with a faint odor of bitter almonds. The odor threshold is 0.2 to 5.1 ppm in air; however, if the odor of hydrogen cyanide is detected, it has already exceeded the allowable amount. The immediately fatal concentration is usually 250 to 300 ppm in air. Hydrogen cyanide blocks the uptake of oxygen by the cells. It is soluble in water, with a specific gravity of 0.69, which is lighter than water. Hydrogen cyanide is highly toxic by inhalation, ingestion, and skin absorption. The TLV is 4.7 ppm in air, and the IDLH is 50 ppm. The target organs are the central nervous system, cardiovascular system, kidneys, and liver. The four-digit UN identification number is 1051 for anhydrous (without water) and 1614 when it is absorbed in a porous material. The NFPA 704 designation is health 4, flammability 4, and reactivity 2. The primary uses are in the manufacture of acrylonitrile, acrylates, cyanide salts, rodenticides, and pesticides. The structure and molecular formula for hydrogen cyanide are shown in Figure 8.21.

Methyl hydrazine, CH_3NHNH_2, is a colorless, hygroscopic liquid with an ammonia-like odor. It is soluble in water, with a specific gravity of 0.87, which is lighter than water. Methyl hydrazine is toxic by inhalation and ingestion, and is a suspected human carcinogen. The TLV ceiling is 0.2 ppm in air, and the IDLH is 50 ppm. The target organs are the central nervous system, respiratory system, liver, blood, eyes, and cardiovascular system. The four-digit UN identification number is 1244. The NFPA 704 designation is health 4, flammability 3, and reactivity 2. The primary uses are as a missile propellant and a solvent. The structure and molecular formula for methyl hydrazine are shown Figure 8.22.

Methyl isocyanate (MIC), CH_3NCO, is a colorless liquid. It is water-reactive, with a specific gravity of 0.96, which is lighter than water. This is the chemical that was released in Bhopal, India, that killed over 3000 people in 1984. Methyl isocyanate is toxic by skin absorption and a strong irritant. The TLV is 0.02 ppm in air,

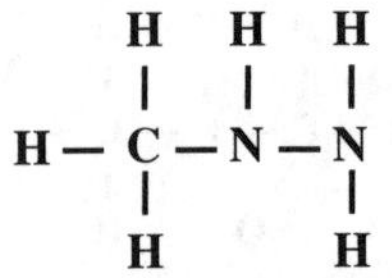

```
     H   H   H
     |   |   |
H — C — N — N
     |       |
     H       H
```

Methyl hydrazine
CH_3NHNH_2

Figure 8.22

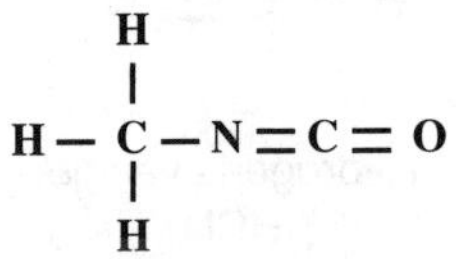

```
     H
     |
H — C — N = C = O
     |
     H
```

Methyl isocyanate
CH_3CNO

Figure 8.23

and the IDLH is 20 ppm. The target organs are the respiratory system, eyes, and skin. The four-digit UN identification number is 2480. The NFPA 704 designation is health 4, flammability 3, and reactivity 2. The white section at the bottom of the diamond has a W with a slash through it, indicating water reactivity. The primary use of methyl isocyanate is as a chemical intermediate. The structure and molecular formula are shown in Figure 8.23.

Toluene diisocyanate (TDI), $(OCN)_2C_6H_3CH_3$, is a water-white to pale-yellow liquid with a sharp, pungent odor. It reacts with water to release carbon dioxide. The specific gravity is 1.22, which is heavier than water. TDI is toxic by inhalation and ingestion, and is a strong irritant to skin and other tissue, particularly the eyes. The TLV is 0.005 ppm in air, and the IDLH is 10 ppm. The target organs are the respiratory system and the skin. The four-digit UN identification number is 2078. The NFPA 704 designation is health 3, flammability 1, and reactivity 3. The white section at the bottom of the diamond has a "W" with a slash through it, indicating water reactivity. The primary uses of TDI are in the manufacture of polyurethane foams, elastomers, and coatings. The structure and molecular formula for TDI are shown in Figure 8.24.

PESTICIDES

The EPA estimates that there are 45,000 accidental pesticide poisonings in the United States each year, where more than 1 billion pounds are manufactured annually. Pesticides can be found in manufacturing facilities, commercial warehouses, agricultural chemical warehouses, farm supply stores, nurseries, farms, supermarkets, discount stores, hardware stores, and other retail outlets. A pesticide is a chemical or mixture of chemicals used to destroy, prevent, or control any living thing considered

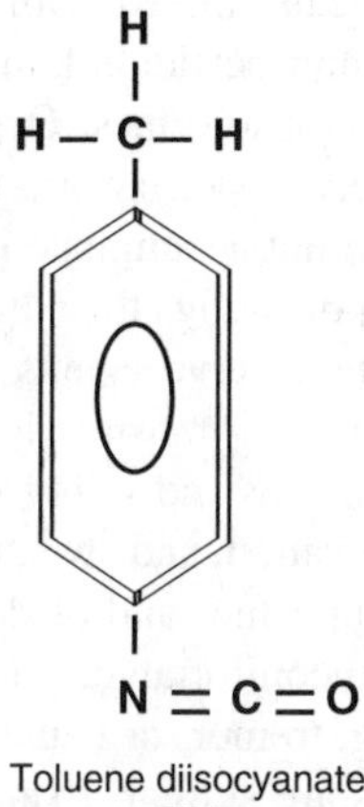

Toluene diisocyanate
TDI
$CH3C_6H_3(NCO)_2$

Figure 8.24

to be a pest, including insects (insecticides), fungi (fungicides), rodents (rodenticides), or plants (herbicides). The definition of a pesticide from the Federal Insecticide, Fungicide, and Rodenticide Act (FIFRA) is "a chemical or mixture of chemicals or substances used to repel or combat an animal or plant pest. This includes insects and other invertebrate organisms; all vertebrate pests, e.g., rodents, fish, pest birds, snakes, gophers; all plant pests growing where not wanted, e.g., weeds; and all microorganisms which may or may not produce disease in humans. Household germicides, plant-growth regulators, and plant-root destroyers are also included."

Agricultural spray plane used to apply pesticides to food crops.

More than 1000 basic chemicals, mixed with other materials, produce about 35,000 pesticide products. Common pesticide families are organophosphates, carbamates, chlorophenols, and organocholorines. **Organophosphates** are derivatives of phosphoric acid and are acutely toxic, but are not enduring. They break down rapidly in the environment and do not accumulate in the tissues. Organophosphates are associated with more human poisonings than any other pesticide and are closely related to some of the most potent nerve agents. Organophosphates function by overstimulating, then inhibiting, neural transmission, primarily in the nervous, respiratory, and circulatory systems. Signs and symptoms of exposure include pinpoint pupils, blurred vision, tearing, salivation, and sweating. Pulse rate will decrease, and breathing will become labored. Intestines and bladder may evacuate their contents. Muscles will become weak and uncomfortable. Additional symptoms include headache, dizziness, muscle twitching, tremor, or nausea.

Symptoms of most pesticides are similar. Examples include malathion, methyl parathion, thimete, counter, lorisban, and dursban. The chemical formulas of the organophosphates contain carbon, hydrogen, phosphorus, and at least one sulfur atom, and some may contain at least one nitrogen atom. **Carbamates** are derivatives of carbamic acid, and are among the most widely used pesticides in the world. Most are herbicides and fungicides, such as 2,4,D, paraquat and dicamba, and function by inhibiting nerve impulses. The formula will contain carbon, hydrogen, nitrogen, and sulfur. Other examples include furadan, temik, and sevin. **Organochlorines** are chlorinated hydrocarbons. The formula contains carbon, hydrogen, and chlorine. They are neurotoxins, which function by overstimulating the central nervous system, particularly the brain. Examples are aldrin, endrin, hesadrin, thiodane, and chlordane. The best-known organochlorine is DDT, which has been banned for use in the United States because of its tissue accumulation and environmental persistence. Organophosphates do not break down in the environment and accumulate in the tissues, affecting the food chain. Chlorophenols contain carbon, hydrogen, oxygen, and chlorine. They affect the central nervous system, kidneys, and liver.

Pesticide labels contain valuable information for the emergency responder and medical personnel treating a patient exposed to pesticides. This information includes product name, "signal word," a statement of practical treatment, EPA registration number, a note to physician, and a statement of chemical hazards. Other information includes active and inert ingredients. "Inert" does not necessarily mean that the ingredients do not pose a danger; it means only that the inert ingredients do not have any action on the pest for which the pesticide was designed. Many times, the inert ingredient is a flammable or combustible liquid.

The label also contains information about treatment for exposure. This information should be taken to the hospital when someone has been contaminated with a pesticide. Do not, however, take the pesticide container to the hospital. Take the label, or write the information down, take a Polaroid picture of the label, or use a pesticide label book. (Label books are available from agricultural supply dealers.)

Pesticides can be grouped generally into three toxicity categories: high, moderate, and low. You can tell the degree of toxicity by the signal word on the label. Three signal words indicating the level of toxicity of a pesticide are "Danger," Caution," and Warning" (see Figure 8.25). Highly toxic materials bear the word

Pesticide Signal Words

Signal Word	Toxicity	Lethal Dose
Danger/Poison*	Highly toxic	Few drops to 1** teaspoon
Warning	Moderately toxic	1 Teaspoon to 1 Tablespoon
Caution	Low toxicity	1 Ounce to more than a pint

*Skull and crossbones symbol included
**Less for a child or person weighing less than 160 pounds

Figure 8.25

"Danger," with a skull-and-crossbones symbol and the word "Poison" printed on the label. The lethal dose may be a few drops to 1 tsp. Moderately toxic pesticides have the word "Warning," and the lethal dose is 1 tsp to 1 tbsp. Low toxicity pesticides carry the word "Caution," and the lethal dose is 1 oz to 1 pt.

Pesticides may poison or cause harm to humans by entering the body in one or more of these four ways: through the eyes, through the skin, by inhalation, and by swallowing. As with most any chemical, exposure to the eyes is the fastest way to become poisoned. Whenever anyone is exposed to a pesticide, it is important to recognize the signs and symptoms of poisoning so that prompt medical help can be provided. Any unusual appearance or feeling of discomfort or illness can be a sign or symptom of pesticide poisoning. These signs and symptoms may be delayed up to 12 hours. When they occur and pesticide contact is suspected, get medical attention immediately. (The National Pesticide Network, located in Texas, provides emergency information through a toll-free telephone number: 1-800–858–7378, from 8 a.m. to 6 p.m. Central Standard Time.) Information can also be obtained from the National Poison Control Center by calling 1–800–222–1222.

OTHER TOXIC MATERIALS

Ethylene glycol, CH_2OHCH_2OH, an alcohol hydrocarbon derivative, is a clear, colorless, syrupy liquid with a sweet taste. Ethylene glycol is soluble in water and has a specific gravity of 1.1, which is slightly heavier than water. It is toxic by inhalation and ingestion. The lethal dose is reported to be 100 cc, and the TLV is 50 ppm. Ethylene glycol has not been assigned a four-digit UN identification number. The NFPA 704 designation is health 1, flammability 1, and reactivity 0. The primary uses are as coolants and antifreeze, brake fluids, low-freezing dynamite, a solvent, and a deicing fluid for airport runways. The structure and molecular formula for ethylene glycol are shown in Figure 8.26.

Phenol, C_6H_5OH, also known as carbolic acid, is a white, crystalline mass that turns pink or red if not perfectly pure or if exposed to light. Phenol absorbs water from the air and liquefies; it may also be found in transport as a molten material. It

```
     H   H
     |   |
 H — C — C — H
     |   |
     O   O
     |   |
     H   H
```

Ethylene glycol
CH_2OHCH_2OH

Figure 8.26

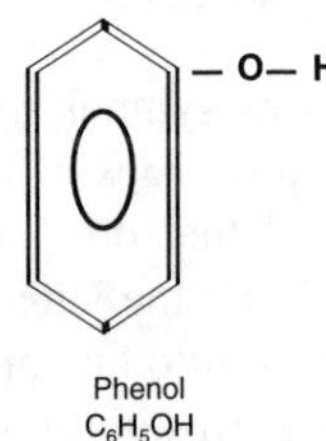

Phenol
C_6H_5OH

Figure 8.27

has a distinctive odor and a sharp, burning taste, but in a weak solution, it has a slightly sweet taste. Phenol is soluble in water, with a specific gravity of 1.07, which is heavier than water. It is toxic by ingestion, inhalation, and skin absorption, and is a strong irritant to tissues. The TLV is 5 ppm in air, and the IDLH is 250 ppm. The target organs are the liver, kidneys, and skin. The four-digit UN identification number is 1671 for solids, 2312 for molten materials, and 2821 for solutions. The NFPA 704 designation is health 4, flammability 2, and reactivity 0. The primary uses are phenolic resins, epoxy resins, 2-4-D herbicides, solvents, pharmaceuticals, and as a general disinfectant. The structure and molecular formula for phenol are shown in Figure 8.27.

Caprolactam, $\mathbf{CH_2CH_2CH_2CH_2CH_2NHCO}$, is a solid material composed of white flakes. Caprolactam is soluble in water and has a specific gravity (in a 70% solution) of 1.05, which is heavier than water. It may also be encountered as a molten material. Caprolactam is toxic by inhalation, with a TLV of (vapor) 5 ppm in air and (dust) 1 mg/m^3 of air. The primary uses are in the manufacture of synthetic fibers, plastics, film, coatings, and polyurethanes. The structure and molecular formula for caprolactam are shown in Figure 8.28.

Nitrobenzene is a greenish-yellow crystal or yellow, oily liquid, and is slightly soluble in water. The primary hazard of nitrobenzene is toxicity; however, it is also combustible. The boiling point is about 410°F, the flash point is 190°F, and the ignition temperature is 900°F. The specific gravity is 1.2, which is heavier than water, and the material will sink to the bottom. The vapor density is 4.3, which is heavier than air. Nitrobenzene is toxic by ingestion, inhalation, and skin absorption, with a

```
  H  HH  H
   \ /  \ /
H   C — C — C = O
 \ /        |
  C         |
 / \        |
H   C — C — N — H
   / \  / \
  H  H H  H
```

Caprolactam
$CH_2CH_2CH_2CH_2CH_2NHCO$

Figure 8.28

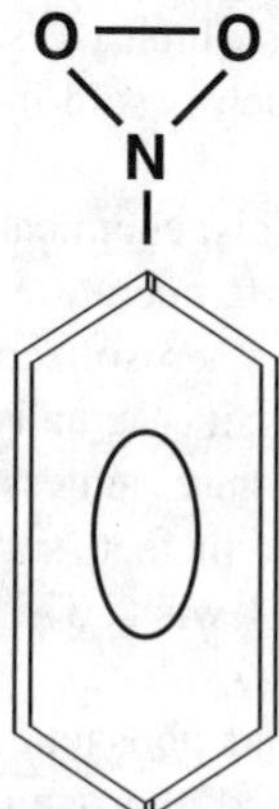

Nitrobenzene
$C_6H_5NO_2$

Figure 8.29

TLV of 1 ppm in air. The four-digit UN identification number is 1652. The NFPA 704 designation is health 3, flammability 2, and reactivity 1. Nitrobenzene is a nitro hydrocarbon derivative, but it is not very explosive. The primary uses are as a solvent, an ingredient of metal polishes and shoe polishes, and in the manufacture of aniline. The structure and molecular formula are shown Figure 8.29.

MILITARY AND TERRORIST CHEMICAL AGENTS

Chemical Agents

Chemical agents can be divided into groups, which include nerve agents, vesicants (blister agents), blood agents, choking agents, and riot-control agents. **Nerve agents** function by inhibiting the enzyme acetylcholinesterase, resulting in an excess of acetylcholine in the body. This excess results in the characteristic uncontrolled muscle movements associated with exposures to nerve agents. Nerve agents presented in this section will include GA (tabun), GB (sarin), GD (soman), GF, and VX. **Vesicants,** or blister agents, include mustard agents; sulfur mustard (H); (HD)(Agent "T"); nitrogen mustards (HN), (HN-2), and (HN-3); lewisite (L); and phosgene oxime (CX). Vesicants produce vesicles (blisters) when in contact with the human body, hence, their name, blister agents. Mustard agents also cause damage to the eyes and respiratory system by direct contact and inhalation. **Blood agents** (cyanide compounds) at high concentrations kill quickly. Common forms of cyanide include hydrocyanic acid (AC) and cyanogen chloride (CK). **Choking agents** (lung-damaging agents) include phosgene (CG), diphosgene (DP), chlorine, and chloropicrin (PS). HC smoke (a smoke that contains zinc), and oxides of nitrogen (from burning munitions) also produce lung hazards when exposed. **Riot-control agents** are incapacitating compounds. CS, which is used by law enforcement agents and the military, and CN (Mace 7), which is sold in devices for self-protection, are the primary types of riot-control agents.

Like all other hazardous materials, chemical agents can exist as solids, liquids, or gases, depending on the existing temperatures and pressures. The only exception is riot-control agents, which exist as aerosolized solids at normal temperatures and pressures. Tear gas really is not a gas; it is actually a solid that is aerosolized. Mustard and nerve agents likewise are not gases unless boiled above 212°F, which is the boiling point of water at sea level. In fact, sulfur mustard has a boiling point of 422°F, nitrogen mustard 495°F, and lewisite 375°F. These agents vaporize much like boiling water.

Chlorine, hydrogen cyanide, and phosgene are common industrial chemicals. Chlorine and phosgene are gases and hydrogen cyanide is a liquid at normal temperatures and pressures. Nerve and mustard agents, and hydrogen cyanide are liquids under the same conditions in which chlorine and phosgene are gases. Some mustard agents are frozen solid at 57°F. When in the liquid state, they evaporate at a rate similar to that of water. Chemical agent evaporation occurs not only because of its chemical makeup, but also because of the temperature, air pressure, wind velocity, and nature of the surface the agent comes in contact with. Water, for example, evaporates at a slower rate than gasoline, but at a faster rate than motor oil at a given temperature and pressure. Mustard is less volatile than the nerve agent sarin, but more volatile than the nerve agent VX. Volatility is the liquid's ability to produce a vapor at normal temperatures and pressures. Thus, a liquid that is said to be volatile is producing a lot of vapor; one that is not volatile will not be producing much vapor.

Evaporation rates of all chemicals mentioned are accelerated by increases in temperature and wind speed or when they are resting on a smooth surface rather than a porous one. Volatility has an inverse relationship to persistence. The more volatile a substance is, the more quickly it evaporates, and the less it tends to stay or persist as a liquid. Because of the relative low volatility of persistent-liquid chemical agents, the liquid hazard is generally more significant than the danger from the small amounts of vapor that may be generated. The reverse is true of the nonpersistent agents. They evaporate quickly enough so as not to present a liquid hazard for an extended period of time. Generally, the division between persistence and nonpersistence is related to the amount of the material left after 24 hours. The nonpersistent agents are usually gone after 24 hours.

The military has developed specialized toxicology terms for exposures to chemical agents in addition to the LD_{50} and LC_{50} already discussed. They are the **ED_{50}** and the **ID_{50}**. ED_{50} is the dose (D) of a liquid agent that will predictably cause effects (E) to anyone exposed. ID_{50} is the dose (D) that will cause the person to become incapacitated (I). When applying the LD_{50} to military agents, the lower the LD_{50} value, the less the amount of agent that is required to cause harm and the more potent is the agent. There is a difference in absorption rates for chemical agents and therefore, the ED_{50} and LD_{50} for a particular agent are specific to the site of entry into the body. For example, the LD_{50} for mustard absorbed through dry, intact skin is much higher than agent absorbed through the eyes.

The military uses a term called the concentration-time product, or **Ct**, which is the concentration of the agent present (usually expressed in terms of milligrams per cubic meter in air (mg/m^3) multiplied by the time (usually expressed in minutes) of exposure to the agent. For example, exposure to a concentration of soman (GD) vapor for 10 minutes results in a Ct of 40 $mg\text{-}min/m^3$. Exposures of 8 mg/m^3 for 5 minutes results in the same Ct (40 $mg\text{-}min/m^3$). This result is true for most of the agents except for cyanide.

The Ct associated with a biological effect remains relatively constant even though the concentration and length of time of exposure may vary within certain limits. For example, a 10-minute exposure to 4 mg/m^3 of soman causes the same effects as a 5-minute exposure to 8 mg/m^3 of the agent or to a 1-minute exposure to 40 mg/m^3. When the exposure threat is in the form of a vapor or gas, the "E" for effect is attached to the "Ct," which is the agent concentration and time of exposure. The result is the effect that will occur from the given time/concentration (ECt_{50}) exposure by inhalation of the vapor or gas to 50% of those exposed. The same holds true for the "I" indicating incapacitation. Fifty percent of those exposed to a certain concentration/time by inhalation will be incapacitated by the exposure, which is expressed by the value ICt_{50}. Lethal concentrations over a given time to 50% of those exposed by inhalation are represented by LCt_{50}. When the exposure is to a liquid agent, the terms used to identify the exposure are ED_{50} for the effects-dose by ingestion or skin absorption, ID_{50} for the incapacitation effect by ingestion or skin absorption, and LD_{50} for the lethal dose by ingestion or skin absorption.

Nerve Agents

The most common nerve agents are Tabun (GA), Sarin (GB), Soman (GD), GF, and VX. "G" agents are volatile, and penetrate the skin well, in terms of seconds or minutes. "V" agents are less volatile, and penetrate the skin well in terms of minutes and hours. All nerve agents are related to organophosphate pesticides and in pure form are colorless. However, nerve agents may also be light brown in color when contaminated and vary in their degree of volatility. Some of the agents have the volatility of motor oil, while others have volatility similar to water. Compared to other liquids that are considered to be volatile, such as gasoline, none of the nerve agents are significantly volatile. When comparing the nerve agents as a group, sarin would be considered the most volatile. VX, on the other hand, is the least volatile of the nerve agents, but is the most toxic. G-agents, such as sarin, are considered to be nonpersistent, and V-agents, like VX, are considered persistent agents. Thickened nonpersistent agents may present a hazard for a longer period of time.

Most nerve agents are odorless, however, some may have a faint fruity odor when contaminated. Nerve agents are highly toxic and quick-acting. They enter the body through inhalation or skin absorption. Poisoning can occur also with ingestion of the agents placed in food or drink. Nerve agents work the quickest if inhaled. Inhalation of high concentrations can cause death within minutes. The LC_{50} for sarin is 100 mg/min/m^3, while VX is 50 mg/min/m^3. Primary target organs for nerve agents are the respiratory and central nervous systems. Entry into the body through skin absorption requires a longer period of time for symptoms to develop. First symptoms from skin contact may not appear for 20 to 30 minutes after exposure. However, if the dose of the nerve agent is high, the poisoning process may be rapid. Eye exposure is extremely dangerous, and eyes should be flushed immediately with copious amounts of water if exposed. Liquid nerve agent splashed into the eyes is absorbed faster into the body than through skin contact.

Acetylcholine is an important neurotransmitter, which is essential to complete the transmission of neural impulses from one neuron (fibers that convey impulses to the nerve cell) to another. Without acetylcholine, the body cannot function normally. When a message is sent from the brain for a muscle to move or some other bodily function to activate, acetylcholine is released. It then binds to the postsynaptic membrane, which starts and continues the movement or action. When it is time for the movement to stop, acetylcholinesterase is released to remove the acetylcholine from the synapse, so it can be used again.

Nerve agents are acetylcholinesterase-enzyme inhibitors and can affect the entire body. Acetylcholinesterase enzymes normally act upon acetylcholine when it is released. Nerve agents inhibit this action, and an accumulation of acetylcholine occurs. Initially, it results in an overstimulation of the nervous system. The system then becomes fatigued and paralysis results. Paralysis of the diaphragm muscle is the primary cause of death from nerve agent poisoning. This results in the cession of breathing.

Accumulation of acetylcholine causes increased nerve and muscle activity, over-functioning of the salivary glands, secretory glands, and sweat glands. Muscular twitching, fatigue, mild weakness, cramps, and flaccid paralysis, accompanied by dyspnea and cyanosis, result from the excess acetylcholine. Accumulation of excess acetylcholine in the brain and spinal cord result in central nervous system symptoms. Unless the accumulation of acetylcholine is reversed by the use of antidotes, the effects become irreversible and death will occur.

Symptoms of nerve agent poisoning resulting from a low dose include increased saliva production, runny nose, and a feeling of pressure on the chest. Pupils of the eyes exhibit pinpoint constriction (miosis), short-range vision is impaired, and the victim feels pain when trying to focus on a nearby object. Headache may follow with tiredness, slurred speech, and hallucinations. Exposures to higher doses present more dramatic symptoms. Bronchioconstriction and secretion of mucous in the

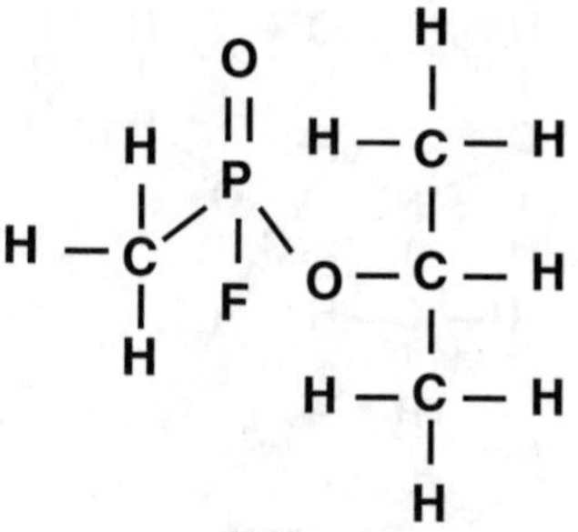

GB, Sarin
$CH_3PO(F)OCH(CH_3)_2$

Figure 8.30

respiratory system leads to difficulty in breathing and coughing. Discomfort in the gastrointestinal tract may develop into cramps and vomiting. Involuntary discharge of urine and defecation may also occur. Saliva discharge is powerful and may be accompanied by running eyes and sweating. Muscular weakness, local tremors, or convulsions may follow in cases of moderate poisoning. Exposure to a high dose of nerve agent may lead to more pronounced muscular symptoms. Convulsions and loss of consciousness may occur. Those individuals most sensitive to nerve agents will experience a lethal dose at about 70 mg/min/m^3. More resistant persons require about 140 mg/min/m^3. The amount of nerve agent for a dermal exposure that would result in a lethal effect is small. In fact, if you were to look at the back of a Lincoln penny and find the Lincoln Memorial, the amount of liquid that it would take to cover one of the columns on the memorial would be a toxic dermal exposure!

Sarin (GB) is a fluorinated organophosphorous compound with the chemical name of phosphonofluoridic acid, methyl-, isopropyl ester. The chemical formula is $\mathbf{CH_3PO(F)OCH(CH_3)_2}$. The structure and molecular formula of the compound are shown in Figure 8.30.

Sarin has a vapor density that is about five times heavier than air and about the same specific gravity as water. It is a colorless and odorless liquid in the pure form. To date, OSHA has not identified a permissible exposure concentration for sarin. Sarin is listed by the American Conference of Governmental Industrial Hygienists (ACGIH) and OSHA as a carcinogen. Sarin is stable when in the pure state. While sarin will burn, it has a high flash point and would be difficult to ignite under normal circumstances. Sarin reacts with steam or water to produce toxic and corrosive vapors.

Tabun (GA) is an organophosphorus compound with a chemical name of ethyl N, N-dimethylphosphoramidocyanidate. The chemical formula is $\mathbf{C_2H_5OPO(CN)N(CH_3)_2}$. The structure and molecular formula of the compound are shown in Figure 8.31.

Tabun is a colorless to brown liquid with a faint fruity odor. In its pure form, it does not have any odor. The boiling point of tabun is 475°F, which is approximately 159 degrees higher than sarin. Tabun has a vapor density higher than sarin and about

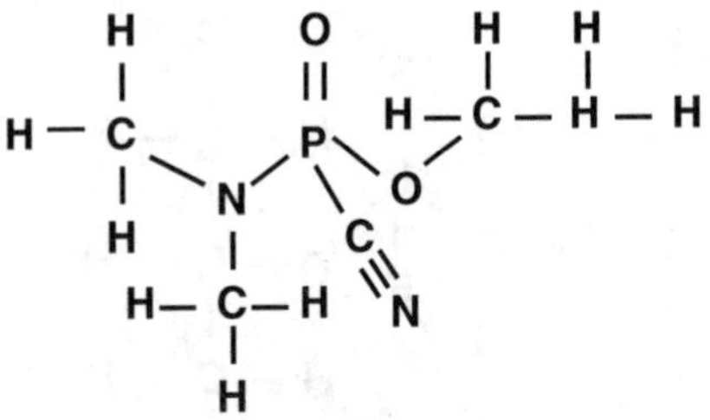

GA, Taban
$C_2H_5OPO(CN)N(CH_3)_2$

Figure 8.31

5.5 times heavier than air. Its specific gravity is slightly heavier than water. Tabun has a flash point lower than sarin. There are no explosive limits available. Contact with the agent liquid or vapor can be fatal. Tabun is a lethal cholinesterase-inhibitor similar to sarin in the way it affects the human body. It is only about half as toxic by inhalation as sarin, but in low concentrations, is more irritating to the eyes. Symptoms presented by GA depend on the concentration and rate of entry into the body. Small dermal exposures may cause local sweating and tremors, with few other effects. Symptoms for larger doses are much the same as for sarin, regardless the route of exposure to the body. They include, in order of appearance: runny nose; tightness of the chest; dimness of vision and pinpoint pupils (miosis); difficulty breathing; drooling and excessive sweating; nausea; vomiting; cramps and involuntary defecation and urination; twitching, jerking, or staggering; headache; confusion; drowsiness; coma; and convulsion. These symptoms are followed by cessation of breathing and death. First symptoms appear more slowly from skin contact than through inhalation. Skin absorption of a dose great enough to cause death can occur in 1 to 2 minutes; however, death may be delayed for 1 to 2 hours. The inhalation lethal dose can kill in 1 to 10 minutes, and liquid splash in the eyes is almost as fast.

Soman (GD) is a fluorinated organophosphorus compound with a chemical name of pinacolyl methyl phosphonofluoridate, and it is a lethal nerve agent. The chemical formula is **$CH_3PO(F)OCH(CH_3)C(CH_3)_3$**. The structure and molecular formula of the compound are shown in Figure 8.32.

Soman has a vapor density more than six times heavier than air. It is a colorless liquid with a fruity odor, when pure, and an amber or dark brown color with an oil of camphor odor, when impure. Doses of soman that can cause death may be only slightly larger than those that produce symptoms. The median incapacitation dose for soman is unknown. In contact with water, it will hydrolyze to form hydrogen fluoride (HF).

Lethal Nerve Agent (VX) is a sulfonated organophosphorous compound with a chemical name of O-ethyl-S-(2-iisopropylaminoethyl) methyl phosphonothiolate. The chemical formula is **$(C_2H_5O)(CH_3O)P(O)S(C_2H_4)N[(C_2H_2)(CH_3)_2]_2$**. The structure and molecular formula of the compound are shown in Figure 8.33.

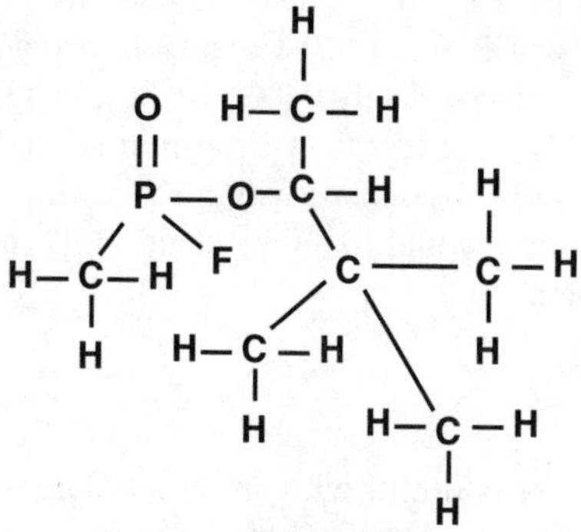

GD, Soman

$CH_3PO(F)OCH(CH3)C(CH3)3$

Figure 8.32

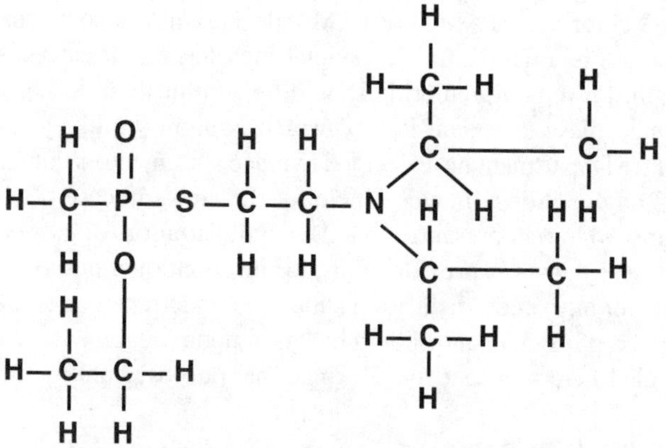

$(C_2H_5O)(CH_3O)P(O)S(C2H4)N[(C2H2)(CH3)2]2$

Figure 8.33

VX has a vapor density more than nine times heavier than air. VX is a colorless to straw-colored liquid with little, if any, odor. It has the consistency and appearance of motor oil. VX is a lethal cholinesterase inhibitor, much more toxic than sarin or any of the other nerve agents. Life-threatening doses may be only slightly larger than those producing symptoms. A drop of VX the size of a pinhead on the skin will kill in 5 to 15 minutes.

Lethal Nerve Agent (GF) is a fluorinated organophosphate compound. Limited information is available on this nerve agent. The chemical formula is **$CH_3PO(F)OC_6H_{11}$**. GF has a sweet, musty odor of peaches or shellac. It has a boiling point of 239°C (463°F); a vapor pressure of 0.044 mm Hg at 20°C; a vapor density of 6.2, which is over six times heavier than air; a volatility of 438 mg/m^3 at 20°C; a specific gravity of 1.1327 at 20°C; and a freezing/melting point of –30°C. The flash point of GF is approximately 94°C (200°F), and the flammable limits are not available. GF has a MCt_{50} of less than 1 mg-min/m^3 and an LD_{50} on the skin of 30 mg. All nerve agents are 6.1 poisons, with a UN identification number of 2810. Their NFPA 704 designation would likely be health 4, flammability 1, reactivity 1, and no special information.

Antidotes

No chemical warfare agent is useful without an antidote to protect your own personnel. Atropine is the universal antidote for chemical nerve agent exposures. It functions by binding to the acetylcholine receptors without causing excitement. This keeps the excess of acetylcholine from reaching the receptors. Atropine treats the symptoms of nerve agent exposure rather than the cause. This is why atropine is administered in conjunction with a group of chemicals called oximes. They treat the cause of the problem by restoring the acetylcholinesterase to operation, by breaking the enzyme-nerve agent bond.

Protopam® chloride, 2-PAMCl (2-PAM-chloride), may also be used in conjunction with atropine. Its effect is on the skeletal muscles, and it causes an increase in muscle strength. Protopam® chloride should be administered in all cases of nerve agent poisoning unless the agent is GD, which is highly unlikely. Officials at the Philadelphia Fire Department have decided to place autoinjectors containing atropine and 2-PAMCl on every department vehicle for the crews in case of chemical nerve agent exposure. Material presented on the administration of atropine and other antidotes for nerve agents is provided only for informational purposes. Administration of atropine or any other medication should be undertaken only upon the advice and guidance of a physician, and should be based upon local medical protocols. The material provided here is meant for informational purposes only

Mustard Agents (Vesicants)

Blister agents, or vesicants as they are also known, include four basic types: sulfur mustard, nitrogen mustard, lewisite, and phosgene oxime. Like the nerve agents, mustard agents are sometimes incorrectly referred to as gases, when, in fact, they are not. All of the blister agents have the same basic characteristics and will be referred to as mustard, blister agents, or vesicants, interchangeably. Mustard agents are persistent by nature. They are made in varying formulations, including distilled, nitrogen, sulfur, and thickened mustards. Military designations for mustard agents are as follows: mustard (H), distilled mustard, (HD) or (HS), and thickened mustard (HT). Nitrogen mustards have the military designation of (HN_1), (HN_2), and (HN_3). Unlike the nerve agents, mustard agent exposure is rarely fatal. Symptoms for mustard agents do not appear for several hours after exposure. Mustard is designed to incapacitate, not to kill. It can, however, be fatal if a large enough dose is administered or a person has other health problems that will be compounded by mustard exposure.

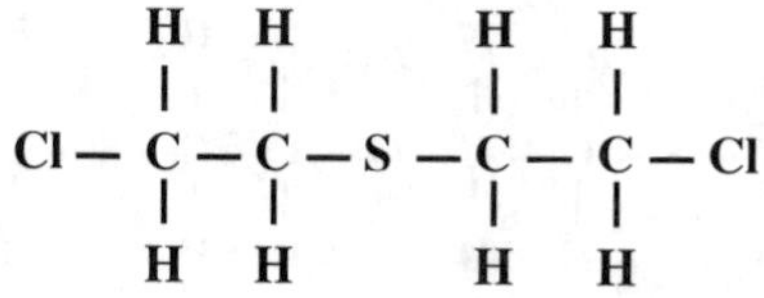

```
       H   H       H   H
       |   |       |   |
Cl  —  C — C — S — C — C — Cl
       |   |       |   |
       H   H       H   H
```

Sulfur Mustard
$(ClH_2CH_2)_2S$

Figure 8.34

Sulfur mustard (H), **Distilled mustard (HD)** is a liquid that has a consistency of motor oil and appears in colors from light yellow to brown and black, based upon the number and type of impurities present. Pure mustard is clear like water. The odor of mustard ranges from onions or garlic to horseradish or mustard, which is where it gets its name. It is a chlorinated sulfur compound. H or HD mustard has a chemical name of Bis-(2-chloroethyl)sulfide. Mustard is only slightly soluble in water and is miscible with organophosphate nerve agents. It has a freezing point of approximately 57°F and, if stored outside, would remain frozen for a large part of the year. Mustard has a specific gravity that is heavier than water. The vapor density of mustard is about 5.5 times heavier than air. While mustard is flammable, it has a fairly high flash point. Explosive charges may be used to ignite mustard agents. Flammable limits or explosive ranges for mustards have not been identified. The chemical formula for sulfur/distilled mustard is $\mathbf{(ClH_2CH_2)_2S}$, and the structure and molecular formula are shown in Figure 8.34.

Sulfur mustard (HT), or Agent T, has a chemical name of Bis-(2-(2-chloroethylthio)ethyl)ether. It has a yellow color with a garlic-like odor, similar to other mustard agents. The specific gravity of HT is heavier than water and similar to distilled mustard. It has a vapor density more than nine times heavier than air and four times heavier than distilled mustard. No airborne exposure limits have been identified for sulfur mustard. Flash point or flammable limits have not been established, and it has a boiling point slightly lower than distilled mustard. Specific health-hazard data, such as airborne exposure limits, have not been established for Agent T; however, it has a toxicity similar to that of distilled mustard, and under no circumstances should anyone be allowed to be exposed to direct vapor or skin or eye contact. The median lethal dose (LCt_{50}) of Agent T in laboratory animals is 1650–2250 mg-min/m^3 based on 10 minutes of exposure. The chemical formula for Agent T sulfur mustard is $\mathbf{C_2H_4S_2C_2H_4OC_2H_4C_2H_4Cl_2}$, and the structure and molecular formula are shown in Figure 8.35.

Nitrogen mustard (HN) was developed in three formulations: HN-1, HN-2, and HN-3. HN-1 was the first to be produced in the late 1920s and early 1930s. Originally, it was developed as a pharmaceutical and used to remove warts before it became a military agent. Agent H-2 was developed as a military agent and became a pharmaceutical. HN-3 was designed as a military mustard agent and is the only one that remains in military use. Therefore, this section will only cover the characteristics of HN-3 mustard agent. HN-3 is colorless to pale yellow with a butter-almond odor.

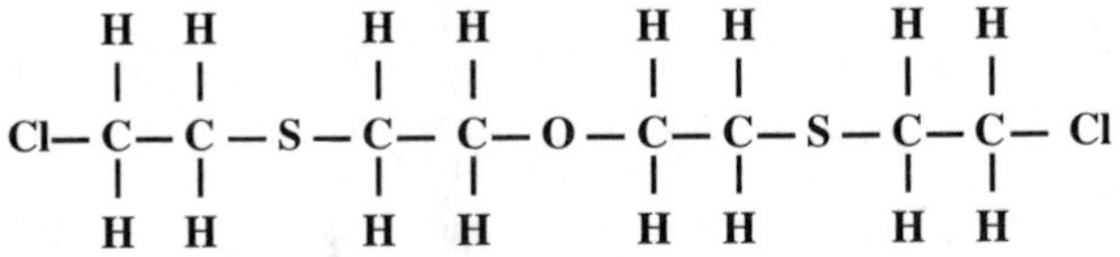

```
    H   H       H   H       H   H       H   H
    |   |       |   |       |   |       |   |
Cl— C — C — S — C — C — O — C — C — S — C — C — Cl
    |   |       |   |       |   |       |   |
    H   H       H   H       H   H       H   H
```

Agent "T" Sulfur Mustard
$C_2H_4S_2C_2H_4OC_2H_4C_2H_4Cl_2$

Figure 8.35

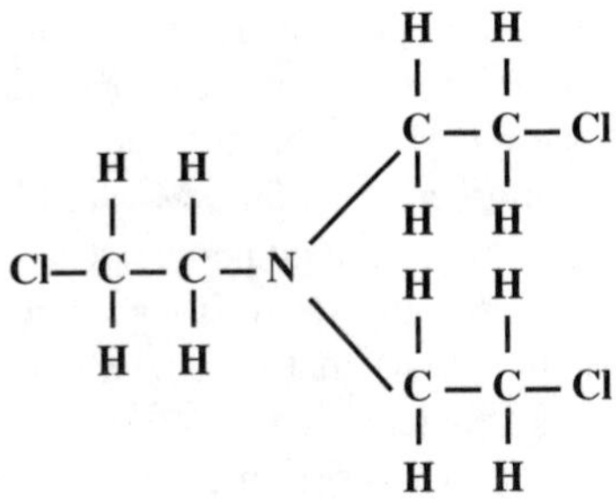

```
                      H   H
                      |   |
                      C — C — Cl
         H   H       /|   |
         |   |      / H   H
    Cl— C — C — N
         |   |      \ H   H
         H   H       \|   |
                      C — C — Cl
                      |   |
                      H   H
```

$N(CH_2CH_2Cl)_3$

Figure 8.36

The chemical formula for nitrogen mustard agent HN-3 is $\mathbf{N(CH_2CH_2Cl)_3}$, and the structure and molecular formula are shown in Figure 8.36.

Mustard is effective in small doses and affects the lungs, skin, and eyes. Symptoms are likely to appear first on delicate tissues, such as the soft membranes surrounding the eyes. This will be followed by tissues of the throat, lungs, nose, and mouth. Mustard agents have the greatest effects on warm, moist areas of the body, such as eyes, respiratory tract, armpits, groin, buttocks, and other skin folds. Because of the delayed onset of symptoms, those exposed often do not know they have contacted mustard. Mustard agents can be an inhalation or skin-absorption hazard. The symptoms of the agents do not appear for 2 to 24 hours after the exposure, and the usual onset in most people is between 4 to 8 hours. Liquid exposures to the skin and eyes produce an earlier onset of symptoms than do vapor exposures. When mustard agent contacts the skin, it does not produce pain or sensation. Those exposed would not likely be aware of the contact. Once mustard contacts the skin, the damage has already occurred. If the exposure is known at that point, decontamination will, at best, only reduce the amount of material still on the skin. It will otherwise be ineffective against stopping the damage to the body.

Because mustard is a persistent agent, it will remain on the clothing of victims and anything else that it touches for a long period of time. Decontamination, while it will not provide much help to the victim, is important to prevent secondary contamination of responders and others. Hypochlorite solution or flooding amounts of water are necessary to remove mustard agents. Unlike nerve agents, there is no known antidote for mustard exposure. It enters the body through the skin or mucous membrane and triggers biochemical damage within seconds to minutes, and no known medical procedures can stop or minimize the damage.

First symptoms from mustard exposure involve erythema, which is a redness of the skin similar to sunburn, followed by itching, burning, or stinging pain. The time between exposure and onset of symptoms may be affected by the dose of mustard, the air temperature, the amount of moisture on the skin, and area of exposure on the skin. Within the vicinity of the reddened skin, vesicles form and eventually become blisters filled with a yellowish, translucent fluid. The eyes are sensitive to mustard agent, and exposure will produce the most rapid onset of symptoms. Irritation, accompanied by a "gritty" sensation of something in the eye, is followed in some cases by inflammation of the eyes (conjunctivitis). Larger doses in the eyes produce edema of the eyelids (fluid buildup), which may be followed by damage to the cornea. The eyelids may become swollen, accompanied by pain in the eye, causing contraction of the muscles surrounding the eye and, finally, complete closure of the eyelids. Liquid mustard causes the most severe eye damage; however, blisters do not form in the eyes. Almost all of the eye exposure symptoms are self-reversing.

Respiratory exposure results in damage to the mucosa, or cell lining, followed by cellular damage and cellular death. The extent of the damage is dependent on the amount of the dose during exposure. Mustard enters the respiratory system through inhalation, and the greater the amount inhaled, the more severe the symptoms and damage. Small amounts of mustard result in nasal irritation and possibly bleeding. Sinus passages may be irritated along with the throat, which will produce a scratchy sore throat-type of pain. Laryngitis may result from larger exposures and can include complete loss of voice for a period of time. Shortness of breath and productive coughs may be present in the respiratory system resulting from large doses. When respiratory symptoms begin within 4 hours of the exposure, it is usually an indication of a large dose through inhalation.

When a large dose of mustard agent has entered the body, it may be carried to the bone marrow. Damage occurs here to the precursor cells, which is followed by a decrease in the white blood cell count. Over a period of days, red blood cells and platelets are also diminished. The gastrointestinal (GI) tract is also a target organ for mustard agent. Nausea and vomiting are not unusual within 12 to 24 hours after exposure. Last, the central nervous system may be affected by exposure to mustard agents. Generally, the symptoms are not well recognized and may involve sluggish, apathetic behaviors. Symptoms have been recorded for as much as a year after exposure. While death from mustard exposure is rare, lethal doses will produce convulsions and severe pulmonary damage that is usually associated with infection. In large dose exposures, death can occur within 24 to 36 hours.

Lewisite (L) is a vesicant from the arsenical (vesicant) chemical family. Lewisite causes many of the same types of damage to skin, eyes, and the respiratory system as do the mustard agents. The chemical name is dichloro-(2-chlorovinyl) arsine. It has not knowingly been used on the battlefield, and human exposure data is limited. The major difference between lewisite and mustard is that lewisite causes pain

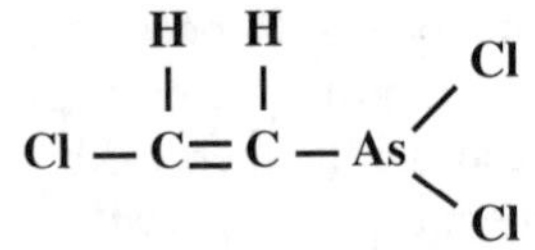

H H Cl
Cl – C=C – As
Cl

Lewisite
$C_2H_2AsCl_3$

Figure 8.37

immediately upon exposure, whereas the mustards have a delayed onset. Visible tissue damage will occur quickly in the form of a grayish-appearing area of dead skin. Blisters also develop more quickly than with mustard, but may take up to 12 hours for full blistering effects to develop.

Lewisite is also a systemic poison, which can result in pulmonary edema, diarrhea, restlessness, weakness, subnormal temperature, and low blood pressure. Severity of symptoms, in order of appearance, are: blister agent, toxic lung-irritant, tissue absorption, and systemic poison. If inhaled in high enough concentrations, lewisite can cause death in as little as 10 minutes. Common routes of exposure into the body are through the eyes, skin, and inhalation. Lewisite is an oily, colorless liquid when pure. "War gas" is amber to dark brown in color with a geranium-like odor; pure lewisite has little, if any, odor. Lewisite is much more volatile than mustard agents. Often, lewisite is mixed with mustard to lower the boiling point of the mixture. The military designation for the mustard/lewisite mixture is (**HL**). Lewisite is insoluble in water and has a specific gravity, which makes it heavier than water. It has a vapor density more than seven times heavier than air. Lewisite has a low level of flammability, with no flash point or flammable range identified. The molecular formula for lewisite is $\mathbf{C_2H_2AsCl_3}$, and the structure and molecular formula are shown in Figure 8.37.

Lewisite is a vesicant and toxic lung-irritant that is absorbed into tissues. If inhaled in high concentrations, it can be fatal in as little as 10 minutes; the body is unable to detoxify itself from lewisite exposure. Routes of entry into the body include the eyes, skin absorption, and inhalation. Eye contact results in pain, inflammation, and blepharospasm (spasms of the muscles of the eyelid), which leads to closure of the eyelids, corneal scarring, and iritis (inflammation of the iris). If decontamination of the eyes occurs quickly after exposure, damage may be reversible; however, permanent injury or blindness can occur within one minute of exposure.

Lewisite is irritating to the respiratory tract, producing burning, profuse nasal discharge, and violent sneezing. Prolonged exposure results in coughing and large amounts of froth mucus. Vapor exposure to the respiratory tract produces much the same symptoms as mustard. The main difference is that edema of the lungs is more exceptional and may be accompanied by pleural fluid. Lewisite does not affect the bone marrow, but does cause an increase in capillary permeability, with ensuing plasma leakage into tissues. This results in sufficient fluid loss to cause hemoconcentration, shock, and death. Exposures that do not result in death can cause chronic conditions, such as sensitization and lung impairment. Lewisite is a suspected carcinogen.

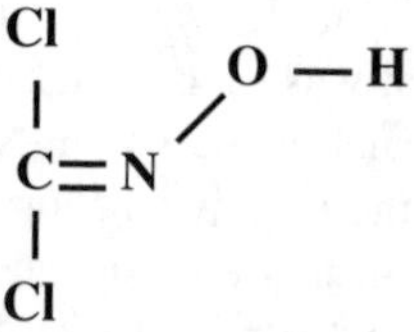

Cl
|
C=N—O—H
|
Cl

Phosgene Oxime
CCl_2NOH

Figure 8.38

An antidote is available for lewisite exposure. BAL (British-Anti-Lewisite; dimercaprol) was developed by the British during World War II. The antidote is produced in oil diluent for intramuscular administration to counter the systemic effects of lewisite. There is no effect, however, on the skin lesions (eyes, skin, and respiratory system) from the antidote. Mustard agents (H), (HD), (HS), and (HT), like nerve agents, would be classified as Class 6.1 poisons by the DOT and would have NFPA 704 designations of health 4, flammability 1, reactivity 1, and special 0. The UN 4-digit identification number for mustard agents would be 2810, as it also is for all of the mustard and nerve agents.

Phosgene oxime (CX) is also a vesicant, or blister agent, but it should not be confused with phosgene, which is a lung-damaging agent. Phosgene oxime causes urticaria (eruptions on the skin), rather than the fluid-filled blisters that occur with other vesicants. CX is a solid (powder) at temperatures below 95°F, but can be considered a flash-point solid, because of the high vapor pressure of 11.2 at 25°C. The boiling point is 53 to 54°C, and it is not considered flammable. Not much is known about the mechanical action or biological activity of phosgene oxime. It is absorbed through the skin, and contact with the material causes extreme pain immediately, much like lewisite, and tissue effects occur quickly. Irritation begins to occur in about 12 seconds at a dose of 0.2 mg-min/m^3. When the agent has been in contact with the skin for 1 minute or more, the irritation becomes unbearable at a dose of 3 mg-min/m^3. It is also irritating to the respiratory system and causes the same types of eye damage as lewisite. Phosgene oxime remains persistent in the soil for about 2 hours and is considered nonpersistent on other surfaces. The agent is known to be corrosive to most metals. There is no known antidote for phosgene oxime, and exposures should be treated symptomatically after decontamination, which should occur immediately. The structure and molecular formula are shown in Figure 8.38.

Blood Agents (Cyanogens)

Blood agents are common industrial chemicals that have been used on the battlefield to produce casualties. Two primary types of blood agents used by the military are hydrogen cyanide (AC) and cyanogen chloride (CK). Poisonous effects of cyanide have been well known since ancient times. Cyanide was the first blood agent used as a chemical warfare agent. Cyanides are salts with the metals potassium, sodium, and calcium most commonly used to form the compounds. The term "blood agent" was used because at the time cyanide was introduced as a warfare agent, it was thought to be the only agent that was transported via blood to the target organ. It is now known that other agents, such as nerve and vesicants, are also carried by the blood. Therefore, realistically, the term has become obsolete.

Cyanide is found naturally in many foods, such as lima beans, cherries, apple seeds, and the pits of peaches and similar fruits. It is reported to have an odor of burnt almonds or peach pits. However, 40% of the population cannot detect the odor of cyanide. Blood agents function by interfering with the body's ability to use oxygen at the cellular level. They are considered chemical asphyxiants, because death occurs from a lack of oxygen in the body. Inhalation is the primary route of exposure for blood agents. Unlike nerve agents and vesicants, blood agents are gases or volatile liquids. They produce vapors easily, which dissipate quickly in air and are, therefore, considered nonpersistent agents. Antidotes are available for cyanide poisoning, but must be administered quickly after exposure to be effective.

Hydrogen cyanide (HCN), (AC), also known as hydrocyanic acid or prussic acid, is a colorless to water-white or pale-blue liquid at temperatures of below 80°F. It has an odor of peach kernels or bitter almonds at 1 to 5 ppm. HCN may not have a detectable odor in lethal concentrations. The acute toxicity of hydrogen cyanide is high. There is rarely a reported chronic exposure to HCN because you either get better quickly from an exposure or you die. It is toxic by inhalation, skin absorption, and ingestion, with a TLV of 10 ppm in air. The LD_{50} through ingestion is 10 mg/kg of body weight. Skin absorption LD_{50} is estimated to be 1500 mg/kg of body weight. Inhalation of HCN has a LC_{50} of 63 ppm for 40 minutes. OHSA has established a PEL of 10 ppm (11 mg/m^3) for skin contact. TLV-TWA values are listed by ACGIH at 10 ppm (11 mg/m^3) for skin absorption. Inhalation of 18 to 36 ppm over a period of several hours can produce weakness, headache, confusion, nausea, and vomiting. Inhalation of 270 ppm can cause immediate death, and 100 to 200 ppm over a period of 30 to 60 minutes can also be fatal. Absorption of 50 mg through the skin can be fatal. Ingestion of 50 to 100 mg of HCN can also be fatal.

HCN is highly soluble and stable in water. It is extremely flammable, with a flash point of 0°F and an explosive range of 6–41% in air. Hydrogen cyanide has a boiling point of 25.7°C (78°F). Its vapor density is 0.990, which is slightly lighter than air, and it has a liquid density of 0.687, which is lighter than water. The autoignition temperature for HCN is 1000°F. Liquid HCN contains a stabilizer (usually phosphoric acid) and, as it ages, may explode if the acid stabilizer is not maintained in the solution at a sufficient concentration. HCN can polymerize explosively if heated above 120°F or if contaminated with any alkali materials. The structure and molecular formula of hydrogen cyanide are shown in Figure 8.39.

Hydrogen cyanide is volatile, and the DOT lists it as a 6.1 poisonous liquid with an inhalation hazard from the vapor. The NFPA 704 designation for AC is health 4, flammability 4, reactivity 2, and special 0. Vapors from hydrogen cyanide are highly toxic. The 4-digit UN identification number is 1613 for less than 20% hydrogen cyanide and 1051 for greater than 20% hydrogen cyanide. The Chemical Abstract Service (CAS) number is 74-90-8. Hydrogen cyanide is thought to act on the body by blending with cytochrome oxidase (an enzyme essential for oxidative processes of the tissues) and blocking the electron carrier system. This results in loss of cellular

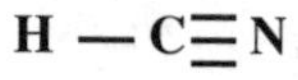

H — C≡N

Hydrogen Cyanide
HCN

Figure 8.39

oxygen use. The central nervous system, and particularly the respiratory system, are notably sensitive to this effect, and respiratory failure is the usual cause of death.

Cyanogen chloride (CK) is a colorless compressed gas or liquefied gas with a pungent odor, and is slightly soluble in water. It is not flammable; however, containers exposed to radiant heat or fire may explode and give off toxic or irritating fumes. CK is highly toxic by ingestion or inhalation, and is an eye and skin irritant. The TLV is 0.3 ppm, and the ceiling is 0.75 mg/m^3 in air. Because it is a gas, it is considered a nonpersistent agent. The vapor density is 2.16, which is heavier than air, and the specific gravity is 1.186, which is heavier than water. It has a boiling point of approximately 58°F, and rapid evaporation can cause frostbite. Cyanogen chloride has a freezing/melting point of –6.9°C.

CK may polymerize violently if contaminated with hydrogen chloride or ammonium chloride. Upon heating, it decomposes, producing toxic and corrosive fumes of hydrogen cyanide, hydrochloric acid, and nitrogen oxides. Hazardous polymerization can occur. Cyanogen chloride will react slowly with water to form hydrogen chloride gas. It acts on the body in two ways. Systemic effects of CK are much like those of hydrogen cyanide. Additionally, it causes irritation of the eyes, upper respiratory tract, and lungs. Eye irritation results in tearing. CK, like AC, stimulates the respiratory system and rapidly paralyzes it. Exposure is followed by immediate, intense irritation of the nose, throat, and eyes, with coughing, tightness in the chest, and lacrimation. This is followed by dizziness and increasing difficulty breathing. Unconsciousness comes next, with failing respiration and death within a few minutes. Convulsions, retching, and involuntary urination and defecation may occur. If these effects are not fatal, the signs and symptoms of pulmonary edema may develop. There may be repeated coughing, with profuse foamy sputum, rales in the chest, severe dyspnea, and distinct cyanosis. Recovery from the systemic effects is usually as prompt as in AC poisoning. However, a higher incidence of residual damage to the central nervous system should be expected. Based upon the concentration of the cyanogen chloride to which the victim has been exposed, the pulmonary effects may evolve instantly or may be delayed until the systemic effects have subsided. Consequently, early prognosis must be cautious. The structure and molecular formula for cyanogen chloride are shown in Figure 8.40.

Cyanogen chloride becomes volatile as temperatures increase, and the DOT lists it as a 2.3 poison gas. The NFPA 704 designation for CK is estimated to be health 4, flammability 0, reactivity 2, and special –0. Cyanogen chloride vapors are highly toxic. It has a 4-digit UN identification number of 1589 (inhibited). Treatment for either AC or CK poisoning is to follow the treatment protocols for Airway, Breathing,

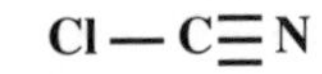

$$Cl - C \equiv N$$

Cyanogen Chloride
ClCN

Figure 8.40

and Circulation (ABCs) and administer oxygen to assist breathing. Instructions for administration and dosage should be based on local protocols and with the advice of a physician. Sodium nitrate is administered to produce methemoglobin, thus seizing the cyanide on the methemoglobin. The sodium thiosulfate combines with the confiscated cyanide to form thiocyanate, which is then excreted from the body.

Choking Agents (Lung-Damaging Agents)

Choking agents are known to cause pulmonary edema, which is the accumulation of fluid in the lungs. Examples of lung-damaging agents include phosgene (CG), diphosgene (DP), chlorine, and chloropicrin (PS). Phosgene is the best known of the choking agents. It is produced by heating carbon tetrachloride, which was once used as a fire-extinguishing agent. It was discontinued as a fire-extinguishing agent because when it came in contact with hot surfaces, it released phosgene gas. Choking agents irritate the bronchi, trachea, larynx, pharynx, and nose, which may result in pulmonary edema and contribute to a choking sensation.

Phosgene (CG), (carbonyl chloride), is produced as a liquid or liquefied gas, which is colorless to light yellow. At ordinary temperatures and pressures, CG is a colorless gas. Odors range from strong and stifling, when concentrated, to the smell of freshly mowed hay in lower concentrations. Phosgene has a boiling point of 8.2°C (45.6°F) and is noncombustible. The vapor density is 3.4, which is heavier than air, and the specific gravity is 1.37 at 20°C, which is heavier than water. The primary route of exposure is through inhalation, by which it is highly toxic. It is also a strong eye irritant, and has a TLV of 0.1 ppm or 0.40 mg/m^3 in air. General population limits are 0.0025 mg/m^3. In addition to mild conjunctival irritation, direct effects of exposure to phosgene result in damage to the lungs. The primary effect of exposure is pulmonary edema (fluid in the lungs). Death can occur within several hours after an exposure to a high concentration. Most fatalities, however, reach a maximum effect from the pulmonary edema in about 12 hours, and death occurs within 24 to 48 hours after exposure. Symptoms include coughing, choking (thus the term choking agent), tightness in the chest, nausea, and possibly vomiting, headache, and lacrimation. There is no real relationship between the symptoms and the prognosis. The structure and molecular formula for phosgene are shown in Figure 8.41.

The DOT lists phosgene as a 2.3 poison gas. The NFPA 704 designation for CG is health 4, flammability 0, reactivity 1, and special 0. Vapors of phosgene are highly toxic. It has a 4-digit UN identification number of 1076.

Diphosgene (DP), trichloromethyl chloroformate, is a clear, colorless liquid with an odor similar to phosgene. It is noncombustible, a strong irritant to the eyes and

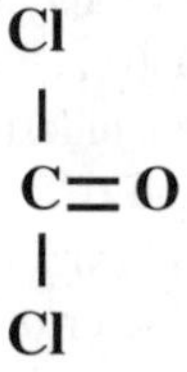

Cl
|
C=O
|
Cl

Phosgene
$COCl_2$

Figure 8.41

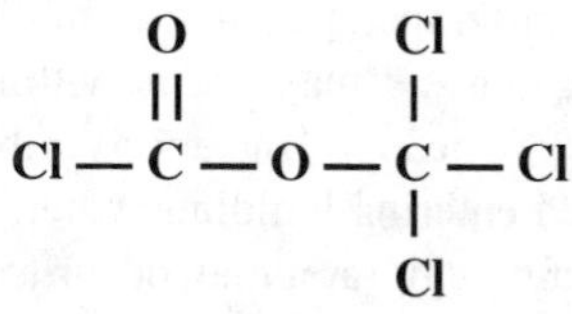

O Cl
|| |
Cl — C — O — C — Cl
|
Cl

Diphosgene
ClCOOCCl

Figure 8.42

tissues, and is toxic by inhalation and ingestion. DP has a boiling point of 127–128°C (263°F) and a vapor pressure of 4.2 at 20°C. The liquid density is 1.65, which is heavier than water, and a melting/freezing point of 157°C. Inhalation LC_{50} is 3600 mg/m^3 for 10 minutes. Effects of exposure are quite similar to phosgene gas. Its molecular formula is **$ClCOOCCl_3$**, and the structure and molecular formula are shown in Figure 8.42.

The DOT lists diphosgene as a 6.1 poison liquid. The NFPA 704 designation for CG is estimated to be health 4, flammability 0, reactivity 1, and special 0. It has a 4-digit UN identification number of 2972.

Riot-Control Agents (Irritant Agents and Vomiting Agents)

Riot-control agents are also referred to as tear gas, irritants, and lacrimators. They are local irritants, which in low concentrations act essentially on the eyes, resulting in intense pain and profuse tearing. Higher concentrations irritate the upper respiratory tract, skin, and sometimes cause nausea and vomiting. Exposure to these materials is rarely serious and may not require medical attention. Tear gases are not really gases at all. They are fine particulate smoke or aerosolized solid materials that can cause contamination of those exposed. Riot-control agents are commonly used by law enforcement officers. **(CS)** and **(CN)** are the primary agents used by law enforcement agencies today. However, it may be possible to encounter an older riot-control agent, **(DM)**, which is a "vomiting gas." CN, chloroacetophenone is also known as Mace. It has been replaced today by CS, O-chlorobenzylidenemalononitrile, which was developed in the 1950s and has become the agent of choice of law enforcement. Other irritant agents include (CNC), chloroacetophenone in chloroform; (CA), bromobenzylcyanide; and (CR), dibenz-(b,f)-1,4-oxazepine. OC is the designation for pepper spray. CK and PS are also considered to be lacrimators.

The biological mechanism by which riot-control agents work has not been well studied or documented compared to the other types of chemical agents. Generally, riot-control agents cause pain, tearing, and conjunctivitis in the eyes, which can be accompanied by spasms of the muscles around the eyes. A burning sensation occurs in the nose and respiratory tract. This is followed by sneezing and a large volume of nasal discharge. The chest may feel tight, and there may be a shortness of breath accompanied by coughing and secretions from the bronchial tubes when inhaled. Contact with the skin produces tingling, a burning feeling, and redness similar to sunburn. If conditions of high temperature and humidity are present along with a high concentration of agent, blisters may occur within 8 to 12 hours, which are similar to those produced by vesicants. Blisters have been reported by firefighters resulting from exposure when entering buildings where law enforcement personnel have discharged CS. Death, although rare, may occur from riot-control agents when exposure occurs in confined areas for an extended period of time. However, there are no reported deaths from open-air use.

Agent CS, O-chlorobenzylidenemalononitrile, is a white, crystalline solid with a boiling point of 310° to 315°C, a flash point of 197°C, and a melting point of 194°F. It has a vapor pressure of 3.4x 10^{-5} at 20°C and a vapor density several times heavier than air. CS is immiscible in water. Volatility is 0.71 mg/m^3 at 25°C. It has an odor that is biting and similar to pepper spray. Clouds produced by CS are white near the source for several seconds following the release. Usually, CS is disseminated by burning, exploding, or by forming an aerosol. It may also be used in liquid form if dissolved into a solvent and aerosolized. CS acts rapidly compared to CN, and it is 10 times more potent, while it is also much less toxic. The LD_{50} for CS is approximately 200 mg/kg of body weight, or 14 grams for a 70-kg person. CS has a TLV-TWA of 0.4 mg/m^3, an ICt_{50} of 10–20 mg-min/m^3, and an LCt_{50} of 61,000 mg-min/m^3. The chemical formula for CS is $\mathbf{C_{10}H_5ClN_2}$, and the structure and molecular formula are shown below:

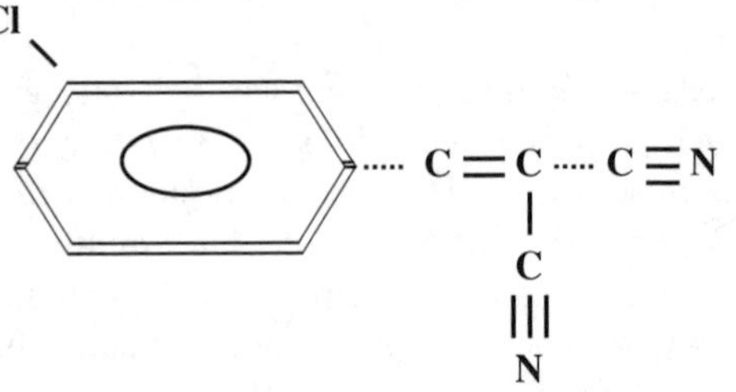

Cl
C=C ····· C≡N
C
N

Riot Control Agent CS
$C_{10}H_5ClN_2$

Figure 8.43

Agent CR, dibenz-(b,f)-1,4-oxazepine, is a pale-yellow, crystalline solid, which has a melting point of 163°F. It has a pepper-like odor and is only used in solution for dissemination in liquid dispensers. Solutions consist of 0.1% of CR in 80 parts of propylene glycol and 20 parts water. CR is an eye irritant in organic solutions at concentrations of 0.0025% or lower. Agent CR is less toxic when inhaled, but has more profound skin effects, which are longer lasting. It is a persistent agent when released in the environment and deposited on clothing.

Agents CN, chloroacetophenone, and **CA**, bromobenzylcyanide, are both white, crystalline solids with boiling points of 478°F and 468°F, respectively, and freezing points of 129°F and 77°F, respectively. Vapor density for CN is 5.3, and for CA it is 4.0, both of which are heavier than air by four and five times, resepctively. CN has an odor similar to apple blossoms, and CA smells like sour fruit. Agent CA is normally found as a liquid, and CN can also be used as a liquid in suitable solvents. When released, these agents produce a bluish-white cloud at the time of release.

Riot-control agents CA and CN are dispersed as minute particulate smoke and as a vapor resulting from burning munitions, such as lacrimator candles and grenades. Liquid agents may be dispersed from aircraft spray or exploding munitions. The OSHA PEL for agent CN is 15 mg/m^3, and the TLV is 10 mg/m^3. ICt_{50} for CN is 30 mg-min/m^3, and the LCt_{50} is estimated to be 8000 to 11,000 mg-min/m^3. For CA, the ICt^{50} is 80 mg/m^3, and the LCt_{50} is 7000 mg-min/m^3 from a solvent and 14,000 mg-min/m^3 from a grenade. CN has a chemical formula of **C_6H_5COCH2Cl**, and CA has a chemical formula of **C_8H_6BrN**. The structures and molecular formulas for CA and CN are shown in Figure 8.44.

Chloropicrin (PS), nitrotrichloromethane, trichloronitromethane, nitrochloroform, is a slightly oily, colorless, pale to transparent liquid that is nearly stable. It is nonflammable, with a boiling point of approximately 235°F and slight water-solubility. The vapor density is 5.7, which is heavier than air. PS is a strong irritant, highly toxic by inhalation and ingestion, with a TLV of 0.1 ppm (0.7 mg/m^3) in air. When heated, it decomposes to form hydrogen chloride, phosgene, carbon monoxide,

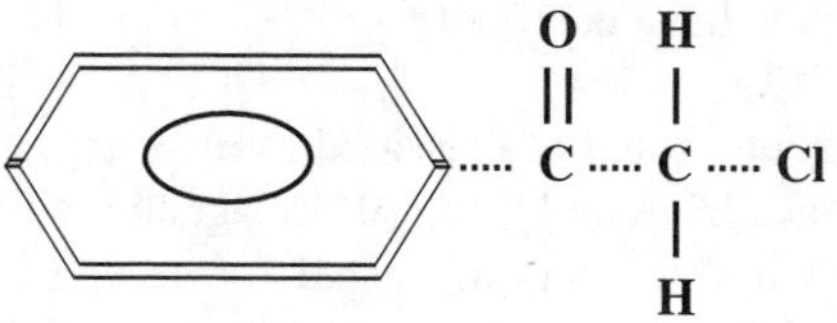

Riot Control Agent CN
$C_6H_5COCH_2Cl$

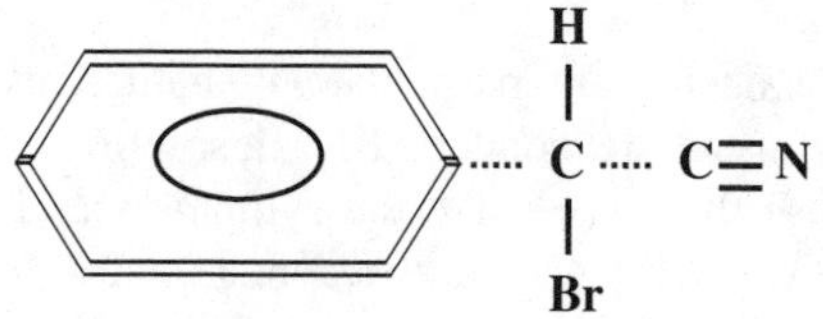

Riot Control Agent CA
C_8H_6BrN

Figure 8.44

and oxides of nitrogen. Decomposition can be explosive and it can become shock-sensitive. The LC_{50} for inhalation is 66 mg/m^3 over 4 hours, and the oral dose, LD_{50}, is 250 mg/kg of body weight. Chloropicrin is incompatible with strong oxidizers, alcohols, sodium hydroxide, and aniline. The DOT lists chloropicrin as a 6.1 poison liquid, and it has a UN identification number of 1580. The NFPA 704 designation for chloropicrin are flammability 0, health 4, reactivity 3, and special 0. The CAS number is 76-06-2.

Vomiting Agents

Vomiting agents create a strong pepper-like irritation of the upper respiratory tract along with irritation of the eyes and profuse tearing. This may be accompanied by intense, uncontrollable sneezing, coughing, nausea, vomiting, and an overall feeling of malaise (nonspecific feeling of illness). Agents primarily associated with this group include (DA) diphenylchloroarsine, (DM) diphenylaminochloroarsine (Adamsite), and (DC) diphenylcyanoarsine. Inhalation of the fine aerosol mist is the principal route of entry into the body, along with irritation and symptoms resulting from direct eye contact. Unlike other riot-control agents, vomiting agents generally do not produce any discomfort when in contact with skin.

These agents are all crystalline solids, which are released as smoke when heated. Smoke produced by DM is a light or canary yellow, while DA and DC produce a white smoke. These agents are highly effective in low concentrations, which may not be detectable during exposure. Symptoms produced include a feeling of pain and sense of fullness in the nose and sinuses, accompanied by a severe headache, intense burning in the throat, and tightness and pain in the chest. Coughing is uncontrollable, and sneezing is violent and persistent. Nasal secretion is greatly increased, and quantities of ropy saliva flow from the mouth. Nausea and vomiting are prominent. Mental depression may occur during the progression of symptoms. There are usually no long-term effects from vomiting agent exposures, and symptoms generally disappear within 20 minutes to 2 hours. Like other riot-control agents, deaths have occurred from exposure to high concentrations in enclosed areas.

Most riot-control agents would be classified as irritants under DOT hazard class 6.1. No information is available on NFPA 704 designations for riot-control agents. UN four-digit identification numbers are listed for several types of tear gas and containers. Grenades and tear gas candles are poison 6.1 and flammable solids. Tear gas devices are assigned the number 1700; devices and the remaining types of tear gas are listed as 1693.

Because riot-control agents are particulate in nature, exposures will require decontamination of victims and responders. Bleach solutions should not be used for decontamination, because they will make the symptoms and effects of the agents worse. Plain soap and water are the best decontamination solutions. Emergency decon can be performed using the water from hose lines. Clothing must be removed. This process will remove the majority of the agent particles off of the person.

A tank car of organophosphate pesticide, the same type of material used as a military nerve agent.

Miscellaneous Chemical Agents

Psychedelic Agent 3 (BZ), 3-quinuclidinyl benzilate, also known as "agent buzz," is a potent psychoactive chemical that affects the central nervous system as well as the circulatory, digestion, salivation, sweating, and vision systems. Those exposed experience hallucinations, and it acts as a sedative. The experience is much the same as that encountered by the narcotics amphetamines and cocaine. Three to four days after exposure, full recovery is expected from BZ intoxication. The TLV-TWA for BZ is 0.004 mg/m^3, and the general population limits are 0.0001 mg/m^3. BZ has an ICt_{50} value of 101 mg-min/m^3 (15 1/min) and an estimated LCt_{50} of 200,000 mg-min/m^3.

INFECTIOUS SUBSTANCES

Infectious substances are microorganisms or toxins derived from living organisms that produce death or disease in humans, animals, or plants. There are both good and bad bacteria present in the environment and living organisms. The bad bacteria are referred to as pathogens, because they can cause death to a living organism. Disease-causing microorganisms (pathogens) are classified as 6.2 Infectious Substances under the DOT hazard class system. This section will also cover potential terrorist biological agents, including bacteria, viruses, and toxins.

Toxic materials are the primary reason decontamination is conducted at HazMat incidents.

Toxins, which are poisons produced by microorganisms or plants, are classified as 6.1 Poisons. DOT defines an infectious substance or etiologic agent as "a viable (live) microorganism, or its toxin, that is capable of producing disease in humans." Included in the DOT regulation are agents listed in 42 CFR 72.3 of the regulations of the DSHS or any other agent that causes or may cause severe, disabling, or fatal disease.

Many materials shipped as infectious substances are diagnostic specimens. A diagnostic substance is any human or animal material, including blood, and its components, tissue, and tissue fluids being shipped for purposes of diagnosis. Medical wastes are also included in this hazard subclass.

The main reason this group is included with poisons is that the living organisms produce toxins (poisons). These toxins are biological wastes produced by the microorganisms. In small doses, the body's defense systems can handle the toxins. However, as the volume of toxin increases, it overwhelms the body's ability to defend against it. Toxins travel through the body until they reach a susceptible tissue and cause damage. Examples of typical infectious substances include *Bacillus anthracis* (anthrax), *Botulinum toxin* (botulism), *Francisella tularensis* (tularemia/rabbit fever/deerfly fever), *Coxiella burnetii* (rickettsia/Q fever), Venezuelan equine encephalitis (VEE), *Brucella suis* (brucellosis/undulant fever/Bang's disease), smallpox, ricin, micotoxins, and *Staphylococcal enterotoxin B*. Infectious substances also include bloodborne pathogens HIV and HBV, as well as any other substance that meets the DOT definition.

Biological agents can be subdivided into several related groups. These include **bacteria** and **rickettsia**, **viruses**, and **toxins**. Bacteria and rickettsia are single-celled, microscopic organisms that can cause disease in plants, animals, and humans. Some of the diseases caused by bacteria include anthrax, botulism, plague, cholera, diphtheria, tuberculosis, typhoid fever, typhus, Legionnaire's disease, Lyme disease, and strep infections. Bacterial organisms have a nucleus, intracellular nonmembrane-bound organelles (a specialized cellular part that resembles an organ), and a cell wall. Rickettsia are pleomorphic (come in varying sizes), parasitic microorganisms that live in the cells of the intestines of arthropods (invertebrate animals), such as insects, spiders, and crabs, which have segmented bodies and jointed limbs. Some are pathogenic to mammals and man, where they are known to cause the typhus group of fevers. **Rickettsia** are smaller than bacteria, but larger than viruses. Like the viruses, rickettsia are obligate (they cannot exist on their own or in any other form); they are considered intracellular parasites. **Viruses** are submicroscopic organisms, smaller than bacteria, and are unable to live on their own. They must invade a host cell and make use of its reproductive mechanism to multiply. **Toxins** are poisons produced by living organisms, including plants, bacteria, and animals. They would be classified as 6.1 Poisons rather than 6.2 Infectious Substances because they are toxic materials rather than disease-causing agents.

Bacterial Agents

Bacteria are single-celled organisms that range in shape and size from cocci (spherical cells), with a diameter of 0.5–1.0 m (micrometer), to long rod-shaped organisms known as bacilli, which may be from 1–5 m in size. Chains of bacilli have been known to exceed 60 m in size. Some bacteria have the ability to turn into spores. In this form, the bacteria are more resistant to cold, heat, drying, chemicals, and radiation than the bacterial form would be. When in the spore form, bacteria are inactive, or dormant, much like the seeds of a plant. When conditions are favorable, the spores will germinate just like seeds.

Bacteria have two methods by which they can cause disease in humans and animals. The first is by attacking the tissues of the living host. Second, all living organisms produce waste. Bacteria may produce a toxic or poisonous waste material that causes disease in the host. Some bacteria attack using both methods.

Anthrax spores are highly persistent and resist adverse environmental elements and remain viable for hundreds of years in soil and in dried or processed hides. Spores are resistant to drying, heat, and sunlight. They have been known to survive in milk for up to 10 years, on dried filter paper for 41 years, on dried silk threads for up to 71 years, and in pond water for 2 years.

Anthrax is a disease that is present in the soil and occurs naturally in livestock, including pigs, horses, goats, cattle, and sheep. Farmers who work with these animals have been known to contract anthrax. Human-to-human contact has not been documented and is thought to be unlikely. Infection occurs from **skin contact** with infected animal tissue and possibly from biting fleas feeding on the animals. The bacterium enters a cut or abrasion on the skin while handling contaminated products from infected animals. Most commonly, the disease appears on the hands and forearms of people working with infected animals. Symptoms include the formation of carbuncles (inflammation of hair follicles and surrounding subcutaneous tissue) and swelling at the location of the infection. Scabs form over the lesion and turn a coal-black. Anthrax is the Greek term for coal, and so the name is derived from the coal-black scabs on the lesions. This localized infection can also become systemic and transformed to the inhalation form of the disease.

Inhalation exposure occurs from spores in contaminated soil areas or from fried or processed skins and hides of infected animals that become airborne. Symptoms from inhalation exposure, depending on the concentration and length of time, will present in two distinct phases. First, spores are carried to the lungs, specifically the avolai within the lungs. This is followed by a pus-producing infection with edema (fluid buildup) and hemorrhage in the lungs. Symptoms are flu-like, including mild fever, malaise, fatigue, myalgia (muscle pain), a dry cough, and a feeling of pressure in the chest. It is believed that the anthrax exposure can be successfully treated at this stage if diagnosed in time. Depending on the dose, the first phase can last for several days or as little as 24 hours. The first phase can be followed by a period when the victim is symptom-free.

Onset of the second phase can be sudden, with the evolution of shortness of breath and cyanosis. **Ingestion** of contaminated meat may also cause infection, although this is a rare occurrence. Symptoms from ingestion exposure to anthrax include acute inflammation of the intestinal tract, nausea, loss of appetite, vomiting, and fever, which is followed by abdominal pain, vomiting of blood, and severe diarrhea. Approximately 25 to 60% of those victims who ingest anthrax will die.

The primary cause of human outbreaks is from skin contact with contaminated hides, leather, and animals. Natural infection from inhalation or ingestion is rare. The Canadian Center for Disease Control reports that the infectious dose in humans is 1300 organisms through inhalation. This equates to one-billionth of a gram (about the size of a speck of dust), which would be lethal to a single person. The United States military reports that the lethal dose is 8000 to 10,000 spores. Once infected, the incubation period for anthrax is 7 days and, on average, symptoms begin within 2 to 5 days. When inhaled at the proper dose, the disease runs a swift sequence of events and death can occur within 24 to 48 hours. Spore size is also an important factor in determining their effectiveness. Particles from 2 to 5 microns (one-millionth of a meter, 25,400th of one inch) in size are considered to be the most efficient in causing infection through inhalation. Particles larger than 5 microns would tend to be filtered out by the upper airway. It is expected that heavy smokers might be more susceptible to large particles.

Skin exposure and ingestion cases have been successfully treated when antibiotic therapy begins quickly. Penicillin, tetracycline, ciprofloxacin, doxycycline or other broad-spectrum antibiotics have proven effective, although inhalation exposure is almost always fatal, even with treatment. Penicillin has been the antibiotic of choice in the past, and dosage is usually 2 million units given intravenously every 2 hours. The FDA has approved Cipro as the antibiotic of choice, but the drug has significant side effects. Naturally occurring strains of anthrax may be resistant to penicillin. Victims who do recover from the cutaneous form of the disease may develop immunity. Injections of anthrax vaccine occur in six steps: it is administered at 0, 2, and 4 weeks, followed by 6, 12, and 18 months, followed by an annual booster. It is believed that after the first three doses, protection against cutaneous anthrax is achieved. Little data is available on the length of protection against inhalation exposure, although tests in primates suggest good protection can be attained after 2 doses, with protection up to 2 years. It is possible that the vaccine could be overwhelmed by an extremely high dose of spores.

Plague (*Yersinia pestis*) is a zoonotic bacterium that is normally spread among rodents by infected fleas. Zoonotic bacterium are capable of being transmitted from lower animals to man under natural conditions. Three forms of the disease can affect humans: bubonic, pneumonic, and primary septicemic. Another type of plague, called pharyngeal, resembles acute tonsillitis. Periodic outbreaks of the plague occur naturally in rodent populations, which may result in a high death rate. Fleas that have lost their usual hosts pursue alternative sources of blood. When this happens, the risk to humans and other animals is increased. Epidemics of plague in humans commonly involves house rats and their fleas. The disease is passed on to humans and other animals when they are bitten by a flea that has bitten an infected rat or other living thing. Animals prone to be carriers in the United States include the rock squirrels, prairie dogs, and other burrowing rodents. During 1924 and 1925, the last epidemic in the United States occurred. Since that time, there have only been isolated cases reported, usually in rural areas from wild rodents.

Plague cases in the United States during the 1980s averaged around 18 per year, mostly in the Southwestern states of New Mexico, Arizona, Colorado, and California. Highest rates of infection occur among Native Americans, particularly the Navajos. Others at risk include hunters, veterinarians, pet owners, campers, and hikers. Of those, most cases involved persons that were under the age of 20, with a fatality rate of one in seven. Death rates from bubonic plague can reach as high as 50 to 60% if not treated. When treated, the death rate is reduced to about 15%. If treatment is not begun within 24 hours after symptoms develop, pneumonic plague has a near 100% death rate.

Plague can also be transmitted when the organism enters the body through a break in the skin. This type of exposure occurs from direct contact with tissue or body fluids of a plague-infected animal, such as skinning a rabbit or other animal. This is, however, a rare occurrence. Plague can also be transmitted through inhalation by contacting infected droplets from a person or domestic animal coughing. Plague that develops from this type of exposure is called pneumonic. This infection involves the lungs as a result of inhalation of organisms, which results in primary pneumonic plague. Secondary pneumonic plague results from septicemia (blood infection) when the organisms spread to the lungs.

Symptoms from plague exposure usually develop within 2 to 6 days following the exposure. Pneumonic plague occurs a little faster, from 1 to 3 days, which is also dependent on the amount of organisms inhaled. Symptoms of bubonic plague include enlarged lymph nodes, fever, chills, and prostration. Pneumonic plague symptoms are similar to bubonic and include high fever and chills, which are accompanied by cough and difficulty breathing, production of a bloody sputum, and toxemia. Symptoms may be followed by rapid shock and death if treatment is not begun early. Death results from respiratory failure, circulatory collapse, and a predisposition toward bleeding. Bubonic plague can progress spontaneously to septicemic plague accompanied by fever, chills, prostration, abdominal pain, shock, and bleeding into the skin and other organs. It can also affect the central nervous system, lungs, and other parts of the body. As many as 80% of bubonic plague victims have positive blood cultures for septicemia. Approximately 25% of the patients will also have various types of skin lesions. Pustules, vesicles, eschars (dead tissue separating from living tissue), or papules (small elevation of the skin) containing leukocytes and bacteria may also be present near the site of the fleabite.

Antibiotic treatment should begin as soon as possible. Streptomycin is the drug of choice for treatment, but others, such as tetracyclines, chloramphenicol, gentamicin, or one of the sulfonamides, may also be effective. A vaccine for the plague is available, however, it is only effective as a preventative measure; once someone is exposed, the vaccine will not help. The initial dose of vaccine is followed by a second one 1–3 months later and a third one 3–6 months later. Booster shots are administered at 6, 12, and 18 months and then every 1 to 2 years. As with *all vaccines*, the level of protection is related to the size of the dose; vaccination defense could be overwhelmed by extremely high doses of the bacteria.

Tularemia (francisella tularensis), also known as rabbit fever, deerfly fever, and Ohara's disease, like the plague, is a bacterial infection that can occur naturally from the bite of insects, usually ticks and deerflies. The disease can also be acquired from contact with infected rabbits, muskrats, and squirrels, ingestion of contaminated food, or inhalation of contaminated dust. Once contracted, it is not directly spread from human to human. Tularemia remains infectious in the blood for about 2 weeks and in lesions for a month. It remains ineffective in deerflies for 14 days and ticks throughout their lifetime (about 2 years). The disease can occur at anytime of the year, but is most common in the early winter during rabbit hunting season and in the summer when tick and deerfly activity is at its peak. Tularemia contracted naturally has a death rate of approximately 5%.

Tularemia can appear in several different forms in humans, depending on the route of exposure. The usual presentation is ulceroglandular, typhoidal, or septicemic. In humans, as few as 10 to 50 organisms can cause disease if inhaled or injected, but over 108 would be required for the disease if ingested. Ulceroglandular tularemia is acquired naturally from dermal or mucous membrane exposures of blood or tissue fluids of infected animals. The typhoidal form makes up 5 to 15% of naturally occurring cases, which result from inhalation of infectious aerosols. Pneumonia can result from any of the forms of tularemia, but is most prominent in typhoidal. Incubation periods range from 2 to 10 days, depending on the dose and route of exposure. The average incubation period occurs within 3 days.

Symptoms of ulceroglandular disease include lymphadenopathy, fever, chills, headache, and malaise. About 90 to 95% of patients may present cutaneous ulcers. When ulcers are absent, it is referred to as glandular tularemia. When symptoms are confined to the throat, it is called primary ulceroglandular disease. Oculoglandular tularemia results from contact to the eyes from an infected fluid or blood. Typhoidal or septicemic tularemia produces fever, prostration, and weight loss, without adenopathy (any disease of the gland, especially a lymphatic gland). After exposure, the usual treatment is two weeks of tetracycline. Streptomycin is given for more severe exposures. Aminoglycosides, genatamycin, kanamycin, and chloramphenicol are also effective antibiotics. Once a person recovers, he or she has permanent immunity from the disease. A vaccine is under development and has been successful during tests on more than 5000 persons, without significant adverse reactions.

Cholera (vibrio cholera) is a bacterial disease that is contracted by ingestion of contaminated water or food. However, it does not spread easily from person to person. Cholera occurs naturally in many underdeveloped countries and has caused widespread outbreaks in South America, with over 250,000 cases reported just in Peru. It can be spread through ingestion of food or water contaminated with feces or vomitus of patients, by dirty water, hands contaminated with feces, or flies. Cholera is an acute infectious disease, represented by a sudden onset of symptoms. Victims may experience nausea, vomiting, profuse watery diarrhea with "rice water" appearance, and the rapid loss of body fluids, toxemia, and frequent collapse. Not everyone exposed will show symptoms. In some cases, there may be as many as 400 people without symptoms for every patient showing symptoms. Where cases go untreated, the death rate can be as high as 50%. With treatment, the death rate drops to below 1%.

Cholera itself is not lethal, but the breakdown of medical treatment systems in large outbreaks can result in many deaths from dehydration, hypovolemia (loss of body fluid), and shock. Fluid loss can be as much as 5 to 10 liters per day, and IV fluids used to replenish fluids can be in short supply. The incubation period varies from 12 to 72 hours, depending on the dose of ingested organisms. An infectious dose is greater than 108 organisms to a healthy individual through ingestion. Treatment involves antibiotic and IV fluid therapy using tetracycline (500 mg every 6 hours for 3 days). Doxycycline (300 mg once, or 100 mg every 12 hours for 3 days), IV solutions of 3.5g NaCl, 2.5g $NaHCO_3$, 1.5g KCl, and 20g glucose per liter are also appropriate treatments. Cholera shows a significant resistance to tetracycline and polymyxin antibiotics. Ciprofloxacin (500 mg every 12 hours for 3 days) or erythromycin (500 mg every 6 hours for 3 days) can be used as substitutes. A vaccine is available for prevention, however, it has not proven very effective. It offers only 50% protection for a period of up to 6 months.

Brucella (brucellosis), also known as undulant fever or Bang's disease, is a bacterial disease caused by any one of four species of coccobacilli. They are naturally occurring diseases in cattle, goats, pigs, swine, sheep, reindeer, caribou, coyotes, and dogs. The organisms can be contracted by humans through ingestion of unpasturized milk and cheese, or from inhalation of aerosols generated on farms and in slaughterhouses. Skin lesions on persons who have come in contact with infected animals can also spread infection. If these bacteria were used by terrorists, it would have to be aerosolized for inhalation by victims or be used to contaminate food supplies. Brucella is present worldwide among the animal populations, but is especially prevalent in the Mediterranean countries of Europe and in Africa, India, Mexico, and South America. The disease is also common in populations who eat raw caribou. When the animal population has a high rate of infection, the rate of disease occurrence is also higher in humans. It is unknown what the infectious dose of brucella is. Symptoms are nonspecific and insidious upon onset (the disease is well established when the symptoms appear). Symptoms include intermittent fever, headache, weakness, profuse sweating, chills, and arthralgia, and localized suppurative (pus-forming) infections are frequent. Incubation periods vary from 5 to 30 days, and in some cases, many months. Evidence of human-to-human transmission of the disease has not been documented. Side effects include depression and mental status changes. Osteoarthritic complications involving the axial skeleton are also common. Death is uncommon even in the absence of treatment. Treatment of brucellosis involves the administration of the antibiotics tetracycline and streptomycin or TMP-SMX. The disease is resistant to penicillin and cephalosporin. There is no vaccine available for use in humans.

Q fever (*Coxiella burnetii*), also known as Query fever and rickettsia, is a bacterial disease that occurs naturally in sheep, cattle, and goats. It is present in high concentrations in the placental tissues of these animals. Incidence of the disease is worldwide, and it is likely that more cases occur than those reported. Many epidemics occur in stockyards, meat-packing plants, and medical labs using sheep for research. Transmission occurs from airborne dissemination of rickettsiae in dust from contaminated premises. Organisms can be carried in the air over half a mile downwind. Infections are also contracted from contact with infected animals, their birth products (especially sheep), wool from sheep, straw, fertilizer, and laundry of exposed persons. The disease has also been traced to unpasturized milk from cows. Transmission from human to human is rare. Several varieties of ticks may also carry the disease and transmit it from animal to animal. Mortality rates from this disease are low, from 1 to 3%. It would, however, be an effective incapacitating agent because it is highly infectious when delivered through inhalation; as little as one organism can cause clinical symptoms.

The usual infectious dose is considered to be 10 organisms through inhalation. Symptoms are not specific to the disease, and it may be mistaken for a viral illness or atypical pneumonia. The incubation period is from 10 to 20 days. Patients may experience fever, cough, and chest pain as soon as 10 days after exposure. Although somewhat rare, other symptoms that may appear include chills, headache, weakness, malaise, severe sweats, hepatitis, endocarditis, pericarditis, pneumonitis, and generalized infections. Patients are not critically ill and, in most cases, the illness lasts from 2 days to 2 weeks.

Q fever is generally a self-limiting illness and will clear up without treatment. Antibiotics given during the illness can shorten the period of incapacitation. Tetracycline is the antibiotic of choice and when given during the incubation period may delay the onset of symptoms. Usual dosage is 500 mg every 6 hours or doxycycline, 100 mg every 12 hours. Antibiotic treatment should be continued for 5 to 7 days. The disease is remarkably resistant to heat and drying, and is stable under diverse environmental conditions. Vaccines for humans are still in the development stages, although tests have shown promise.

Salmonella spp. (Salmonellosis) is a naturally occurring bacteria that is present in a wide variety of animal hosts and environmental sources. There are over 2000 strains of bacteria in the salmonella family. Ten strains are responsible for most of the reported salmonella infections. We will limit our discussions here to the four most common strains: *Salmonella spp.*, *S. typhi*, *S. paratyphi*, and *S. choleraesuis*. Some strains may make humans sick and not animals, while others make animals sick and not humans. The bacterium that causes salmonella infections is a single-celled organism that cannot be seen with the naked eye, touched, or tasted.

Salmonella bacteria occur naturally in the intestines and waste of poultry, dogs, cats, rats, and other warm-blooded animals. When live salmonella bacteria enter the body, which is usually the result of contaminated food, a salmonella infection occurs. Salmonellosis is the most common bacterial foodborne illness. Salmonella infections can be prevented by cooking food at the proper temperature and cleaning food utensils and preparation areas effectively. There are approximately 40,000 salmonella infections reported to public health officials each year in the United States. Experts believe that between 500,000 and 4,000,000 cases actually occur that go unreported. It is estimated that 2 of every 1000 cases results in the death of the patient, over 500 annually. Salmonella affects those that are young, those that are older, and those whose body has already been weakened by some other illness. Many of those who contract the illness believe they have the flu and never go to see a doctor.

While there are many strains of salmonella, with a few exceptions, they produce similar symptoms. Incubation periods range from 6 hours to many weeks, depending on the strain and the amount of bacteria ingested. The infectious dose varies widely, and there is no rhyme or reason for why certain people are affected by the bacteria. Some become ill with ingestion of as little as 10 organisms, while others have ingested food contaminated with millions of bacteria and experienced no adverse effects. There is not any particular type of food that contains salmonella. Infection can occur from any undercooked food, eating raw foods, or eating food contaminated by people preparing the food.

Salmonella spp. (excluding *s. typhi*, *s. choleraesuis*, and *s. paratyphi*) causes acute gastroenteritis (inflammation of the gastrointestinal tract). After exposure and the appropriate incubation period, the onset of symptoms are sudden, with abdominal pain, diarrhea, nausea, and vomiting. Dehydration can occur and be severe in infants. Hosts range from humans to domestic and wild animals. The infectious dose is 100 to 1000 organisms by ingestion. Symptoms can appear in 6–72 hours, and normally within 12–36 hours. Transmission occurs from eating food made from infected animals or food contaminated from feces of an infected animal or person. Infected animal feeds and fertilizers prepared from contaminated meat scraps, and fecal-oral transmission from person to person also occurs. Salmonella bacteria is sensitive to antibiotic therapy using ampicillin, amoxicillin, TMP-SMX, and chloramphenicol.

Salmonella chloeraesuis produces acute gastroenteritis (infection of the gastrointestinal tract), sudden headache, abdominal pain, diarrhea, nausea, and sometimes vomiting. This bacterial illness may develop into enteric fever, with septicemia or focal infection in any tissue of the body. The infectious dose is 1000 organisms by ingestion. Modes of transmission and incubation periods are similar to *Salmonella spp*. Carriers are contagious throughout the infection, which can last for several days to several weeks. Antibiotic treatment can prolong the period in which the disease is contagious.

Salmonella typhi is a generalized systemic infection, which produces the following symptoms: fever, headache, malaise, anorexia, enlarged spleen, rose spots on the trunk of the body, and constipation. These symptoms are followed by more serious ones, including ulceration of Peyer's patches in the ileum, which can produce hemorrhage or perforation. Mild and atypical infections can also occur, with a death rate of 10% if not treated with antibiotics. Drug-resistant strains are appearing in several parts of the world. The infectious dose is 100,000 organisms by ingestion. Transmission occurs by ingesting food or water that have been contaminated with the feces or urine of a patient or carrier. Infection can also occur from food handlers who do not practice proper hygiene, or flies contaminating foods. Incubation periods depend on the strength of the dose, but are usually 1 to 3 weeks. Antibiotic treatment with chloramphenicol, ampicillin, or amoxicillin is usually effective.

Salmonella paratyphi is a bacterial enteric (intestinal) infection with an abrupt outbreak, which produces the following symptoms: continued fever, headache, malaise, enlarged spleen, rose spots on the trunk of the body, and diarrhea. These symptoms are similar to those of typhoid fever, but the death rate is much lower. Mild and asymptomatic infections may also occur upon exposure. Outbreaks and locations are similar to those of the other salmonella bacteria. The infectious dose is 1000 organisms by ingestion. Transmission occurs by direct or indirect contact with feces or, in rare cases, urine of patients or carriers. It is spread by food, especially milk and dairy products, shellfish, and in some isolated cases, water supplies. Incubation depends on the strength of the dose, but usually 1–3 weeks for enteric fever and 1–10 days for gastroenteritis. Antibiotic treatment with chloramphenicol, ampicillin, or TMP-SMX is usually effective.

Rickettsia are pleomorphic (come in varying sizes), parasitic microorganisms. A number of strains of these bacteria exist naturally. **Rickettsia Canada**, also known as louse-borne typhus fever and classical typhus fever, occurs in areas of poor hygiene that are also louse-infected. Outbreaks generally occur in Central America, South America, Asia, and Africa. It is primarily a disease of humans and squirrels. The body louse, *Pediculus humanus*, is the primary carrier of the disease. It feeds on the blood of an infected patient with acute typhus fever and becomes infected. Once this occurs, the lice excrete rickettsiae in their feces, which is defecated as they feed. Infection occurs from feces left on the skin by the lice, which is rubbed in at the site of the bite or other breaks already existing in the skin, or through inhalation of infected dust. Squirrels become infected from the bite of the squirrel flea.

The incubation period ranges from 1 to 2 weeks, with an average of 12 days. Rickettsia Canada cannot be directly transmitted from human to human, but it is a bloodborne pathogen, and universal precautions should be practiced. Victims are infective for lice from the time in which the febrile (fever) illness is present and for 2 to 3 days after the body temperature returns to normal. Infection remains in the louse for 2 to 6 days after biting the source, although it may occur quicker if the louse is crushed. Symptoms include headache, chills, fever, prostration, and general pains. On the 5th or 6th day, a macular eruption (unraised spots on the skin) occurs on the upper trunk and spreads to the entire body (except for the face, palms of the hands, and soles of the feet). The illness lasts for approximately 2 weeks. Without treatment, the fatality rate is about 10 to 40%. Treatment involves antibiotic therapy with tetracyclines and chloramphenicol.

Glanders, *Burkholderia* (formerly *Pseudomonas*) *mallei*, is a gram-negative bacillus found naturally in horses, mules, and donkeys. Man is rarely infected even during frequent and close contact with infected animals. No naturally occurring cases in man have been reported in the United States in over 60 years. Sporadic cases do continue to occur in Asia, Africa, the Middle East, and South America. The disease exists in four basic forms in man and horses. Acute forms occur primarily in mules and donkeys, and result in death in 3 to 4 weeks. Chronic forms occur in horses, and result in lymphadenopathy (enlargement of the lymph nodes), skin nodules that ulcerate and drain, and induration (increase in fibrous tissue). Transmission to man occurs through the nasal, oral, and conjunctival mucous membranes, by inhalation into the lungs, and breaks in the skin. Aerosols have been reported to be highly infectious in laboratory exposures and resulted in over 46% of the cases becoming severe. There is no vaccine or effective treatment available. Incubation periods are from 10 to 14 days. Symptoms include fever, rigors, sweats, myalgia, headache, pleuritic chest pain, cervical adenopathy, splenomegaly, and generalized papular/pustular eruptions. The disease is almost always fatal without treatment. Sulfadiazine may be an effective treatment in some cases. Ciprofloxacin, doxycycline, and rifampin have also been shown to be effective. Most antibiotic sensitivities are based upon animal studies because of the low incidence of human exposure.

Viruses

Viruses are the simplest type of microorganisms and the smallest of all living things. They are much smaller than bacteria and range in size from 0.02 to 1 m (m = 1,000 mm). One drop of blood can contain over 6 billion viruses! Viruses were first discovered in 1898 and observed for the first time in 1939. Virus is a Latin word meaning "poisonous slime." Every living entity is composed of cells, except for viruses. They are, in fact, totally inert until they come in contact with a living host cell. Hosts can include human, animal, plants, or bacteria. The infection point created from the virus occurs at the cellular level. There must be an exact fit between the virus and the cell, or the invasion of the cell cannot occur. After a virus attaches to a cell, it begins to reproduce itself, resulting in an acute viral infection. Once a virus takes hold of the cell, it can cause the host cell to die. Common examples of viral agents include measles, mumps, meningitis, influenza, and the common cold. Viruses that most people are familiar with today are HIV (the virus that causes AIDS), HBV (the virus that causes hepatitis B), and HCV (the virus that causes hepatitis C). These viruses are not airborne and are usually difficult to transmit. Very specific actions have to take place to transmit the viruses. HIV, HBV, and HCV are transmitted by contact with blood and body fluids. They are more commonly referred to as bloodborne pathogens. Viral hemorrhagic fevers are a group of viruses that include Ebola, Marburg, Arenaviridae, Lassa fever, Argentine and Bolivian, Congo-Crimean, Rift Valley, Hantavirus, Yellow Fever, and Dengue.

HIV is caused by the human immunodeficiency virus. There are more than 1 million people in the United States today that are HIV-positive. Exposure to HIV occurs through direct contact with blood or blood components; cerebrospinal fluid; synovial fluid; peritoneal fluid; amniotic fluid; or blood in saliva, semen, vaginal secretions, or any other body fluid that contains blood. Infection may occur from unprotected sex with an infected person, sharing needles, needle and other sharp-object sticks, and contact with blood or blood products. The incubation period for

HIV varies from 2 months to over 10 years. Treatments are available, which may extend the incubation period. Symptoms include an acute, self-limited mono-like illness lasting for 1 to 2 weeks. Those infected then remain symptom-free for many years. Full-blown AIDS then rears its ugly head. Symptoms are somewhat nonspecific and vary from person to person. HIV destroys the body's immune system, allowing infections, some which are rare, to develop. Common infections include pneumocystic carinii pneumonia and cancers, such as Kaposi's sarcoma. Victims also may develop a wasting syndrome, extrapulmonary tuberculosis, or neurologic diseases, like HIV dementia. Symptoms are insidious at onset. They may include swollen glands, anorexia, chronic diarrhea, weight loss, and fever. Treatments and vaccines are under development, which have offered the promise of reducing HIV to a chronic illness.

HBV and HCV are diseases of the liver, sometimes referred to as hepatitis or inflammation of the liver. Both HBV and HCV are bloodborne pathogens, like HIV. The virus is most concentrated in blood, serum, and wound exudates. The virus is also present in smaller concentrations in semen, vaginal fluid, and breast milk. Low concentrations exist in urine, feces, sweat, tears, and saliva. The incubation period for HVB is from 45 to 160 days, and for HCV 2 to 26 weeks, with a fatality rate of 1 to 1.4%. Like HIV, symptoms are insidious at onset. They include anorexia, malaise, nausea, vomiting, abdominal pain, jaundice, skin rashes, arthralgies, and arthritis. A vaccine is available for HBV.

Smallpox is a lethal infection caused by the variola virus, which has at least two strains, variola major and variola minor. Cases of smallpox date back over 2000 years, and it is the oldest-known human pathogen. Naturally occurring smallpox was declared eradicated from the earth in 1980 by the World Health Organization (WHO), a branch of the UN. The last reported case in the world occurred in Somalia in 1977. Two laboratories in the world still hold the last-known stocks of variola virus: the Centers for Disease Control (CDC) in Atlanta and VECTOR in Novizbersk, Russia. Clandestine stocks could exist in other parts of the world, but are as yet unknown. The WHO's governing body recommended the total destruction of the remaining stockpiles by the year 1999. An effective vaccination is available for smallpox and has been used for years for the general population. Since it is primarily a children's disease, vaccinations were given during early childhood and were only effective for about 10 years. Vaccination of civilians in the United States was discontinued in the early 1980s. Children who are no longer vaccinated would be at great risk from exposure to smallpox.

Monkey pox and cowpox are closely related to variola and might be genetically manipulated to produce a smallpox-like virus. Once exposure to the smallpox virus occurs, the incubation period is approximately 12 days. Those who may have contacted exposed persons are quarantined for a minimum of 16 to 17 days following the exposure. Symptoms of smallpox include malaise, fever, rigors, vomiting, headache, and backache, and about 15% of the patients develop delirium (hallucinations). In approximately 2 to 3 days, an enanthem develops concomitantly with a particular rash on the face, hands, and forearms. This is followed by eruptions on the lower extremities and the trunk of the body, which occurs over a week's time. Lesions progress from discolored spots flush with the surface of the skin, to raised spots on the skin, and finally to an inflamed swelling on the skin containing pus (skin blisters). Lesions are more abundant on the extremities and face, which is important in the diagnosis of the disease. Within 8 to 14 days, scabs form on the skin blisters. Once the scabs fall off, a discolored depression is left behind. As long as the scabs are in place, the patient is considered contagious and should be isolated. Transmission occurs from close person-to-person contact, and it is unknown if an airborne dispersion would be effective. Most anitviral drugs for smallpox are experimental at the present time and would not be available for large numbers of victims. Vacinia-immune globulin (VIG) has shown to be an effective prevention following an exposure to smallpox.

Venezuelan equine encephalitis (VEE) is a virus that is naturally transmitted from horse to horse by mosquitoes. Humans can also get the virus from the infected mosquito. Each year, thousands of persons acquire the disease naturally from mosquito bites. Human outbreaks usually follow an epidemic among the horse population. Humans infected can infect mosquitoes for up to 72 hours. Once a mosquito becomes infected, it remains so for life.

VEE is considered a bloodborne pathogen. Universal precautions for bloodborne pathogens should be taken by emergency workers when around the VEE patient. Human-to-human transmission through inhalation of respiratory droplets can theoretically cause infection, but has not been proven. VEE is rarely fatal (less than 1%) and acts as an incapacitating agent. Nearly 100% of those exposed acquire the disease, however, only a small number actually develop encephalitis. Usually young children are the most vulnerable for developing encephalitis.

VEE is characterized by convulsions, coma, and paralysis. Encephalitis is characterized by inflammation of the meninges (surrounding membranes) of the brain and the brain itself, which produces central nervous system symptoms. Onset of symptoms is sudden, following an incubation period of 1 to 5 days. Symptoms are flu-like and may include malaise, spiking fevers, rigors (chills and severe shivering), severe headache, photophobia (sensitivity to light), and myalgias (muscle pain). Symptoms of VEE may include nausea, vomiting, cough, sore throat, and diarrhea. Complete recovery requires 1 to 2 weeks. Diagnosis of the disease is difficult, as physical symptoms are nonspecific. White blood cell counts may show a striking leukopenia (abnormally low leukocytes in the blood) and lymphopenia (reduction in number of lymphocytes circulating in the blood). The virus can be isolated from the serum (fluid that forms blood clots). Because there are no drugs or specific treatments for the disease, treatment is supportive. Analgesics may be given to relieve headache and myalgia. Victims that develop encephalitis can be treated with anticonvulsants and fluid therapy for electrolyte balance. Two vaccines are available, but are still in the investigation phase of development. TC-83, a live vaccine, is given in a single dose of 0.5 ml subcutaneously. A second vaccine, C-84, is used to boost those who do not respond to the TC-83. It is given as 0.5 ml subcuntaneously in three doses at two- to four-week intervals. Research is also underway using antiviral drugs that have shown some promise with laboratory animals. However, no human clinical data is available.

Viral hemorrhagic fevers (VHFs) are an assorted group of human diseases originated by viruses from several different families. Included are the Filoviridae family, which includes Marburg and Ebola, and the Arenaviridae family, which encompasses Lassa fever and Argentine and Bolivian hemorrhagic fevers. Additionally, the Bunyaviridae family, which involves various members of the Hantavirus genus, the Congo-Crimean hemorrhagic fever family from the Nairovirus genus, and the Rift Valley fever from the Phlebovirus genus are family members. The last family is the Flaviriridae, which includes Yellow Fever and the Dengue hemorrhagic fever virus.

The virus spreads through close personal contact with a person who is infected with the disease and can also be spread through sexual contact. Universal bloodborne pathogen precautions should be practiced when treating victims. It is not known what the natural host is for the virus or how a person contracts the disease initially. The incubation period is 2 to 21 days.

Mortality rates in Africa from **Ebola** range from 50 to 90% of those infected. Another strain of Ebola, the **Reston**, was reported among monkeys in the Philippines in 1989. This strain is not yet known to transmit to humans. African strains, on the other hand, have caused severe disease and death. Why the disease has only shown sporadic outbreaks is unknown.

Marburg virus has been reported to cause infection in man on four previous occasions. Three occurrences were in Africa and one in Germany, where the virus was named. Outbreaks first occurred in Germany and Yugoslavia involving 31 reported cases, which occurred from exposure to African green monkeys. Seven people died. The incubation period for Marburg is 3 to 7 days. Methods of transmission of these diseases are not well known. The disease seems to be spread by direct contact with infected blood, secretions, organs, or semen.

Argentine hemorrhagic fever (AHF) is caused by the Junin virus, which first appeared in 1955 among corn harvesters in Argentina. The virus is spread naturally from contact with the infected excreta of rodents. Somewhere between 300 and 600 cases of AHF occur each year in the Pampas region of Argentina. A similar disease, **Bolivian hemorrhagic fever** (BHF), which is caused by the Machupo virus, appeared in northeastern Bolivia following the appearance of AHF. Another closely related virus is Lassa, which occurs over much of western Africa.

Congo-Crimean hemorrhagic fever (CCHF) is a disease carried by ticks and transmitted by bites to humans. Infections occur primarily in the Crimea and other parts of Africa and in Europe and Asia. The incubation period is 3 to 12 days. It is stable in blood up to 10 days at 105°F. **Rift Valley fever** also occurs only in Africa and results in sporadiç, widespread epidemics of the disease.

Hantavirus was first identified prior to World War II in Manchuria along the Amur River. Incubation periods vary from 5–42 days, with 12–16 days being the average.

Yellow fever and **dengue** fever are two diseases that are transmitted by infected mosquitos. Yellow fever has an incubation period of 3 to 6 days. Yellow fever is the only one of the VHFs that does have a vaccine available. Dengue fever virus has an incubation period from 3–14 days, with 7–10 being the average. The virus is stable in dried blood and exudates up to 2 days at room temperature. All VHFs, except for Dengue fever, are capable of being spread by aerosolization and skin-crawling insects. Patients infected with VHFs other than hantavirus will have the virus in the blood, and it can be transmitted through contact with blood or body fluids containing blood. Bloodborne pathogen precautions should be undertaken when exposure to blood is a possibility. Routes of infection for the filoviruses in humans is not well understood at this time.

VHFs are feverish illnesses that are complicated by easy bleeding, petechiae (bleeding under the skin), hypotension (abnormally low blood pressure), shock, flushing of the face and chest, and edema (excess fluid in tissues). Not all infected humans actually develop VHFs. The reason for this is not well understood. Congenital symptoms, such as malaise, myalgias, headache, vomiting, and diarrhea may appear in association with any of the hemorrhagic fevers. Treatment is supportive of the symptoms that are presented. Ribavirin may be an effective antiviral therapy for Lassa fever, Rift Valley fever, and Congo-Crimean hemorrhagic fever viruses. During recovery, plasma may be effective in Argentine hemmorrhagic fever.

Toxins

Biological toxins are defined as any toxic substance occurring in nature produced by an animal, plant, or microbes (pathogenic bacteria), such as bacteria, fungi, flowering plants, insects, fish, reptiles, or mammals. Under the DOT classification system, these materials would be classified as Class 6.1 Poisons. Unlike chemical agents, such as sarin, cyanide, or mustard, toxins are not manmade. Generally, toxins are not volatile and not considered a dermal exposure hazard (except for mycotoxins). Toxins as a family are much more toxic than any of the chemical agents, including VX and sarin. Ricin, a biological toxin produced from the castor bean plant, is 10,000 times more toxic than sarin; as little as a milligram (1/1000 of a gram) can kill an individual. Botulinum toxin is the most toxic material known to man. One ounce of botulinum toxin (.12 micrograms or 1200-millionths of a gram) is enough to kill 60 million people. Mycotoxins are the least toxic of the toxins (thousands of times less than botulism).

Routes of exposure also have a bearing on the level of toxicity. Some toxins are much more lethal when aerosolized and inhaled than when taken orally. Ricin, saxitoxin, and T2 mycotoxins are examples of these types of materials. Botulism, for example, has a lower toxicity through aerosolization and inhalation than through ingestion. When looking at the potential toxicity of toxins, the lower the LD_{50}, in micrograms per kilogram of body weight, the less agent that would be required to be toxic. Conversely, some agents such as ricin, would require great quantities (tons) for an aerosol attack out in the open. Toxins can either be used for their lethality or as an incapacitating agent. Some toxins are incapacitating at lower doses and would cause serious illness.

Toxins can be divided into groups based upon the mechanism by which they function. **Protein toxins** are created by bacteria. Protein toxins include botulinum (seven related toxins), diphtheria, tetanus, and staphylococcal enerotoxins (seven different toxins). They function by paralyzing the respiratory muscles. Staphylococcal enterotoxins can incapacitate at levels at least 100 times lower than the lethal level.

Bacterial toxins can be classified as membrane-damaging. This group includes escherichia coli (hemolysins), aeromonas, pseudomonas, and staphylococcus alpha (cytolysins and phospholipases). Many of these toxins function by interfering with bodily functions and kill by creating pores in cell membranes.

Marine toxins may be developed from marine organisms. Examples include saxitoxin, tetrodotoxin, palytoxin, brevetoxins, and microcystin. Saxitoxin is a sodium-channel blocker and is most toxic by inhalation compared to the other routes of exposure. Saxitoxin and tetrodotoxin are similar in mechanical action, toxicity, and physical attributes. They can be lethal within a few minutes when inhaled. It has not yet been chemically synthesized efficiently, or easily created in large quantities from natural sources. Palytoxin is produced from soft coral and is highly toxic. It is, however, difficult to produce or harvest from nature.

Trichothecene mycotoxins are created by numerous species of fungi. Over 40 toxins are known to be produced by fungi. T2 is a stable toxin even when heated to high temperatures. However, unlike other toxins, the mycotoxins are dermally active. Once absorbed into the body, it would become a systemic toxin.

Plant toxins are derived from plants or plant seeds. One of the premier plant toxins is ricin. It is a protein taken from the castor bean. Approximately 1 million tons of castor beans are grown annually worldwide for the production of castor oil. Castor beans to grow the plants can be purchased through seed catalogs, from lawn and garden centers, and from agricultural co-ops. Waste mash resulting from the production of castor oil contains 3 to 5% ricin by weight. Videos and books are available giving detailed step-by-step instructions for producing ricin.

Animal venoms can contain protein toxins as well as nontoxic proteins. Many venom toxins can be cloned, created by molecular biological procedures, and manufactured by simple chemical synthesis. Venoms can be divided into groups: **ion channel toxins**, like those found in rattlesnakes, scorpions, and cone snail; **presynaptic phospholipase A2 neurotoxins**, found in the banded krait, Mojave rattler, and Australian tiapan snake; **postsynaptic neurotoxins**, found in the coral, mamba, cobra, sea snake, and cone snail; **membrane-damaging toxins** of the Formosan cobra and rattlesnake; and the **coagulation/anticoagulation toxins**, found in the Malayan pit viper and carpet viper. Toxins are unlike chemical agents in that they vary widely in their mechanism of action. Time from exposure to onset of symptoms also varies widely.

Saxitoxin acts quickly, and can kill an individual within a few minutes of inhalation of a lethal dose. It acts by directly blocking nerve conduction, and causes death by paralyzing muscles of respiration. At slightly less than the lethal dose, the victim may not experience any effects at all. Botulinum toxin needs to invade nerve terminals in order to block the release of neurotransmitters, which under normal conditions control muscle contraction. The symptoms from botulinum toxin are slow to develop (from hours to days), but are just as lethal, causing respiratory failure. This toxin blocks biochemical action in the nerves, which activate the muscles necessary for respiration, which leads to suffocation. Unlike saxitoxin, toxicity for botulinum is greater through ingestion than inhalation.

Neurotoxins are effective in stopping nerve and muscle function without producing microscopic injury to the tissues, whereas other toxins destroy or damage tissue directly. Microcystin is a toxin produced by blue-green algae. When it enters the body, it binds to an important enzyme inside the liver cells. No other cells in the body are affected. If microcystin is not blocked from reaching the liver within 15 to 60 minutes of receiving a lethal dose, irreversible damage to the liver will occur. Damage to the liver from this toxin is the same, regardless of the route of exposure. With other toxins, the damage that occurs after contact may vary greatly, depending on the route of exposure, even with the same toxin family. Death occurs from ricin because it blocks protein synthesis in many different cells within the body. However, no damage occurs to the lungs unless the route of exposure is inhalation.

For example, the staphylococcal enterotoxins can cause illness at low concentrations, but require large doses to be lethal. Trichothecene mycotoxins are the only biological toxins that are dermally active. Exposure results in skin lesions and systemic illness without being inhaled and absorbed through the respiratory system. Primary routes of exposure are through skin contact and ingestion. Nanogram (one-billionth of a gram) quantities per square centimeter of skin can cause irritation. One-millionth-of-a-gram quantities per square centimeter of skin can cause destruction of cells. Microgram doses to the eyes can cause irreversible damage to the cornea. Because most biological agents are not skin-absorbent hazards, simple washing of contaminated skin surfaces with soap and water within one to three hours of an exposure can greatly reduce the risk of illness or injury.

Botulinum clostridium (botulism) toxin is a deadly illness caused by any one of seven different, but related, neurotoxins (A through G). All seven types have similar mechanisms of action. They each produce similar symptoms and effects when inhaled or ingested, but the length of time to development of symptoms may vary, depending on the route of exposure and dose received. Botulism is not spread from person to person. Botulinum toxins as a group are among the most toxic compounds known to man. Lethal doses in research animals are 0.001 micrograms per kilogram of body weight, which will kill 50% of the animals. Botulinum toxins are 15,000 times more toxic than lethal nerve agent VX, and 10,000 times more toxic than sarin. Botulism occurs naturally in improperly canned foods and infrequently in contaminated fish. Ingestion of the canned food or fish causes the illness. Low-dose inhalation may not produce symptoms for several days. Inhalation or ingestion of high doses would produce symptoms much quicker. Botulism bacterium is commonly found in the soil.

Two types of illness are associated with the botulinum toxin, infant and adult botulism. An adult becomes ill by eating spoiled food that contains the toxin. Infants become ill from eating the spores of the botulinum bacterium. One source of these spores comes from the ingestion of honey. Spores are not normally toxic to adults. Botulinum toxins work by binding to the presynaptic nerve terminal at the neuromuscular junction and at cholinergic autonomic sites. They then act to stop the release of acetylchloline presynaptically, thus blocking neurotransmission.

This function is quite unlike the action of the nerve agents, where there is too much acetylcholine due to the inhibition of acetylcholinesterase. What occurs with botulism is a lack of the neurotransmitter in the synapse. Therefore, using atropine as an antidote would not be helpful and could even provoke symptoms. When contaminated food is ingested by adults, the toxin is absorbed from the intestines and attaches to the nerves, causing the signs and symptoms of botulism poisoning.

Symptoms include blurred vision, dry mouth or sore throat, difficulty in swallowing or speaking, impairment of the gag reflex, general muscular weakness, dilated pupils, and shortness of breath. Paralysis of the skeletal muscles follows, with a proportional, downward, and growing weakness, which may result in sudden respiratory failure. The time from the beginning of symptoms to respiratory failure can occur in as little as 24 hours when the toxin is ingested. One-third of patients die within 3 to 5 days. An antitoxin for botulism can be effective if administered quickly after the onset of symptoms. The bacteria that produces botulism survives well in soil and agricultural products. Botulism toxin can be destroyed by boiling for 10 minutes.

Staphylococcal enterotoxin B (SEB) is a common cause of food poisoning. Cases have usually been isolated to a group of people exposed to contaminated food at some public event or through airline travel. While it can cause death, it is thought of as an incapacitating agent rather than a lethal agent. It could, however, make a large number of people ill for an extended period of time. The primary route of exposure is ingestion. Symptoms from inhalation exposure and ingestion are completely different. Either route of exposure may produce fatalities. Symptoms appear within 3 to 12 hours after aerosol exposure. They include sudden onset of fever, chills, headache, myalgia, and cough. Some patients may also exhibit shortness of breath and retrosternal (behind the breastbone) chest pain. Fevers of between 103° and 106°F generally last for 2 to 5 days, and cough may persist for several weeks. Ingestion produces nausea, vomiting, and diarrhea. Very large doses may result in pulmonary edema, septic shock, and possibly death. No antitoxin has been developed for this illness, so treatment remains supportive. No preventative vaccines are available. Naturally occurring food poisoning cases would not present with respiratory symptoms.

SEB infection has a tendency to develop quickly due to a somewhat unchanging clinical condition. Respiratory difficulties occur much later with SEB inhalation. Laboratory testing will provide limited data for diagnosing the disease. SEB toxin is difficult to detect in serum when symptoms develop; however, a baseline specimen for antibody detection should be drawn anyway as early as possible after exposure. Additional specimens should be drawn during recovery. SEB can be detected in the urine, and a sample should be taken and tested. Test results may be helpful retrospectively in developing a diagnosis. High concentrations inhibit kidney function. Disinfectant solutions include 0.5% sodium hypochlorite for 10 to 15 minutes or soap and water.

Ricin (*Ricinus communis*) is a protein toxin that is produced from the castor bean and functions as a cellular poison. Ricin is widely available in any part of the world. It is highly toxic and can enter the body by ingestion, inhalation, and injection. Inhalation from aerosol dispersion produces symptoms based upon the dose that was inhaled. During the 1940s, humans were accidentally exposed to sublethal doses, which produced fever, chest tightness, cough, dyspnea, nausea, and arthralgias within 4 to 8 hours. After several hours, profuse sweating occurred, which signaled the end of the symptomatic phase.

Little data are available on human inhalation exposure, but victims would be expected to develop severe lung inflammation with a progressive cough, dyspnea, cyanosis, and pulmonary edema. When ricin enters the body through routes other than inhalation, it is not a direct lung irritant. Ingestion causes gastrointestinal hemorrhage with hepatic, splenic, and renal necrosis. Intramuscular injection causes severe localized necrosis of muscle and regional lymph nodes, with moderate visceral organ involvement.

The toxicity of ricin compared to botulinum and SEB, based upon LD_{50} values, is much less. Natural intoxication from ricin can occur by the ingestion of the castor bean. This produces severe gastrointestinal symptoms, vascular collapse, and death. When exposure to ricin occurs through inhalation of small particles, pathogenic changes can occur in as little as 8 hours. This is followed by severe respiratory symptoms and acute hypoxic respiratory failure in 36 to 72 hours. Intravenous injection may result in disseminated intravascular coagulation, microcirculatory failure, and multiple organ failure. Ricin is toxic to the cells in the body and acts by inhibiting protein synthesis. During tests conducted on rodents, ricin was more toxic through inhalation than ingestion. A vaccine for ricin is under development, but not currently available. A vaccine would provide the best protection against ricin poisoning.

Trichothecene mycotoxins are toxins produced by several types of fungi (mold). They are the only group of biological agents that enter the body through skin absorption. Other routes of exposure include inhalation and ingestion. Most mycotoxins act by inhibiting protein synthesis and respiration. Fungi toxins most likely to be used by terrorists include diacetoxyscirpenol (DAS), Nivalenol, 4-Deoxynivalenol (DON), and T2. Of those listed, T2 is the most likely candidate for terrorist use because of its stability. T2 could be aerosolized or used to contaminate food supplies.

Mycotoxins are fast-acting and may produce symptoms within minutes of exposure. Initial symptoms include burning skin pain, redness, tenderness, blistering, and progression to skin necrosis (tissue death), with leathery blackening and sloughing of large areas of skin in lethal cases. When inhaled, the symptoms include itching and pain, sneezing, epistaxis (bleeding from the nose), and rhinorrhea. Other symptoms include pulmonary/tracheobronchial toxicity by dyspnea, wheezing, and cough. Mouth and throat exposures are characterized by pain and blood-tinged saliva and sputum. When the toxin reaches the gastrointestinal tract, anorexia, nausea, vomiting, watery or bloody diarrhea, and abdominal cramp pain may occur.

SEB and ricin can cause similar systemic symptoms, however, neither of them produce eye or skin symptoms. If the eyes are exposed, eye pain, tearing, redness, foreign-body sensation, and blurred vision may result. Irrespective of the route of exposure, when the toxin reaches the rest of the body's systems, it may cause weakness, prostration, dizziness, ataxia, and loss of coordination. When victims have been exposed to lethal doses, tachycardia, hypothermia, and hypotension follow. Death may occur in minutes, hours, or days. No antidotes are known for mycotoxins. Treatment is supportive and symptomatic.

CHEMISTRY OF CLANDESTINE DRUG LABS

According to the DEA, "Clandestine drug labs are illicit operations consisting of chemicals and equipment necessary to manufacture controlled substances." Production of some substances, such as methamphetamine, PCP, MDMA, and methcathinone, requires little sophisticated equipment or knowledge of chemistry. Synthesis of other drugs, such as fentanyl and LSD, requires much higher levels of expertise and equipment. The clandestine drug problem continues to grow across the United States. No part of the country is spared as a potential site. Once-tranquil rural areas have become targets because of the remote locations. No emergency responders can safely say they will not find a clandestine drug lab in their jurisdiction. Unlike other hazardous materials locations, drug labs, because of their clandestine nature, do not have the usual signs or hints to the presence of hazardous materials. Some hints to the presence of illicit drug activity include:

- Mixing of unusual chemicals in a house, garage, or barn by persons not involved in the chemical industry
- Late-night secretive activity in a rural/farm area
- The possession of chemical glassware by someone not involved in the chemical field
- Possession of unusual chemicals, such as large quantities of methyl ethyl ketone (MEK), Coleman Fuel 7, toluene/paint thinner, acetone, alcohol, benzene, freon, chloroform, starting fluid, anhydrous ammonia, "Heet," white gasoline, phenyl-2-propane, phenylacetone, phenylpropanolamine, iodine crystals, red phosphorous, black iodine, lye (Red Devil Lye), muriatic/hydrochloric acid, battery acid/sulfuric acid, Epsom salts, batteries/lithium, sodium metal, wooden matches, propane cylinders, bronchodialators, rock salt, diet aids, energy boosters, or cold/allergy medications

Response personnel need to be alert to strong unusual odors, such as cat urine, ether, ammonia, acetone, or other chemicals; fast-burning fires; chemical containers and apparatus in places they should not be expected; houses with windows blacked out; excessive trash, including large amounts of items such as antifreeze containers, lantern fuel cans, red chemically stained coffee filters, drain cleaner, and duct tape. Clandestine drug labs have been found in houses, apartments, trailers, motels, mountain cabins, rural farms, and other occupancies.

Under non-fire conditions, health effects of drug labs that responders could be exposed to can be varied. They can range from respiratory problems, skin and eye irritation, headaches, nausea, and dizziness. Little is known about long-term effects of exposure. Cleanup of an abandoned lab is a hazardous materials team function, usually by a private contractor, but emergency services HazMat teams could be called upon to assist in evidence preservation and collection.

Anhydrous ammonia is a key ingredient in the illegal production of methamphetamines. Drug makers often steal ammonia from farms and agricultural supply companies. Leaks and releases of anhydrous ammonia have occurred as a result of the thefts from valves being left open. Ammonia has been transferred to inappropriate makeshift containers, such as propane tanks used on barbeque grills. These leaks add to the risks of emergency responders who are called upon to deal with the releases.

SUMMARY

Class 6.1 Poisons, when spilled, usually do not affect large segments of the population, as do the Class 2.3 Poison Gases. However, they can still present serious dangers to emergency responders. Several protective measures may minimize the effects of toxic materials. Antidotes are available for a small number of toxic materials, but they must be administered immediately after exposure. Your body has the ability to filter out some toxic materials through the normal process of eliminating wastes. Subclass 6.2 Infectious Substances are transported, used, and stored in small quantities. They can, however, create a significant hazard to responders and the public if mishandled. Protect yourself from toxic materials by wearing protective clothing and avoiding contact with toxic materials. Practice contamination prevention. Establish zones, deny entry, and provide protection to responders and to the public.

REVIEW QUESTIONS CHAPTER 8

1. A one-time, short-duration exposure to a toxic material is referred to as which of the following?
   A. Chronic exposure
   B. Subacute exposure
   C. Acute exposure
   D. None of the above
2. Multiple exposures, including 8 hours per day, 40 hours per week, are referred to as which of the following?
   A. Chronic
   B. Subacute
   C. Acute
   D. Mutagenic
3. List the four routes of exposure by which toxic materials may enter the body.
4. Long-term effects of exposure to toxic materials may cause which of the following?
   A. Cancer
   B. Skin rash
   C. Eye irritation
   D. Birth defects
5. Etiologic agents are which of the following?
   A. Chemicals
   B. Pesticides
   C. Living organisms
   D. None of the above
6. Which of the following is a rate of exposure to toxic materials?
   A. RADs
   B. Isomers
   C. Curies
   D. TLV-TWA

7. Protective measures against toxic materials include all but which of the following?
   A. Internal
   B. Vaccinations
   C. External
   D. Antidotes
8. Provide the names, structures, and hazards for the following toxic hydrocarbon and hydrocarbon-derivative compounds.

   C_7H_8 C_2H_5F CH_3OH $HCOOH$ $t\text{-}C_4H_9OH$

9. What are the three "signal words" associated with pesticide labels?
10. List four terms that express concentrations of toxic materials.

CHAPTER 9

Radioactive Materials

HISTORY OF RADIATION

Radiation is a phenomenon characterized more by its ability to cause biological effects than where it originates. Radiation was first discovered by German scientist Antoine Henri Becquerel, who received the Nobel Prize of Physics in 1903 for his work. Many of the terms associated with radioactivity come from those early pioneers in radiation physics: Wilhelm Conrad Roentgen (1845–1923) and Pierre (1859–1906) and Marie Curie (1867–1934), who also received the Nobel Prize in Physics in 1903 for their work on radiation. Ernest Rutherford (1871–1937) is considered the father of nuclear physics. He developed the language that describes the theoretical concepts of the atom and the phenomenon of radioactivity. Particles named and characterized by him include the alpha particle, beta particle, and proton. Rutherford won the Nobel Prize for Chemistry in 1909 for his work.

The effects of radiation have been studied for over 100 years. Scientists know a great deal about how to detect, monitor, and control even the smallest amounts of radiation. More is known about the health effects of radiation than any chemical or biological agent. Radiation is naturally occurring and is a part of everyday life. It is present in the earth's crust, travels from outer space, and is in the air and rocks. Radiation can also be manmade. Our way of life is characterized by the many uses of manmade radiation. Radioisotopes are used in medicine, scientific research, energy production, manufacturing, mineral exploration, agriculture, and consumer products. Radioisotopes can come from three sources. They are naturally occurring, such as radon in the air or radium in the soil. Linear accelerators and cyclotrons can produce radioisotopes, as can a nuclear reactor. Forty-seven nuclear reactors are licensed in research facilities around the United States by the NRC. They are primarily located in colleges and universities.

Radioactivity is caused by changes in the nucleus of the atom. Radioactivity is not a chemical activity, but rather it is a nuclear event. Chemical activity involves electrons orbiting around the nucleus of an atom, particularly the outer-shell electrons. It is within these outer-shell electrons that chemical reactions and chemical bonding take place. Radioactivity, on the other hand, involves the nucleus of the atom. There is normally a "strong force" that holds the nucleus of an atom together.

Determination of Radioactive Placarding and Labeling

Radioactive I:	≤0.5 Millirem per hour
Radioactive II:	≥0.5 – ≤50 Millirem per hour
Radioactive III:	≥50 – ≤200 Millirem per hour

All measurements at the surface of the package

Figure 9.1

There are some nuclei of elements that the force cannot hold together, and the nuclei begin to disintegrate. A basic law of nature says that unstable materials may not exist naturally for long. Unstable materials must do whatever they can to achieve stability. Radioactive elements throw off particles from the nucleus to reach stability. This throwing-off of particles is called radioactivity; the process is known as nuclear decay. This decay process is a random, spontaneous occurrence. There is no way that it can be shut off, nor is there any way to predict when a particular atom will begin to decay.

According to the DOT, a radioactive material is "any material having a specific activity greater than 0.002 microcuries per gram." Specific activity of a radionuclide includes "the activity of the radionuclide per unit mass of that nuclide." Simply stated, a microcurie is a measurement of radioactivity. When a radioactive material emits more than 0.002 microcuries per gram of material, which is a term of weight, the material is then regulated in transportation by the DOT.

Radiation is ionizing energy spontaneously emitted by a material or combination of materials. A radioactive material, then, is a material that spontaneously emits ionizing radiation. There are three types of DOT labels used to mark radioactive packages. They are Radioactive I, II, and III. These radioactive materials are determined by the radiation level at the package surface (see Figure 9.1). Radioactive III materials are the only radioactives that require placarding on a transportation vehicle.

Elements above lead (atomic numbers 83 and above) on the Periodic Table are radioactive (see Figure 9.2). Other elements may have one or more radioactive isotopes. Some elements occur naturally, while others are manmade. Each symbol on the Periodic Table represents one atom of that element. An atom is made up of a nucleus with varying numbers of electrons in orbits circling around the nucleus (see Figure 9.3). Located inside the nucleus are protons and neutrons. Protons in the nucleus of an atom represent the atomic number of that element. Neutron numbers may vary within the same type of element or from one element to another, but the number of protons must stay the same. The atom is the smallest part of an element that normally exists; so any particle of an element that is smaller than an atom is commonly referred to as a subatomic particle.

TYPES OF RADIATION

There are two types of radiation: ionizing and nonionizing. Ionizing radiation involves particles and "waves of energy" traveling in a wave-like motion. Examples

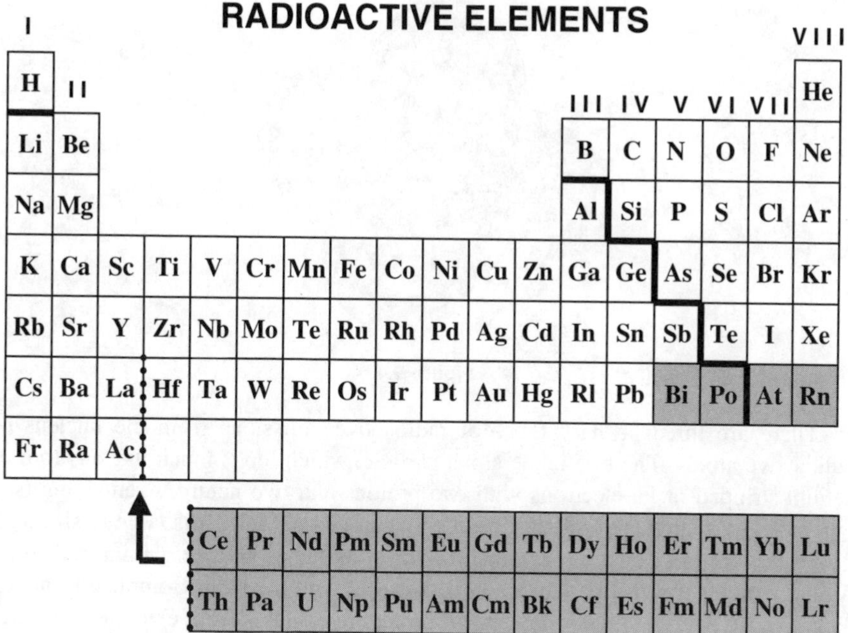

Figure 9.2

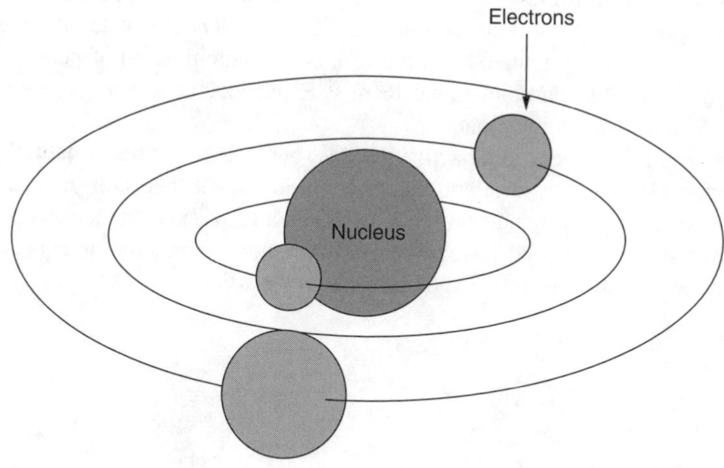

Figure 9.3

are alpha, beta, gamma, x-ray, neutron, and microwave. Nonionizing radiation is also made up of "waves of energy." Examples include ultraviolet, radar, radio, visible light, and infrared light. While all radioactive waves travel in a wave-like motion, all radioactivity travels in a straight line.

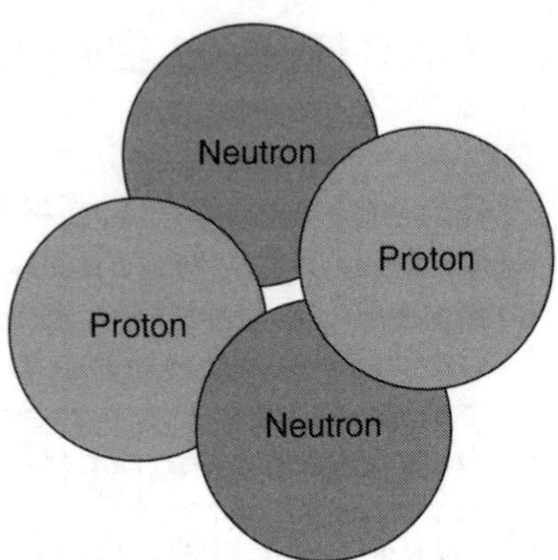

Figure 9.4

There are three primary types of radioactive emissions from the nucleus of radioactive atoms. The first is the alpha particle, which looks much like an atom of helium stripped of its electrons with two protons and two neutrons remaining (see Figure 9.4). An alpha particle is a positively charged particle. It is large in size and, therefore, will not penetrate as much or travel as far as beta or gamma radiation. Alpha particles travel 3 to 4 in and will not penetrate the skin. Complete turnouts, including SCBA, hood, and gloves, will protect responders from external exposure. However, if alpha emitters are ingested or enter the body through a broken skin surface, they can cause a great deal of damage to internal organs.

Beta particles are negatively charged and smaller, travel faster, and penetrate farther than alpha particles. A beta particle is 1/1800 the size of a proton, or roughly equal to an electron in mass (see Figure 9.5). Beta particles will penetrate the skin and travel from 3 to 100 ft. Full turnouts and SCBAs *will not* provide full protection from beta particles. Particulate radiation results in contamination of personnel and equipment where the particles come to rest. Electromagnetic energy waves, like gamma, do not cause contamination.

There is a third type of radioactive particle, but it does not occur naturally. The neutron particle is the result of splitting an atom in a nuclear reactor or accelerator, or it may occur in a thermonuclear reaction. When an atom is split, neutron particles are thrown out. You would have to be inside a nuclear reactor or experience a thermonuclear explosion to be exposed to neutron particles.

Figure 9.5

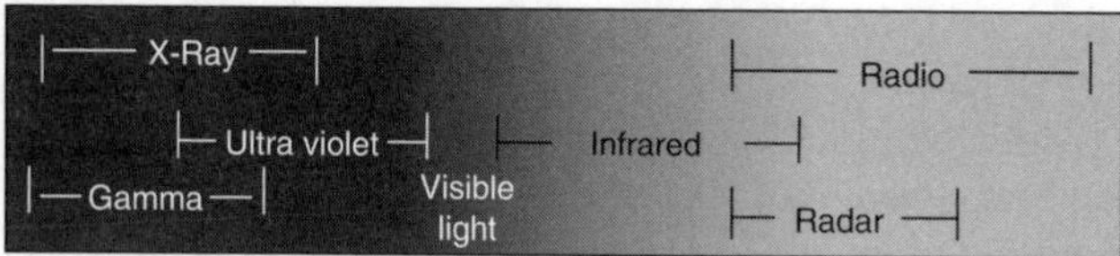

- Speed = 186,000 miles per second
- The energy of gamma is greater than visible light

Figure 9.6

Gamma radiation is a naturally occurring, high-energy electromagnetic wave that is emitted from the nucleus of an atom. It is not particulate in nature and has high-penetrating power. Gamma rays have the highest energy level known and are the most dangerous of common forms of radiation. Gamma rays travel at the speed of light, or more than 186,000 miles per second, and will penetrate the skin, can injure internal organs, and pass through the body. No protective clothing can protect against gamma radiation. Shielding from gamma radiation requires several inches of lead, other dense metal, or several feet of concrete or earth. Gamma radiation does not result in contamination because there are no radioactive particles, only energy waves. Examples of other electromagnetic energy waves include ultraviolet, infrared, microwave, visible light, radio, and x-ray (see Figure 9.6). There is little difference between gamma rays and x-rays; x-rays are produced by a cathode ray tube. To be exposed to x-rays, however, there has to be electrical power to the x-ray machine and the machine has to be turned on. If there is no electrical power, there is no radiation.

ISOTOPES

When the nucleus of an element contains more or fewer neutrons than a "normal" atom of that element, it is said to be an isotope of that element. All atoms have from 3 to 25 isotopes; the average is 10 per element (see Figure 9.7). *All isotopes are not radioactive.* Hydrogen has three important isotopes: Hydrogen 1, sometimes called protonium, has 1 proton in the nucleus and no neutrons. Hydrogen 2 has 1 proton and 1 neutron in the nucleus and is called deuterium, or heavy water. Hydrogen 1 and Hydrogen 2 are not radioactive. Hydrogen 3 has 1 proton and 2 neutrons in the nucleus and is called tritium. Hydrogen 3 is radioactive. Tritium is used in some "exit" signs, which gives them the glow-in-the-dark ability without batteries or any other electrical source.

Isotopes of Selected Elements

CARBON 12 (Normal)	HYDROGEN 1	Protium
CARBON 13	HYDROGEN 2 (heavy water)	Deuterium
CARBON 14	HYDROGEN 3	Tritium

Figure 9.7

Carbon also has several isotopes. "Normal" carbon is known as carbon 12. Carbon 12 has six protons and six neutrons in the nucleus and is not radioactive. Carbon 13 has six protons and seven neutrons in the nucleus and is not radioactive. Carbon 12 makes up 999 out of 1000 carbon atoms; the other is carbon 13. Carbon-14 has six protons and eight neutrons and is naturally radioactive. Carbon 14 is a beta emitter that is produced in the atmosphere by the action of cosmic radiation on atmospheric nitrogen. In the process, a proton is forced from the nucleus of nitrogen, which then becomes carbon 14. The human body is made up of 1/10,000 of 1% of carbon 14. You inhale carbon 14 every time you take a breath. It decays and is replaced in the body, so there is a constant supply. Carbon 14 is sometimes used to determine the age of organic materials, as it takes years to disappear. The amount of carbon 14 can be used to determine how long a person has been dead, or the age of a body or other object. There are three other carbon isotopes that are all manmade: carbon 11 has six protons and five neutrons, carbon 15 has 6 protons and 9 neutrons, and carbon 16 has 6 protons and 10 neutrons; all three are radioactive.

The DOT heavily regulates radioactive materials during transportation, and the Nuclear Regulatory Commission, which is part of the Department of Energy, regulates them in other instances. Design and construction of packaging for radioactive materials in transportation makes the likelihood of a release small. Packaging undergoes rigorous testing before it is approved for use with radioactive materials. Metal casks used for high-level radioactive materials have never been involved in an accident in which a serious release occurred. Most releases of radioactive materials are low-level in nature and often involve radioactive isotopes used for medical purposes.

INTENSITY OF RADIATION

Several terms are used to express the intensity of radiation (see Figure 9.8). "Radiation level" is a term often substituted for dose rate or exposure rate. It is generally referred to as the effect of radiation on matter; i.e., the amount of radiation that is imparted from the source and absorbed by matter due to emitted radiation per unit of time. The curie is a radiological term for the physical amount of a radioactive material. A curie consists of 37 billion disintegrations per second. It is a physical amount of material that is required to produce a specific amount of ionizing radiation:

Radiation Measurements

Curie = A physical amount of radioactive material
1 Megacurie (MCi) = 1,000,000 Curies
1 Kilocurie (kCi) = 1,000 Curies
1 Millicurie (mCi) = 0.0001 Curies
1 Microcurie (uCi) = 0.000001 Curies

Roentgen = Ionization per cm^3 of dry air

RAD = Radiation absorbed dose — Dosage

REM = Biological effects

Figure 9.8

1 millicurie = 0.001 curie and 1 microcurie = 0.000001 curie. Several hundred pounds of one radioactive material may be required to produce the same amount of curies as one pound of another radioactive material.

While a curie is a measure of the physical amount, the roentgen is a measure of the amount of ionization produced by a specific material. It is the amount of x-ray or gamma radiation that produces 2 billion ionizations in 1 cm^3 of dry air. A RAD is the radiation-absorbed dose (roughly equal to a roentgen). The radiation equivalent man (REM), also roughly equal to a roentgen, is a term for how much radiation has been absorbed, or the biological effect of the dose.

When radioactive materials are released, the human senses *cannot* detect radioactivity. The only way responders will know if radioactive materials are present is with the use of instruments specially designed to detect radioactivity. During the Cold War of the 1950s and 1960s, America's civil defense agencies distributed radiation meters around the country to measure radiation fallout from a thermonuclear detonation. Many of those meters have fallen into the hands of emergency responders because they were free and maintained by the civil defense agencies (a program that has since been abandoned).

There are two types of civil defense meters that have been widely available to emergency responders: the CD V-700 and the CD V-715. Neither of these instruments is able to detect alpha radiation. The CD V-700 survey meter has a range of 0 to 50 mR/h. An experienced meter operator can detect beta radiation with the CD V-700 through a process of elimination. Geiger-Mueller (GM) tubes are designed to survey for gamma radiation. If you check for radiation with the tube on the CD V-700 with the window closed, and no radiation exists beyond normal background, no gamma radiation is present. If the window is opened and another reading detects radiation, then it is beta radiation that is present. The CD V-715 survey meter has a range of 0.05 to 500 R/h or 50 to 500,000 mR/h. Radiation is detected through an ionization chamber much like the ionization chamber of a smoke detector.

Dosimeters are used in conjunction with survey meters to monitor the amount of radiation that personnel are exposed to over a given period of time. There are two types of civil defense dosimeters with different monitoring scales. The CD V-138 is used for monitoring relatively low levels of exposure and has a minimum-scale reading of 200 mR. The CD V-742 has a range up to 200 R (200,000 mR) and is used for high levels of personnel exposure. Both meters should be worn by responders to ensure proper protection.

Similar commercial radiation instruments are available, including those sensitive enough to detect alpha radiation. Health effects of exposure to radiation can vary (see Figure 9.9). Nonionizing radiation comes from ultraviolet and infrared energy waves. This type of radiation causes a sunburn type of injury. This is not a major concern for hazardous materials responders. Ionization damage occurs at the cellular level. Four types of short-term effects on the cells can occur:

No damage at all; the ionization passes through the cell
Damage occurs, but the damaged cells can be repaired
Irreparable damage to the cells that does not cause death, but the damage is permanent
Destruction of the cells

Health Effects of Exposure

25 REM	Maximum single lifetime exposure
20–100 REM	Chromosomal damage, alteration of white blood count
100–200 REM	Nausea and vomiting, WBC reduction
200–400 REM	Severe WBC reduction, hair loss, some death from infection
600–1,000 REM	50% Death in 30 days
1,000–2,000 REM	Death within 4–14 days
2,000 or more	Death (immediate)

Figure 9.9

There are also long-term effects from ionizing radiation. Exposures can cause cancer and birth defects of a teratogenic or mutagenic nature. Teratogenic birth defects result from the fetus being exposed to and damaged by radiation. The child is then born with some kind of birth defect as a result of the exposure. Providing no further exposures occur during pregnancy, the mother can have normal children again. Mutagenic damage occurs when the DNA or other part of the reproductive system is damaged by exposure to radiation. The ability to produce normal children is lost; the damage is permanent.

RADIATION EXPOSURE

Routes of entry for radioactive materials are much the same as for poisons. However, the radioactive source or material does not have to be directly contacted for radiation exposure to occur. Exposure occurs from the radiation being emitted from the radioactive source. Once a particulate radioactive material enters the body, it is dangerous because the source now becomes an internal source rather than an external one. You cannot protect yourself by time, distance, or shielding from a source that is inside your body. Contact with or ingestion of a radioactive material does not make you radioactive. Contamination occurs with radioactive particles, but with proper decontamination, these can be successfully removed. After they are removed, they cannot cause any further damage to the body.

Because radiation exposure can be cumulative, there are no truly safe levels of exposure to radioactive materials. Radiation does not cause any specific diseases. Symptoms of radiation exposure may be the same as those from exposure to cancer-causing materials. The tolerable limits for exposure to radiation that have been proposed by some scientists are arbitrary. Scientists concur that some radiation damage can be repaired by the human body. Therefore, tolerable limits are considered acceptable risks when the activity benefits outweigh the potential risks. The maximum annual radiation exposure for an individual person in the United States is 0.1 REM. Workers in the nuclear industry have a maximum exposure of 5 REMs per year. An emergency exposure of 25 REMs has been established by The National Institute of Standards and Technology for response personnel. This type of exposure should be attempted under only the most dire circumstances and should occur only once in a lifetime.

Radiation Sickness

25 REM	No detectable symptoms
50 REM	Temperature, blood count change
100 REM	Nausea, fatigue
200–250 REM	Fatal to some in 30 days, all sick
500 REM	1/2 dead in 30 days
600+ REM	All will die

Figure 9.10

Effects of exposure to radiation on the human body depend on the amount of material the body was exposed to, the length of exposure, the type of radiation, the depth of penetration, and the frequency of exposure. Cells that are the most susceptible are rapidly dividing cells, such as in the bone marrow. Children are more susceptible than adults, and the fetus is the most susceptible. Radiation injuries frequently do not present themselves for quite a long time after exposure. It can be years, or even decades, before symptoms appear. Cancer is one of the main long-term effects of exposure to radiation. Leukemia may take from 5 to 15 years to develop. Lung, skin, and breast cancer may take up to 40 years to develop. Figure 9.10 shows exposure rates and resulting radiation-sickness effects. Varying levels of illness will occur from radiation exposure depending on the dose. No detectable symptoms are the result of up to 25 REM. Elevated temperature and changes in blood count occur from 50 REM. Nausea and fatigue result from 100 REM. Two hundred to two-hundred fifty REM results in sickness to all exposed and death to some within 30 days. Five hundred REM results in the death of half of those exposed in 30 days. Exposure to over 600 REM results in the death of all exposed. Radiation burns are much like thermal burns, although they can be much more severe. First-degree radiation burns result from an exposure of 50 to 200 RADs, second-degree burns result from 500 RADs, and third-degree burns result from 1000 RADs.

Because of the physical characteristics of radioactive materials, protection for emergency responders can be provided by taking a few simple, protective actions, commonly referred to as time, distance, and shielding. **Time** refers to the length of exposure to a radioactive source and the half-life of a radioactive material. A half-life is the length of time necessary for an unstable element or nuclide to lose one half of its radioactive intensity in the form of alpha, beta, and gamma radiation. Half-lives range from fractions of seconds to millions of years. In 10 half-lives, almost any radioactive source will no longer put out any more radiation than normal background.

Distance is the second protective measure against radiation. As previously mentioned, radiation travels in a straight line, but only for short distances. Therefore, the greater the distance from the radioactive source, the less the intensity of the exposure will be. There is a law in dealing with radioactivity known as the "inverse-square law." This means that as the distance from the radioactive source is doubled, the radiation intensity drops off by one quarter. If the distance is increased 10 times, the intensity drops off to 1/100 of the original intensity (see Figure 9.11).

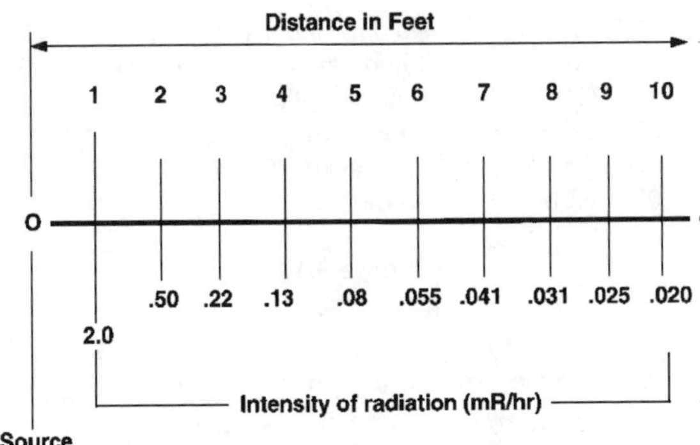

Figure 9.11

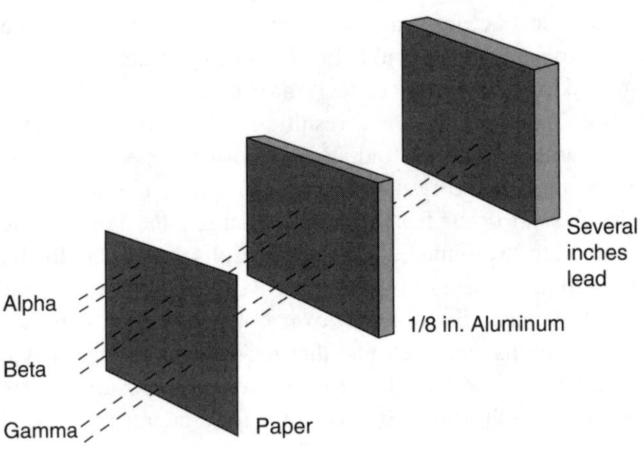

Figure 9.12

Shielding is the third protective measure against radiation. Shielding simply means placing enough mass between personnel and the radiation, which will provide protection from the radiation (see Figure 9.12). In the case of alpha particles, your skin or a sheet of paper will produce enough shielding. Turnouts will provide extra protection. Ingestion is the major hazard of radioactive particles, and wearing SCBA will prevent ingestion. Beta particles require more substantial protection from entering the body. A 1⁄24-in-thick piece of aluminum will stop beta radiation. Turnouts will not provide adequate protection. Gamma radiation requires 3 to 9 in of lead or several feet of concrete or earth.

RADIOACTIVE ELEMENTS AND COMPOUNDS

Radioactive materials are a part of everyday life. **Uranium, U**, is a radioactive metallic element. Uranium has three naturally occurring isotopes: uranium 234

(0.006%), uranium 235 (0.7%), and uranium 238 (99%). Uranium 234 has a half-life of 2.48×10^5 years; uranium 235 has a half-life of 7.13×10^8 years; uranium 238 has a half-life of 4.51×10^9 years. Uranium is a dense, silvery, solid material that is ductile and malleable; however, it is a poor conductor of electricity. As a powder, uranium is a dangerous fire risk and ignites spontaneously in air. It is highly toxic and a source of ionizing radiation. The TLV, including metal and all compounds, is 0.2 mg/m^3 of air. The four-digit UN identification number for uranium is 2979. Uranium is used in nuclear reactors to produce electricity and in the production of nuclear weapons systems.

Uranium Compounds

Uranium compounds are primarily used in the nuclear industry. Uranium has been used over the years for a number of commercial ventures, some successful and others not. Uranium dioxide was employed as a filament in series with tungsten filaments for large incandescent lamps used in photography and motion pictures. Uranium dioxide has a tendency to eliminate the sudden surge of current through the bulbs when the light is turned on, which extends the life of the bulbs. Some alloys of uranium were used in the production of steel; however, they never proved commercially valuable. Sodium and ammonium diuranates have been used to produce colored glazes in the production of ceramics. Uranium carbide has been suggested as a good catalyst for the production of synthetic ammonia. Uranium salts in small quantities are claimed to stimulate plant growth; however, large quantities are clearly poisonous to plants.

Uranium carbide, **UC_2**, is a binary salt. It is a gray crystal that decomposes in water. It is highly toxic and a radiation risk. Uranium carbide is used as nuclear reactor fuel.

Uranium dioxide, also known as yellow cake, has a molecular formula of **UO_2**. It is a black crystal that is insoluble in water. It is a high radiation risk and ignites spontaneously in finely divided form. It is used to pack nuclear fuel rods.

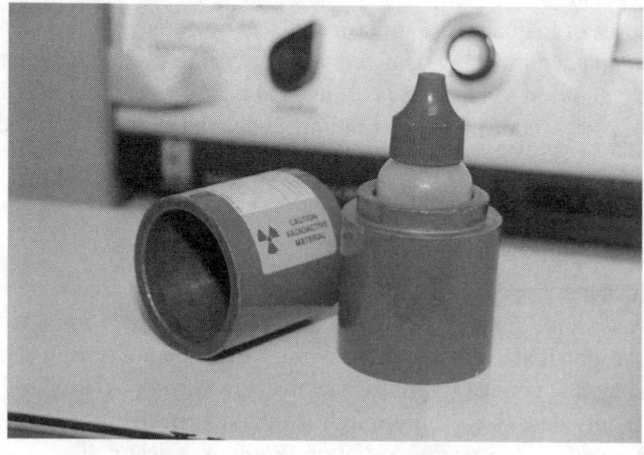

Radioactive-isotope lead container.

A nuclear reactor at a power plant can be a source of highly radioactive materials.

Uranium hexafluoride has a molecular formula of $\mathbf{UF_6}$. It is a colorless, volatile crystal that sublimes and reacts vigorously with water. It is highly corrosive and is a radiation risk. The four-digit UN identification number for fissile material containing more than 1% of uranium 235 is 2977; for lower specific activity, the number is 2978. Uranium hexafluoride is used in a gaseous diffusion process for separating isotopes of uranium.

Uranium hydride has a molecular formula of $\mathbf{UH_3}$ and is a brown-gray to black powder that conducts electricity. It is highly toxic and ignites spontaneously in air.

Uranium tetrafluoride, with the molecular formula of $\mathbf{UF_4}$, is a green, nonvolatile, crystalline powder that is insoluble in water. It is highly corrosive and is also a radioactive poison.

Medical uses of radioactive sources include sterilization, implants using radium, scans using iodine, and therapy using cobalt. X-rays are used in diagnostic medical procedures. In addition to medical facilities, radioactive materials may be found in research laboratories, educational institutions, industrial applications, and hazardous waste sites.

Radium Compounds

Radium, Ra, is a radioactive metallic element. There are 14 radioactive isotopes of radium; however, only radium 226, with a half-life of 1620 years, is usable. It is a brilliant, white solid that is luminescent and turns black upon exposure to air. Radium is water-soluble, and contact with water evolves hydrogen gas. It is in the alkaline-earth metal family and, like calcium, it seeks the bones when it enters the body. It is highly toxic and emits ionizing radiation. Radium is destructive to living tissue.

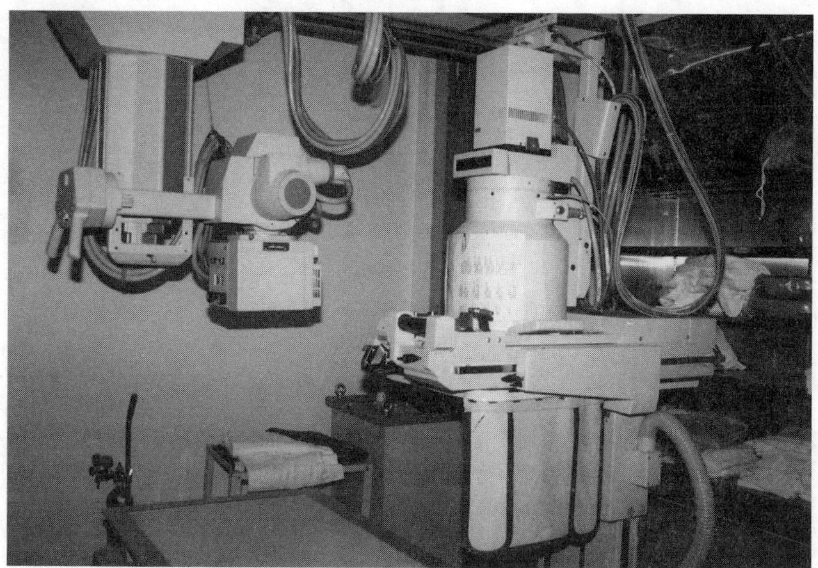

An x-ray machine at a hospital is a source of radiation only when the machine is turned on. There is no radioactive source when the power is off.

It is used in the medical treatment of malignant growths and industrial radiography. Compounds formed with radium all have the same hazards as radium itself. Most are used in the treatment of cancer and for radiography in the medical and industrial fields. The compounds are all solids, and the degree of water solubility varies. **Radium bromide** has a molecular formula of **$RaBr_2$**; it is composed of white crystals that turn yellow to pink. It sublimes at about 1650¡F and is water-soluble. The hazards are the same as for radium. It is used in the medical treatment of cancers.

Radium carbonate, with the molecular formula of **$RaCO_3$**, is an amorphous, radioactive powder that is white when pure. Because of impurities, radium carbonate is sometimes yellow or pink. It is insoluble in water.

Radium chloride, with the molecular formula of **$RaCl_2$**, is a yellowish-white crystal that becomes yellow or pink upon standing. It is radioactive and soluble in water. It is used in cancer treatment and physical research.

Cobalt

Cobalt, **Co**, is a metallic element. Cobalt 59 is the only stable isotope. Common isotopes are cobalt 57, cobalt 58, and the most common, cobalt 60. Cobalt is a steel-gray, shining, hard, ductile, and somewhat malleable metal. It has magnetic properties and corrodes readily in air. Cobalt dust is flammable and toxic by inhalation, with a TLV of 0.05 mg/m^3 of air. It is an important trace element in soils and animal nutrition. Cobalt 57 is radioactive. It has a half-life of 267 days. It is a radioactive poison and is used in biological research. Cobalt 58 is also radioactive and has a half-life of 72 days. It is a radioactive poison, and it is used in biological and medical research. Cobalt 60 is one of the most common radioisotopes. It has a half-life of

Refrigerator with radioactive isotopes inside.

5.3 years, is available in larger quantities, and is cheaper than radium. It is a radioactive poison and is used in radiation therapy for cancer and radiographic testing of welds and castings in industry. Compounds of cobalt are not radioactive.

Iodine

Iodine, **I**, is a nonmetallic element of the halogen family. There is only one natural stable isotope: iodine 127. There are many artificial radioactive isotopes. Iodine is a heavy, grayish-black solid or granules having a metallic luster and characteristic odor. It is readily sublimed to a violet vapor and is insoluble in water. Iodine is toxic by ingestion and inhalation and is a strong irritant to the eyes and skin, with a TLV ceiling of 0.1 ppm in air. Iodine 131 is a radioactive isotope of iodine and has a half-life of 8 days. It is used in the treatment of goiter, hyperthyroidism, and other disorders. It is also used as an internal radiation therapy source. Most iodine compounds are not radioactive.

Krypton

Krypton, **Kr**, is an elemental, colorless, odorless, inert gas. It is noncombustible, nontoxic, and nonreactive; however, it is an asphyxiant gas and will displace oxygen in the air. Krypton 85 is radioactive and has a half-life of 10.3 years. The four-digit UN identification number for krypton is 1056 as a compressed gas and 1970 as a cryogenic liquid. These forms of krypton are not radioactive. Radioactive isotopes of krypton are shipped under radioactive labels and placards as required. Its primary uses are in the activation of phosphors for self-luminous markers, detecting leaks, and in medicine to trace blood flow.

Radon

Radon, **Rn**, is a gaseous radioactive element from the noble gases in family eight on the Periodic Table. There are 18 radioactive isotopes of radon, all of which have short half-lives. For example, radon 222 has a half-life of 3.8 days. Radon is a colorless gas that is soluble in water. It can be condensed to a colorless transparent liquid and to an opaque, glowing solid. Radon is the heaviest gas known, with a density of 9.72 g/L at 32ºF. Radon is derived from the radioactive decay of radium. It is highly toxic and emits ionizing radiation. Lead shielding must be used in handling and storage. Radon has appeared naturally in the basements of homes, causing some concern for the residents. The primary uses are as a cancer treatment, a tracer in leak detection, in radiography, and in chemical research.

Radioactive materials are often found in transportation. They are heavily regulated, and the containers are well constructed. Most radioactive incidents are not handled by local emergency responders. Agencies other than fire, police, and EMS are responsible for response and handling of radioactive emergencies. Emergency responders must, however, be aware of radioactive materials and know how to protect themselves. Each state has radiological response teams for radioactive emergencies. They may be a part of the emergency management agency, the health department, the department of environment, or some other agency. Federal interests are represented by the United States Department of Energy and the Nuclear Regulatory Commission (301) 492–7000. Incidents involving weapons are handled by the Department of Defense Joint Nuclear Accident Center (703) 325–2102.

Radioactive materials may be found in many types of transportation vehicles, including taxi cabs, delivery trucks, and private cars.

Storage areas for radioactive materials are marked with this type of sign.

DEPARTMENT OF ENERGY NUCLEAR EMERGENCY SEARCH TEAM (NEST)

For response to emergencies involving nuclear incidents, the United States Department of Energy (DoE) has created a Nuclear Emergency Search Team (NEST). It is headquartered on a remote corner of Nellis Air Force Base near Las Vegas, Nevada. It is, in reality, the nation's "nuclear fire department," poised to respond to terrorist incidents involving nuclear devices or materials. Its job would be to locate the nuclear device or materials, determine what it is, render it safe, and get rid of it. Resources available to the team include satellites and radiation-sensing planes and helicopters. It is reported it can detect a radioactive particle or debris as small as a grain of salt. On the ground, resources include power generators, secure phone systems, and radiation monitors packed into nondescript vans. These vans can be loaded into a wide-body aircraft and flown anywhere in the country, or the world for that matter, within a moment's notice.

The team was organized in 1975, following a threat by an extortionist in Boston to set off a nuclear device. This team is comprised of over 200 searchers and is a joint effort of the FBI, DoD, and DoE. Since its inception, it has responded to dozens of incidents involving threats or actual use of radiological materials. Some of its responses included stolen plutonium from a Wilmington, North Carolina plant, recovery of radioactive debris from a fallen Soviet satellite, radioactive monitoring after the Three Mile Island Incident in Pennsylvania, and nuclear extortion attempts against Union Oil Company in Los Angeles and Harrah's Club in Reno, Nevada. The team has responded to threats of actual nuclear devices, but has never actually come face to face with one.

REVIEW QUESTIONS FOR CHAPTER 9

1. Radioactivity takes place in which part of the atom?
   A. Outer-shell electrons
   B. Electrons next to the nucleus
   C. In the nucleus
   D. In the protons
2. Which of the following best describes an isotope of an element?
   A. It is located in the nucleus
   B. More or less neutrons than normal
   C. More or less protons than normal
   D. Extra electrons
3. List the three major types of radioactivity.
4. Name the types of radiation that are particulate and create contamination.
5. Radiation always travels in the following manner.
   A. Up
   B. Down
   C. East to west
   D. In a straight line
6. Name the two types of radiation.
7. Match the following radiation terms with the correct statement.

A. Beta	High-energy wave
B. Alpha	Small, travels faster, and penetrates farther
C. Gamma	Large and travels short distances

8. Gamma radiation is considered to be which of the following?
   A. Limited hazard
   B. Particulate in nature
   C. Electromagnetic energy wave
   D. Easy to protect against
9. Which of the following radioactive-labeled materials require placarding on the transport vehicle regardless of the quantity?
   A. Radioactive II
   B. Radioactive I
   C. Radioactive III
   D. All of the above
10. List the three protective measures for radiation.

Chapter 10

Corrosives

Corrosives are the largest class of chemicals used by industry, so it stands to reason that they would frequently be encountered in transportation and at fixed facilities. DOT Class 8 materials are corrosive liquids and solids. There are no DOT subclasses of corrosives. There are, however, two types of corrosive materials found in Class 8: acids and bases. Acids and bases are actually two different types of chemicals that are sometimes used to neutralize each other in a spill. They are grouped together in Class 8 because the corrosive effects are much the same on tissue and metals, if contacted. It should be noted, however, that the correct terminology for an acid is *corrosive* and for a base is *caustic*. DOT, however, does not differentiate between the two when placarding and labeling. The DOT definition for a corrosive material is "a liquid or solid that causes visible destruction or irreversible alterations in human skin tissue at the site of contact, or a liquid that has a severe corrosion rate on steel or aluminum. This corrosive rate on steel and aluminum is 0.246 inches per year at a test temperature of 131°F."

A definition for an acid from the *Condensed Chemical Dictionary* is "a large class of chemical substances whose water solutions have one or more of the following properties: sour taste, ability to make litmus dye turn red and to cause other indicator dyes to change to characteristic colors, the ability to react with and dissolve certain metals to form salts, and the ability to react with bases or alkalis to form salts." It is important to note here that tasting any chemical is not an acceptable means of identification for obvious reasons.

In addition to being corrosive, acids and bases can explode or polymerize; they can also be water-reactive, toxic, flammable (applies to organic acids only, because inorganic acids do not burn), reactive, and unstable oxidizers.

There are two basic types of acid: organic and inorganic (see Figure 10.1). Inorganic acids are sometimes referred to as mineral acids. As a group, organic acids are generally not as strong as inorganic acids. The main difference between the two is the presence of carbon in the compound: inorganic acids do not contain carbon. Inorganic acids are corrosive, but they do not burn. They may, however, be oxidizers and support combustion, or may spontaneously combust with organic material.

Types of Acids

Inorganic	Organic
Sulfuric H_2SO_4	Formic HCOOH
Hydrochloric HCl	Acetic C_2H_5COOH
Nitric HNO_3	Propionic C_2H_5COOH
Phosphoric HPO_4	Acrylic C_2H_3COOH
Perchloric $HClO_4$	Butyric C_3H_7COOH

Figure 10.1

Inorganic-acid molecular formulas begin with hydrogen (H). For example, H_2SO_4 is the molecular formula for sulfuric acid, HCl is hydrochloric acid, and HNO_3 is nitric acid. Organic acids are hydrocarbon derivatives, therefore, they have carbon in the compound, and the name begins with the prefix indicating the number of carbons. For example, the prefix for a one-carbon compound with the organic acids is "form," so a one-carbon acid is called formic acid; a two-carbon acid is acetic acid; a three-carbon acid is propionic acid, etc. Organic acids are corrosive, may polymerize, and may burn.

INORGANIC ACIDS

Acids are materials that release hydrogen ions, H^+, when placed in water. Inorganic acids can generally be identified by hydrogen at the beginning of the formula (see Figure 10.2), because few other compounds begin with hydrogen. The hydrogen ion, H^+, consists of just a hydrogen nucleus, without electrons, and is composed of just one proton. Acids that supply just one H^+ are often referred to as monoprotic acids, e.g., HCl and HNO_3. Acids that supply more than one H^+ are referred to as polyprotic acids; more specifically, H_2CO_3 and H_2SO_4 are referred to as diprotic acids and H_3PO_4 as triprotic acids.

There are two general types of inorganic acids: binary and oxyacids. Binary acids are composed of just two elements: hydrogen and some other nonmetal, e.g., HCl and H_2S. These acids are named by placing the prefix "hydro" before and the suffix "ic" after the nonmetal element; the compound ends with the word "acid." For example, when hydrogen is combined with chlorine, the "ine" is dropped from

Inorganic Acids Begin with Hydrogen in Formula

Binary Acids	Oxy Acids
Hydrofluoric HF	Nitric HNO_3
Hydrochloric HCl	Perchloric $HClO_4$
Hydrobromic HBr	Sulfuric H_2SO_4
Hydrodic HI	Phosphoric H_3PO_4
Hydrosulfuric H_2S	Carbonic H_2CO_3

Figure 10.2

chlorine and the prefix "hydro" and suffix "ic" are added: hydrochloric acid; hydrogen combined with sulfur is called hydrosulfuric acid.

Acids that contain hydrogen, oxygen, and some other nonmetal element are called oxyacids, e.g., H_2SO_4, HNO_3, and $HClO_4$ (note the similarities to the oxyradicals). Like the oxysalts, these acids are named according to the number of oxygen atoms in the compound. The acid with the largest number of oxygen atoms in a series ends with the suffix "ic," and the one with the fewest number of oxygen atoms takes the suffix "ous" (similar to the alternate naming of the transitional metal salts discussed in Chapter 2).

For example, when hydrogen is combined with sulfur, the base state of the compound is SO_4 and the acid, H_2SO_4, is called sulfur*ic* acid. If there is one less oxygen present in the compound, such as SO_3, the ending changes to "ous" and the acid, H_2SO_3, is called sulfur*ous* acid. HNO_3 is nitric acid, HNO_2 is nitrous acid, etc. When halogens are present in the acid, the compound with the most oxygen atoms in the base state ends in "ic," such as chloric acid, $HClO_3$. If the oxygen is increased by one to $HClO_4$, the prefix "per" is added, yielding the name *per*chlor*ic* acid. The acid compound with the least number of oxygen atoms ends with "ous," such as chlorous acid, $HClO_2$. If the oxygen is reduced by one, to HClO, the prefix "hypo" is added, yielding the name *hypo*chlor*ous* acid.

STRENGTH AND CONCENTRATION

Most inorganic acids are produced by dissolving a gas or liquid in water, e.g., hydrochloric acid is derived from dissolving hydrogen chloride gas in water. All acids contain hydrogen. This hydrogen is the form of an ion (H^+) and can be measured by using the pH scale (see Figure 10.3). In simple terms, the pH scale measures the hydrogen-ion concentration of a solution. Concentrated acids and bases measure off the pH scale. The pH scale only measures acids and bases in solutions.

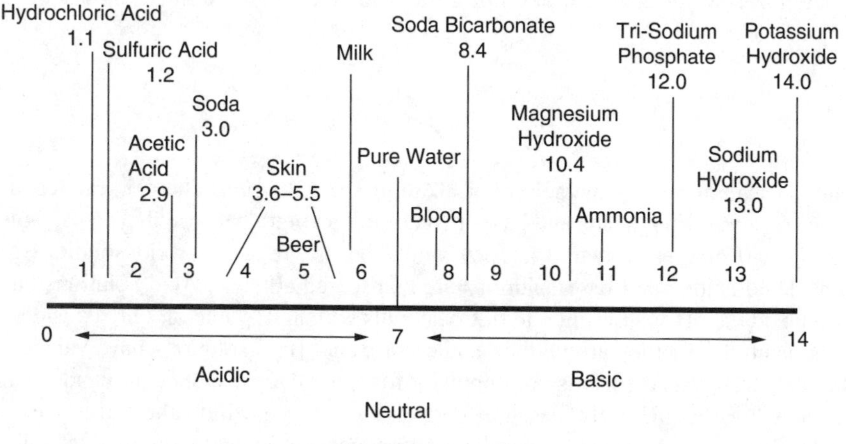

Figure 10.3

Ionization of Common Acids and Bases

Completely Ionized	Moderately Ionized	Slightly Ionized
Nitric	Oxalic	Hydrofluoric
Hydrochloric	Phosphoric	Acetic
Sulfuric	Sulfurous	Carbonic
Hydriodic		Hydrosulfuric
Hydrobromic		(Most others)
Potassium hydroxide		Ammonium hydroxide
Sodium hydroxide		(All others)
Barium hydroxide		
Strontium hydroxide		
Calcium hydroxide		

Figure 10.4

To determine if a concentrated material is an acid or a base, litmus paper is used; however, this does not yield a numerical value. As a group, acids have high hydrogen-ion concentrations. Bases have low hydrogen ion concentrations and high hydroxyl (OH^-) concentrations. The strength or weakness of an acid or base is the amount of hydrogen ions or hydroxyl ions that are produced as the acid or base is created. If the hydrogen ion concentration of an acid is high, the acid is concentrated. If the hydroxyl concentration is high, it is a concentrated base. In both cases, there is almost total ionization of the material dissolved in water to make the strong acid and base (see Figure 10.4). For example, hydrochloric acid is a strong acid because practically all of the hydrogen chloride gas is ionized in the water. Acetic acid is a weak acid because only a few molecules ionize in producing the acid.

Another term associated with corrosives is *concentration.* Concentration has to do with the amount of acid that is mixed with water and is often expressed in terms of percentages. A 98% concentration of sulfuric acid is 98% sulfuric acid and 2% water; a solution of 50% nitric acid is 50% nitric acid and 50% water. In the 50% concentration, the solution has only half the H^+ ions that the 100% concentration would have. A 50% concentration of nitric acid is a solution diluted to 50% of the original acid.

PH

The pH scale measures the acidity or alkalinity of a *solution.* The pH scale cannot measure some strong acids and bases that are full strength because they have values less than 0 or greater than 14. They would be off the scale. Acid solutions are considered acidic, and base solutions are considered alkaline. Acid solutions have a value on the pH scale from 1 to 6.9. Materials with a pH value of 7 are considered to be neutral, i.e., they are neither acidic nor basic. Base solutions have values on the scale from 7.1 to 14. It is not important for emergency responders to understand or know how the pH scale measures corrosivity or the specific values of any given acid or base. It is, however, important for responders to know that numerical values lower than 7 are acids and values higher than 7 are bases.

Exponential Logarithm of pH Values

pH Value	H^+ and OH^- Concentration
1	1,000,000
2	100,000
3	10,000
4	1,000
5	100
6	10
7	1
8	10
9	100
10	1,000
11	10,000
12	100,000
13	1,000,000

Figure 10.5

Usually, when dealing with numerical values, the higher the number, the greater the value that is being measured, and the number 2 is twice the value of 1. When using the pH scale for acids, however, the lower the pH number, the more acidic an acid solution is, and an acid solution with a pH of 1 is 10 times more acidic than an acid solution with a pH of 2, and so on (see Figure 10.5). The ratio and the intervals between the numbers are exponential, e.g., a pH of 5 is 10 times more acidic than a pH of 6, etc. The result of this exponential ratio is that on the full scale, a solution with a pH of 1 is 1,000,000 times more acidic than an acid with a pH of 6.9. So the difference between individual values on the pH scale is great and is one of the reasons why dilution and neutralization are not as simple as they might sound. Those terms will be discussed further under the section Dilution vs. Neutralization.

If the chemical name of a hazardous material is known and it is determined to be a corrosive, looking up the chemical name in reference sources will identify whether the material is an acid or a base. It will not be necessary for responders to get a pH measurement of the material unless it is to verify the reference information. The use of pH measurements can be useful when a material has not been positively identified. The pH measurement can be used to narrow the chemical-family possibilities in the identification process.

There are a number of ways for emergency responders to measure the pH of a corrosive material. First of all, the proper chemical protective clothing must be worn when working around corrosive materials. The simplest and least expensive method of determining pH is the use of pH paper, which changes color based on the type and strength of corrosive material that is present. The colored paper is then compared to a chart on the pH paper container. The chart indicates numerical pH values much the same way as do expensive measuring instruments. Although not as accurate as a pH meter, the numbers will give a "ballpark" measure of the pH of the material.

There are also commercially available pH meters, from handheld to sophisticated laboratory instruments. This equipment can be expensive, and pH paper is accurate enough for emergency response identification purposes. If the only information needed is whether a material is an acid or a base, litmus paper can also be used. Litmus paper turns blue if the corrosive material is a base, red if the corrosive material is an acid. The litmus paper will not give actual pH numerical values.

The definition of a base from the *Condensed Chemical Dictionary* is "a large class of compounds with one or more of the following properties: bitter taste, slippery feeling in solution, ability to turn litmus blue and to cause other indicators to take on characteristic colors, and the ability to react with (neutralize) acids to form salts." It is important to note that while the definition of acid and base mentions the taste and feeling of the materials, these are dangerous chemicals and can cause damage to tissues upon contact. Therefore, it is ***NOT*** recommended that responders come in contact with these materials through taste or touch!

Ionization occurs with the bases, just as with the acids, as they are made. Most bases are produced by dissolving a solid, usually a salt, in water. However, with the bases, the ion produced is the hydroxyl ion (OH^-). The base is considered strong or weak depending on the number of hydroxyl ions produced as a corrosive material is dissolved in water. A large OH^- concentration produces a strong base; a small OH^- concentration produces a weak base. Sodium and potassium hydroxide are strong bases; calcium hydroxide (hydrated lime) is a weak base. Bases will have a pH from 7.1 to 14 on the pH scale. The degree of alkalinity increases from 7.1 to 14, with 7.1 being the least basic and 14 the most basic. The amount of alkalinity between the numerical values on the pH scale is exponential, just as with the acids. A base with a pH of 9 is 10 times more basic than one with a pH of 8, and so on.

Corrosivity is not the only hazard of Class 8 materials. In addition to being corrosive, they may have other hazards, such as toxicity, flammability, or oxidation. Many corrosives, especially acids, can be violently water-reactive. Contact with water may cause splattering of the corrosive, produce toxic vapors, and evolve heat that may ignite nearby combustible materials. Some of the water may be turned to steam by the heat produced in the reaction. This can cause overpressurization of the container. Many corrosives may also be unstable, reactive, may explode, polymerize, or decompose and produce poisons.

Picric acid, $C_6H_2(NO_2)_3OH$, for example, becomes a high explosive when dried out and is sensitive to shock and heat. The hazard class for picric acid is 4.1 Flammable Solid. It is considered a wetted explosive. The name would indicate acid, however, the corrosivity of picric acid is far outweighed by its explosive dangers. The slightest movement of dry picric acid may cause an explosion. Picric acid, when shipped, is mixed with 12 to 20% water to keep it stable. When this water evaporates in storage over time, the material becomes explosive.

Perchloric acid, $HClO_4$, is a colorless, volatile, fuming liquid that is unstable in its concentrated form. It is a strong oxidizing agent and will spontaneously ignite upon contact with organic materials. It is corrosive, with the highest concentration of 70%. Contact with water produces heat; when shocked or heated, it may detonate. The boiling point is 66°F, and it is soluble in water, with a specific gravity of 1.77, which is heavier than water. The vapor density is 3.46, which is heavier than air. Perchloric acid is toxic by ingestion and inhalation. It is used in the manufacture of explosives and esters; in electropolishing and analytical chemistry; and as a catalyst.

Hydrocyanic acid, HCN, is corrosive in addition to toxic. It is also a dangerous fire and explosion risk. It has a wide flammable range of 6 to 41% in air. The boiling point is 79°F, the flash point is 0°F, and the ignition temperature is 1004°F. It is toxic by inhalation, ingestion, and through skin absorption. The TLV of hydrocyanic acid is 10 ppm in air. It is used in the manufacture of acrylonitrile, acrylates, cyanide salts, dyes, rodenticides, and other pesticides.

ORGANIC ACIDS

Organic acids are hydrocarbon derivatives. They are flammable, corrosive, and may polymerize by exposure to heat or sudden shock. Organic acids are "super-duper" polar materials; they are the most polar of the hydrocarbon derivatives. Organic acids have hydrogen bonding and a carbonyl that gives them a double dose of polarity. The functional group is represented by a carbon atom, two oxygen atoms, and a hydrogen atom. The general formula is **R-C-O-O-H**. One radical is attached to the carbon atom of the functional group.

Organic acids use the alternate prefix for one- and two-carbon compounds. When naming them, all of the carbons, including the one in the functional group, are counted to determine the hydrocarbon prefix name. To represent an acid, "ic" is added to the hydrocarbon prefix and the name ends in "acid," e.g., a one-carbon acid uses the alternate prefix name "form"; "ic" is added to "form," making it formic, and "acid" is added to the end: formic acid. A two-carbon acid uses the alternate name for two carbons, which is "acet," plus "ic," and ends with "acid": acetic acid. Naming three- and four-carbon acids reverts back to the normal prefixes for three- and four-carbon radicals, with some minor alterations to make the names flow more smoothly. For example, a three-carbon acid uses the prefix "prop," indicating three carbons; the letters "ion" are then added to make the name flow smoothly. The ending "ic" is added to the radical, and the word "acid" for the compound name: propionic acid. A four-carbon organic acid begins with the radical prefix "but"; the filler letters "yr" are attached; the radical ends with "ic"; acid is added, and the name for a four-carbon organic acid is butyric acid. Structures and molecular formulas for organic acids with one through four carbons are shown in Figure 10.6. Note that the carbon in the functional group is counted when determining which hydrocarbon radical is used in naming it.

It is also possible to add double-bonded radicals to the organic-acid functional group. For example, when the vinyl radical is attached to the carbon atom in the functional group, a three-carbon, double-bonded radical is created. The acryl radical is used for three carbons with a double bond; the ending "ic" is added to the radical, and the word "acid" is added to the end. The compound formed is acrylic acid. The double bond between the carbons can come apart in a polymerization reaction. Generally, materials that have double bonds are reactive in some manner. If polymerization occurs inside a container, an explosion may occur that can produce heat, light, fragments, and a shock wave.

Formic acid, HCOOH, is a colorless, fuming liquid with a penetrating odor. The highest commercial concentration is 90%. Formic acid, as well as all organic

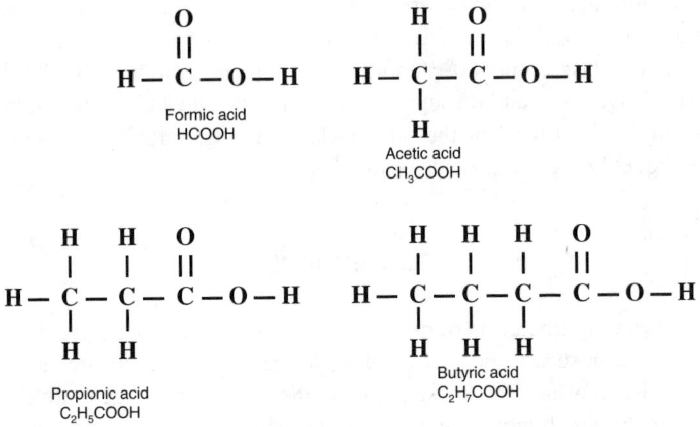

```
        O                H  O
        ||               |  ||
   H— C —O—H        H— C — C —O—H
                         |
                         H
```

Formic acid
HCOOH

Acetic acid
CH_3COOH

```
    H   H   O               H   H   H   O
    |   |   ||              |   |   |   ||
H— C — C — C —O—H      H— C — C — C — C —O—H
    |   |                   |   |   |
    H   H                   H   H   H
```

Propionic acid
C_2H_5COOH

Butyric acid
C_2H_7COOH

Figure 10.6

acids, are polar materials that are soluble in water, and it has a specific gravity of 1.2, which is heavier than water. As with many of the organic acids, formic acid is flammable. The boiling point is 213°F, the flash point is 156°F, and the flammable range is 18 to 57%. The ignition temperature is 1004°F, and the vapor density is 1.6, which is heavier than air. Formic acid is toxic, with a TLV of 5 ppm in air. The four-digit UN identification number is 1779. The NFPA 704 designation is health 3, flammability 2, and reactivity 0. It is used in the dyeing and finishing of textiles, the treatment of leather, and the manufacture of esters, fumigants, insecticides, refrigerants, etc.

Propionic acid, C_2H_5COOH, is a colorless, oily liquid with a rancid odor. It is a polar compound and soluble in water. Propionic acid is flammable, with a flammable range of 2.9 to 12% in air, a boiling point of 286°F, a flash point of 126°F, and an ignition temperature of 955°F. Polar-solvent foam will have to be used to extinguish fires. It is toxic, with a TLV of 10 ppm in air. The four-digit UN identification number for propionic acid is 1848. The NFPA 704 designation is health 3, flammability 2, and reactivity 0. It is used as a mold inhibitor in bread and as a fungicide, an herbicide, a preservative for grains, in artificial fruit flavors, pharmaceuticals, and others.

Butyric acid, also known as butanoic acid, is a colorless liquid with a penetrating, obnoxious odor. It is miscible in water, with a specific gravity of 0.96, which makes it slightly lighter than water. Butyric acid is flammable, with a boiling point of 326°F and a flash point of 161°F. The flammable range is 2 to 10% in air, and the ignition temperature is 846°F. Butyric acid is a strong irritant to skin and eyes. The four-digit UN identification number is 2820. The NFPA 704 designation is health 3, flammability 2, and reactivity 0. The primary uses of butyric acid are in the manufacture of perfume, flavorings, pharmaceuticals, and disinfectants.

Acrylic acid, C_2H_3COOH, is a colorless liquid with an acrid odor. It polymerizes readily and may undergo explosive polymerization. The boiling point is 509°F, the flash point is 122°F, and the ignition temperature is 820°F. The flammable range is

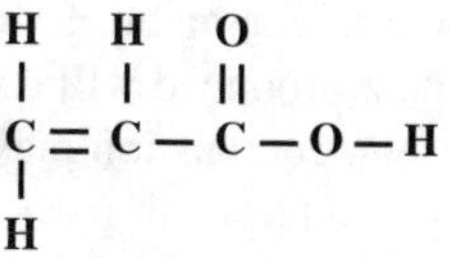

```
H   H   O
|   |   ||
C = C – C – O – H
|
H
```

Acrylic acid
C_2H_3COOH

Figure 10.7

2.4 to 8% in air. Acrylic acid is miscible with water and has a specific gravity of 1.1, which is slightly heavier than water. The vapor density is 2.5, which is heavier than air. It is an irritant and corrosive to the skin, with a TLV of 2 ppm in air. The four-digit UN identification number is 2218. Acrylic acid must be inhibited when transported. The NFPA 704 designation is health 3, flammability 2, and reactivity 2. The primary uses are as a monomer for polyacrylic and polymethacrylic acids and other acrylic polymers. The structure and molecular formula for acrylic acid are shown in Figure 10.7.

Phosphorus trichloride, PCl_3, is a clear, colorless, fuming, corrosive liquid. It decomposes rapidly in moist air and has a boiling point of about 168°F. PCl_3 is corrosive to skin and tissue and reacts with water to form hydrochloric acid. The TLV is 0.2 ppm, and the IDLH is 50 ppm in air. The four-digit UN identification number is 1809. The NFPA 704 designation is health 4, flammability 0, and reactivity 2. The white section at the bottom of the diamond contains a W with a slash through it, indicating water reactivity. The primary uses are in the manufacture of organophosphate pesticides, gasoline additives, and dyestuffs; as a chlorinating agent; as a catalyst; and in textile finishing.

Corrosives in contact with a poison may produce poison gases as the poison decomposes. In responding to an incident involving corrosives, the toxicity of the vapors could be much more of a concern for personnel than the corrosivity. When acids come in contact with cyanide, hydrogen cyanide gas, which is highly toxic, with a TLV of 10 ppm in air, is produced. The structure and molecular formula of phosphorous trichloride are shown in Figure 10.8.

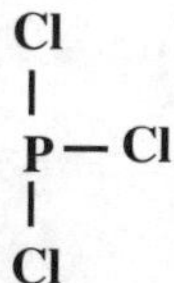

Phosphorus trichloride
PCl_3

Figure 10.8

When strong corrosives contact flammable liquids, the chemical reaction that occurs may produce heat. The heat produced will cause more vapor to be produced and, if an ignition source is present, combustion may occur. Corrosives may also be strong oxidizers. If they come in contact with particulate combustible solids, spontaneous combustion may occur. Once ignition has occurred, the corrosive will act as an oxidizer and accelerate the rate of combustion. Nitric acid in contact with combustible organic materials containing cellulose will produce a chemical reaction. This reaction will produce nitrocellulose, which is a dangerous fire and explosion risk. Toxic vapors may also be produced when the cellulose burns. After flammable liquids and gases, corrosives are the next most-common hazardous material encountered by emergency responders.

Sulfuric acid, H_2SO_4, is a strong corrosive, with a solution pH of 1.2. It is a dense, oily liquid, colorless to dark brown, depending on purity. Sulfuric acid is miscible in water, but violently water-reactive, producing heat and explosive splattering if water is added to the acid. The boiling point is 626°F, and the specific gravity is 1.84, which is heavier than water. Sulfuric acid is highly reactive and dissolves most metals. When in contact with metals, hydrogen gas is released. The vapors are toxic by inhalation, and the TLV is 1 ppm in air. Sulfuric acid is incompatible with potassium chlorate, potassium perchlorate, potassium permanganate, and similar compounds of other light metals. Sulfuric acid has a four-digit UN identification number of 1830. The NFPA 704 designation is health 3, flammability 0, and reactivity 2. The white section contains a W with a slash through it, indicating water reactivity. Sulfuric acid is used in batteries for cars and other vehicles. It is also used in the manufacture of fertilizers, chemicals, and dyes; as an etchant and a catalyst; in electroplating baths and explosives; in or for pigments; and many other uses.

Bulk container of sulfuric acid.

Railcar with liquid sulfuric acid.

Fuming sulfuric acid is also called oleum, which is a trade name. Fuming sulfuric acid is a solution of sulfur trioxide in sulfuric acid. Sulfur trioxide is forced into solution with sulfuric acid to the point that the solution cannot hold any more. As soon as the solution is exposed to air, the fuming begins, forming dense vapor clouds. It is violently water-reactive, as are most acids. The four-digit UN identification number is 1831.

Lime, CaO (also known as calcium oxide, quicklime, hydrated lime, and hydraulic lime), a binary oxide salt, is a white or grayish-white material in the form of hard clumps. It may have a yellowish or brownish tint due to the presence of iron. Lime is odorless and crumbles upon exposure to moist air. It is a corrosive caustic that yields heat and calcium hydroxide when mixed with water, and is a strong irritant, with a TLV of 2 mg/m^3 of air. The four-digit UN identification number for calcium oxide is 1910. The primary uses are in the manufacture of other chemicals, like calcium carbide; pH control; and in the neutralization of acid waste, insecticides, and fungicides.

Sodium hydroxide, NaOH, also known as caustic soda, is a hydroxide salt. Sodium hydroxide is a strong base and is severely corrosive, with a solution pH of 13. It is the most important industrial caustic material. Sodium hydroxide is a white, deliquescent solid found in the form of beads or pellets. It is also found in solutions with water of 50 and 73%. Sodium hydroxide is water-soluble, water-reactive, and absorbs water and carbon dioxide from the air. The specific gravity is 2.8, which is heavier than water. It is corrosive to tissues in the presence of moisture and is a strong irritant to eyes, skin, mucous membranes, and is toxic by ingestion. The TLV ceiling is 2 mg/m^3 of air. The four-digit UN identification number is 1823 for dry materials and 1824 for solutions. The NFPA 704 designation is health 3, flammability 0, and reactivity 1. It is used in the manufacture of chemicals, as a neutralizer in petroleum refining, in metal etching, electroplating, and as a food additive.

Tank car of sodium hydroxide solution, a corrosive material.

Phosphoric acid, **H_3PO_4**, is a colorless, odorless, sparkling liquid or crystalline solid, depending on concentration and temperature. Phosphoric acid has a boiling point of 410°F; at 20°C, the 50 and 75% concentrations are mobile liquids. The 85% concentration has a syrupy consistency, and the 100% acid is in the form of crystals. Phosphoric acid is water-soluble and absorbs oxygen readily, and the specific gravity is 1.89, which is heavier than water. It is toxic by ingestion and inhalation, and an irritant to the skin and eyes, with a TLV of 1 mg/m^3 of air. The four-digit UN identification number is 1805. The NFPA 704 designation is health 3, flammability 0, and reactivity 0. The primary use of phosphoric acid is in chemical analysis and as a reducing agent.

Sodium carbonate, **Na_2CO_3**, also known as soda ash and sodium bicarbonate, is an oxysalt and is a base with a pH of 11.6. It is not particularly hazardous and is used to neutralize acid spills.

Nitric acid, **HNO_3**, an inorganic acid, is a colorless, transparent, or yellowish, fuming, suffocating, corrosive liquid. Nitric acid will attack almost all metals. The yellow color results from the exposure of the nitric acid to light. Nitric acid is a strong oxidizer, is miscible in water, and has a specific gravity of 1.5, which is heavier than water. It may be found in solutions of 36, 38, 40, and 42, degrees B′e (specific gravity) and concentrations of 58, 63, and 95%. Nitric acid is a dangerous fire risk when in contact with organic materials. It is toxic by inhalation and is corrosive to tissue and mucous membranes. The TLV is 2 ppm in air. Nitric acid is incompatible with acetic acid, hydrogen sulfide, flammable liquids and gases, chromic acid, and aniline. The four-digit UN identification number for nitric acid at <40% concentration is 1760. The NFPA 704 designation for nitric acid at <40% concentration is health 3, flammability 0, and reactivity 0. There is not any information in the white area of the diamond for <40% concentrations. Below 40% concentration, nitric acid is not considered an oxidizer.

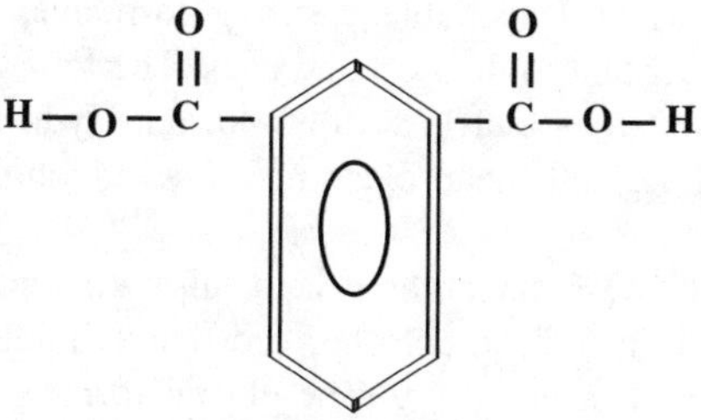

Terephthalic acid
TPA
$C_6H_4(COOH)_2$

Figure 10.9

The four-digit UN identification number for nitric acid at >40% is 2031. It is placarded as a Class 8 Corrosive; however, individual containers are labeled corrosive, oxidizer, and poison. The NFPA 704 designation for nitric acid at >40% concentration is health 4, flammability 0, and reactivity 0. The prefix "oxy" appears in the white section of the diamond. Nitric acid >40% concentration is an oxidizer.

Nitric acid is used in the manufacture of ammonium nitrate fertilizer and explosives, in steel etching, and in reprocessing spent nuclear fuel. There are two types of fuming nitric acid. **White fuming nitric acid** is concentrated with 97.5% nitric acid and less than 2% water. It is a colorless to pale-yellow liquid that fumes strongly. It is decomposed by heat and exposure to light and becomes red in color from nitrogen dioxide. **Red fuming nitric acid** contains more than 85% nitric acid, 6 to 15% nitrogen dioxide, and 5% water. The four-digit UN identification number for red fuming nitric acid is 2032. The NFPA 704 designation is health 4, flammability 0, and reactivity 1. The prefix "oxy" appears in the white section of the diamond. Red fuming nitric acid is considered an oxidizer. Both white and red fuming acids are toxic by inhalation, strong corrosives, and dangerous fire risks that may explode upon contact with reducing agents. They are used in the production of nitro compounds, rocket fuels, and as laboratory reagents.

Terephthalic acid (TPA), $C_6H_4(COOH)_2$, is an organic acid. It is a white crystalline or powdered material that is insoluble in water. It undergoes sublimation above 572°F. In addition to being corrosive, it is also combustible. The primary uses are in the production of polyester resins, fibers, and films; it is also an additive to poultry feeds. The structure and molecular formula for terephthalic acid are shown in Figure 10.9.

The naming of terephthalic acid does not follow any of the rules of naming organic acids under the trivial naming system. However, the formula and the structure indicate an organic compound and the name indicates acid. The hazards of the acids, except for flammability, are similar. The fact that the name indicates acid should lead you to assume flammability and toxicity, in addition to corrosiveness, until other information is known.

Hydrochloric acid, HCl, an inorganic acid, is a colorless or slightly yellow, fuming, pungent liquid produced by dissolving hydrogen chloride gas in water. Hydrochloric acid in solution has a pH of 1.1. The specific gravity is 1.19, which is heavier than water. It is water-soluble, a strong corrosive, and toxic by ingestion and inhalation. It is an irritant to the skin and eyes. The four-digit UN identification number is 1050 for anhydrous and 1789 for solution. Hydrochloric acid is used in food processing, pickling, and metal cleaning, as an alcohol denaturant, and as a laboratory reagent.

Acetic acid, CH_3COOH, an organic acid, also known as ethanoic acid and vinegar acid. Acetic acid is a clear, colorless, corrosive liquid with a pungent odor. In solution, acetic acid has a pH of 2.9. The glacial form is the pure form without water; it is 99.8% pure. Glacial acetic acid is a solid at normal temperatures. It is flammable, with a flash point of 110°F and a flammable range of 4 to 19.9%. The ignition temperature is 800°F. Acetic acid is "super-duper" polar and water-soluble. It will require polar-solvent type foam to extinguish fires. The specific gravity is 1.05, which is slightly heavier than water, but being miscible, it will mix rather than form layers. It is toxic by inhalation and ingestion, with a TLV of 10 ppm in air. Acetic acid is a strong irritant to skin and eyes. It is incompatible with nitric acid, peroxides, permanganates, ethylene glycol, hydroxyl compounds, perchloric acid, and chromic acid. The four-digit UN identification number is 2789. The NFPA 704 designation is health 3, flammability 2, and reactivity 0. It is a food additive at lower concentrations; it is used in the production of plastics, pharmaceuticals, dyes, insecticides, and photographic chemicals. The structure for acetic acid is shown in the Organic Acids section of this chapter.

Caustic potash, KOH, also known as potassium hydroxide and lye, is a hydroxide salt. It is found as a white solid in the form of pieces, lumps, sticks, pellets, or flakes. Potassium hydroxide may also be found as a liquid. It is water-soluble and may absorb water and carbon dioxide from the air. The specific gravity is 2.04,

Hydrofluoric acid in 55-gal drums.

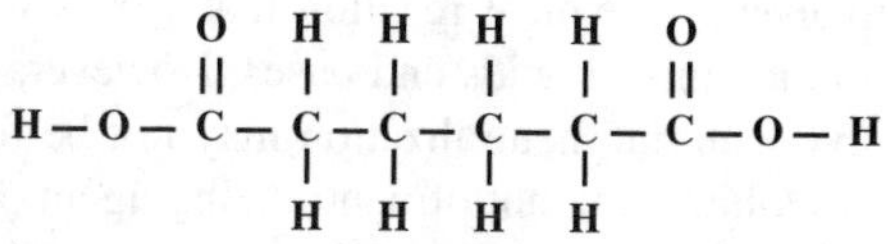

```
      O    H    H    H    H    O
      ||   |    |    |    |    ||
H—O— C — C — C — C — C — C—O—H
           |    |    |    |
           H    H    H    H
```

Adipic acid
(Hexanedioic acid)
$COOH(CH_2)_4COOH$

Figure 10.10

which is heavier than water; however, it is miscible in water, so it will mix rather than form layers. It is a strong base and is toxic by ingestion and inhalation. The TLV ceiling is 2 mg/m^3 of air. The four-digit UN identification number is 1813 for the solid and 1814 for the solution. The NFPA 704 designation is health 3, flammability 0, and reactivity 1. It is used in soap manufacture, bleaching, as an electrolyte in alkaline storage batteries and some fuel cells, as an absorbent for carbon dioxide and hydrogen sulfide, and in fertilizers and herbicides.

Adipic acid, also known as hexanedioic acid, is an organic acid. It is a white, crystalline solid that is slightly soluble in water. In addition to being a corrosive, it is also flammable; however, it is a relatively stable compound. Adipic acid is used in the manufacture of nylon and polyurethane foams. It is also a food additive and adhesive. The structure and molecular formula are shown in Figure 10.10; notice that the structure has two organic-acid functional groups attached.

The naming of adipic acid does not follow any of the rules of naming organic acids under the trivial naming system. However, the formula and the structure indicate an organic acid, and the name indicates acid. The hazards of the acids, except for flammability, are similar. The fact that the name indicates acid should lead you to assume flammability and toxicity, in addition to corrosiveness, until other information is known.

DILUTION VS. NEUTRALIZATION

Dilution and neutralization are often tactics considered when dealing with spills of corrosive materials. **Dilution** involves placing water into the acid to reduce the pH level. The addition of water to a corrosive can create a dangerous chemical reaction. Acids are highly water-reactive, creating vapors, heat, and splattering. With dilution, you must consider the exponential values of the numbers on the pH scale. Just moving the pH from 1 to 2 on the scale will take an enormous amount of water. Dilution may not be a practical approach for large spills. For example, if a 2000-gal spill of concentrated hydrochloric acid occurs, enough water to dilute the material to a pH of 6 would require the following efforts: one 1000-gpm pumper, pumping 24 hours a day, 7 days a week, 365 days a year, for 64 years. This would produce 1,440,000 gal of water per day! A large reservoir would be required to hold the water. As the process proceeds, it would become necessary to stir the mixture of water and acid to ensure uniformity in the dilution process. Dilution may work on small spills, but it will not work well on large spills.

Neutralization involves a chemical reaction that works well under laboratory conditions, using small amounts of acids and bases. However, in the field, facing a large spill of a corrosive material, neutralization may not be feasible. The neutralization reaction requires a large amount of neutralizing agent. For the same spill of 2000 gallons of concentrated hydrochloric acid mentioned in the previous example, it would require 8.7 tons of sodium bicarbonate, 5.5 tons of sodium carbonate, or 4.15 tons of sodium hydroxide to neutralize the spill. The latter would not be recommended, because sodium hydroxide is a strong base and would be dangerous to work with by itself without trying to add it to a concentrated acid. There would also be a need for a method to apply the neutralizing agent. The reaction that occurs will be a violent one, producing heat, vapor, and splattering of product. Neutralization may not work well for emergency responders at the scene of an incident with a large spill. The method of choice may turn out to be one of cleaning up the product by a hazardous waste contractor. They may use vacuum trucks; absorbent, gelling materials; or neutralization to accomplish the task.

The main danger of corrosive materials to responders is the contact of these materials with the body. Corrosive materials destroy living tissue. Destruction begins immediately upon contact. Many strong acids and bases will cause severe damage upon contact with the skin. Weaker corrosives may not cause noticeable damage for several hours after exposure. A chemical burn is nine times more damaging than a thermal burn. There are four basic methods of reducing the chemical action of corrosives on the skin: physical removal, neutralization, dilution, and flushing. Flushing is the method of choice. Removal of a corrosive material is difficult to accomplish and may leave a residue behind. Neutralization is a chemical reaction that may be violent and produce heat. This type of reaction on body tissues may cause more damage than it prevents. Neutralization should not be attempted on personnel wearing chemical suits, for the same reason as mentioned above. The layer of chemical protection is thin, and the heat from the neutralization may melt the suit and cause burns to the skin below the suit.

Dilution takes a large amount of water to lower the pH to a neutral position. While dilution may be similar to flushing, the intended outcome is different. With dilution, the goal is to reduce the pH number to as near neutral as possible. With flushing, the goal is to remove as much of the material as possible with a large volume of water.

Flushing should be started as soon as possible to reduce the amount of chemical damage and should continue for a minimum of 15 min. This also applies to the eyes. Most corrosives are highly water-soluble. Contact lenses should not be worn by personnel at HazMat incident scenes. Contact with acids can "weld" the contact to the eye, which almost always produces blindness. The person being treated may be in a great deal of pain and may have to be restrained during the flushing operation. Treatment after flushing involves standard first-aid for burns.

Corrosives are transported in MC/DOT 312/412 tanker trucks. These trucks have a small-diameter tank with heavy reinforcing rings around the circumference of the tank. The tank diameter is small because most corrosives are heavy. No other type of hazardous material is carried in this type of tanker. The 312/412 is a corrosives tanker regardless of how it is placarded. The placard may indicate a poison, an

MC/DOT 312/412 tanker for heavy corrosive materials.

oxidizer, or a flammable; but do not forget the "hidden hazard": the tank identifies corrosives. Lighter corrosives may also be found in MC/DOT 307/407 tankers and may be placarded corrosive, flammable, poison, and oxidizer. Corrosives may also be found in tank cars, intermodal containers, and varying sizes of portable containers. Portable containers may range from pint and gallon glass bottles to stainless steel carboys and 55-gal drums. Some are also shipped in plastic containers.

INCIDENTS

Emergency responders should have a thorough knowledge of corrosive materials. After flammable liquids and gases, corrosives are the most frequently encountered hazardous material. Responders should have proper chemical protective equipment and SCBA to deal safely with corrosive materials. Firefighter turnouts will not provide protection from corrosives. The most common exposures are contact with the hands and feet, and inhalation of the vapors. Make sure that the chemical suits chosen for use are compatible with the corrosive material. No suit will protect you from chemicals indefinitely; they all have breakthrough times. Make sure personnel are rotated to avoid prolonged exposure, and make sure they do not contact the material unless absolutely necessary. Safety should be your primary concern.

In California, an MC/DOT 312/412 tanker truck developed a leak along an interstate highway. On arrival, responders found a reddish-brown vapor cloud coming from the tank. The shipping papers indicated that the load was spent sulfuric acid; however, the color of the vapor coming from the trailer was in conflict with that information. As it turns out, the driver was hauling spent sulfuric acid, but had room to pick up some nitric acid and put it in the same tank with the sulfuric acid. The nitric acid was not compatible with the tank and ate through it quickly. The entire load of acid was spilled onto the highway when the tank failed. Certain hazardous materials have specific colors, and responders should be aware of these colors.

A tank car placarded "empty," which contained an estimated 800 gal of anhydrous hydrogen fluoride, a corrosive liquid, was found leaking in a rail yard. "Empty" or "residue" placarded tank cars, as they are now called, can have as much as 3000 gallons of product still in the tank if it has not been purged. Responders attempted to control the leak over a 4-hour period. In the meantime, a vapor cloud formed and traveled approximately 2.5 miles downwind. This forced the evacuation of 1500 people from a 1.1-sq-mile radius around the leaking tank car for 9 hours. Local hospitals treated approximately 75 people for minor skin and eye irritations.

REVIEW QUESTIONS FOR CHAPTER 10

1. List the two types of chemicals that make up the DOT corrosive class.
2. List the two types of acids.
3. The strength of an acid is a result of passing a gas through water. Which of the following terms reflects the name of the process?
   A. Radiation
   B. Sublimation
   C. Ionization
   D. Cationazation
4. Concentration of an acid is an expression of the relationship between acid and what other material?
   A. Alcohol
   B. Water
   C. Gas
   D. Dissolved solid
5. A corrosive hazardous material with a pH of 7.0 is considered to be:
   A. Weak
   B. Acidic
   C. Basic
   D. Neutral
6. A corrosive hazardous material with a pH of 2.3 is considered to be:
   A. Acidic
   B. Basic
   C. Neutral
   D. Weak
7. A corrosive hazardous material with a pH of 8.7 is considered to be:
   A. Neutral
   B. Acidic
   C. Basic
   D. Weak
8. List two procedures that can be used to reduce the corrosive effects of an acid.
9. Provide the names and structures for the following organic acids.

   CH_3COOH C_2H_5COOH C_3H_7COOH

10. List the formulas and names for the following organic acid structures.

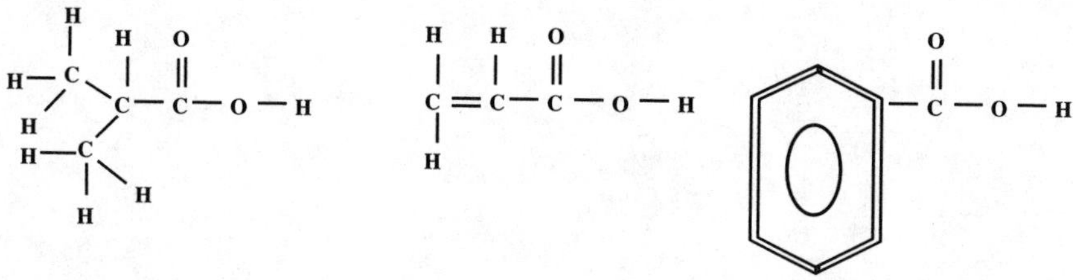

Figure 10.11

CHAPTER 11

Miscellaneous Hazardous Materials

Miscellaneous hazardous materials in DOT/UN Class 9 are defined as "a material which presents a hazard during transportation, but which does not meet the definition of any other hazard class." Other hazards might include anesthetic, noxious (harmful to health), elevated temperature, hazardous substance, hazardous waste, or marine pollutant. They may be encountered as solids of varying configurations, gases, and liquids. Examples include asbestos, dry ice, molten sulfur, and lithium batteries. These materials would be labeled and placarded with the Class 9 Miscellaneous Hazardous Materials placard, which is white with seven vertical black stripes on the top half.

Also included in the Miscellaneous Hazardous Materials class is "Other Regulated Materials ORM-D, Consumer Commodities." They are "materials that present a limited hazard during transportation due to the form, quantity, and packaging." Some of these materials, if they were shipped in tank or box truck quantities, would fit into another hazard class. However, because the individual packaging quantities are so small, the DOT considers the hazard is limited and they are labeled ORM-D. Generally, these ORM-D materials are destined for use in the home, industry, and institutions. The materials are in small containers, including aerosol cans, with a quantity that is usually a gallon or less.

Caution should be observed if fire is involved in an incident, since many small containers can become projectiles as pressure builds up inside from the heat and the containers explode. Aerosol cans may be particularly dangerous because they are already pressurized, and exposure to heat can cause them to explode and rocket from the pressure. Those materials used in industry and institutions are usually service products used in cleaning and maintenance rather than in industrial chemical processes.

Examples of ORM-D materials include low-concentration acids, charcoal lighter, spray paint, disinfectants, and cartridges for small firearms. Even though the container sizes may be small, the products inside can still cause contamination of responders or death and injury if not handled properly.

Intermodal container of miscellaneous hazardous materials.

There is no one specific hazard that can be attributed to Class 9 materials. The hazards will vary, and may include all of the other eight hazard classes. The physical and chemical characteristics mentioned in the first nine chapters of this book may be encountered with Class 9 materials. The difference is that the quantities may be small, or the materials may be classified as hazardous wastes, which can be almost any of the other hazard classes. With miscellaneous hazardous materials, it is important to obtain more information about the shipment to determine the chemical names and the exact hazards of the materials involved.

Class 9 placards on transportation vehicles may include a four-digit UN identification number. The corresponding information in the *North American Emergency Response Guide (NAERG)* may not give detailed names of the materials. There may be generalizations, such as “hazardous substance, n.o.s (not otherwise specified).” When the material is not specifically identified in the *NAERG,* the shipping papers or other sources will have to be consulted to determine the exact hazard of the shipment.

ELEVATED-TEMPERATURE MATERIALS

In addition to the Class 9 placard, a second placard may appear next to it with the word HOT. The word may also appear outside of a placard by itself. It indicates that the material inside has an elevated temperature that may be a hazard to anyone who comes in contact with it. An elevated-temperature material is usually a solid that has been heated to the point that it melts and becomes a molten liquid. The change is in physical state only; the chemical characteristics of the material remain

Electrical transformers were once a primary source of PCBs, a miscellaneous hazardous material.

the same. There may, however, be vapors produced from molten materials that are not present in the solid form. These vapors may be flammable or toxic. Water in contact with molten materials can cause a violent reaction and instantly turn to steam. If this happens inside a container, the pressure buildup from the steam can cause a boiler-type explosion that has nothing to do with the characteristics of the chemical inside. The steam, which is a gas, builds up pressure inside what is usually a nonpressure container. When the container can no longer withstand the pressure, it fails. The molten material inside may be splattered around by the explosion.

There are two molten materials specifically listed in the Hazardous Materials Tables in 49 CFR: molten aluminum and molten sulfur. **Molten aluminum** has a four-digit UN identification number of 9260. The *NAERG* refers to Guide 169 for hazards of the material. Molten aluminum is the only material that refers to this guide. The guide indicates that the material is above 1300°F and will react violently with water, which may cause an explosion and release a flammable gas.

Molten materials in contact with combustible materials may cause ignition if the molten material is above the ignition temperature of the combustible. For example, gasoline has an average ignition temperature of around 800°F. Diesel fuel has an average ignition temperature around 400°F, depending on the blend and additives. In an accident, gasoline or diesel fuel could be spilled. The molten material could be an ignition source for the gasoline or diesel fuel with which it comes in contact. When contacting concrete on a roadway or at a fixed facility, molten materials could cause spalling and small pops. This could cause pieces of concrete to become projectiles. Contact with the skin would cause severe thermal burns. There is no personnel protective clothing that adequately protects responders from contact with molten materials.

Other molten materials are not as hot as molten aluminum. **Sulfur,** in the molten state, refers you to Guide 133 in the *ERG.* Molten sulfur, with the four-digit UN identification number 2448, may ignite combustible materials that it comes in contact with if it is above the ignition temperature of the material. Molten sulfur has a melting point of approximately 245°F. The molten sulfur in transportation would be above that temperature, but not as hot as molten aluminum. Contact would still cause severe thermal burns, and the vapor is toxic.

Hot asphalt in the liquid form can also cause combustion of combustible materials and severe thermal burns. Asphalt refers you to Guide 130 in the *NAERG* for hazard information. Asphalt has a boiling point of >700°F and a flash point of >400°F. The ignition temperature is 905°F. Fires involving asphalt should be fought with care. Water may cause frothing, as it does with all combustible liquids with flash points above 212°F. This does not mean that water should not be used, but be aware that the frothing may be violent and the water contacting the molten material may also cause a reaction. Asphalt has a four-digit UN identification number of 1999 for all forms. The NFPA 704 designation for asphalt is health 0, flammability 1, and reactivity 0.

OTHER MISCELLANEOUS HAZARDOUS MATERIALS

White, gray, green, brown, and blue **asbestos** are impure magnesium silicate minerals that occur in fibrous form. Asbestos is noncombustible and was used extensively as a fire-retardant material until it was found to cause cancer. Asbestos is highly toxic by inhalation of dust particles. The four-digit UN identification number for white asbestos is 2590. The primary uses of asbestos are in fireproof fabrics, brake linings, gaskets, as a reinforcing agent in rubber and plastics, and as cement reinforcement. Many uses of asbestos are banned because of the cancer danger of the material.

Ammonium nitrate fertilizers that are not classified as oxidizers are classified as miscellaneous hazardous materials. This type of fertilizer has other materials in the mixture, and there are controlled amounts of combustible materials. Mixtures of ammonium nitrate, nitrogen, and potash that are not more than 70% ammonium nitrate and do not have more than 0.4% combustible material are included as a miscellaneous hazardous material. Additionally, ammonium nitrate mixtures with nitrogen and potash, with not more than 45% ammonium nitrate, may have combustible material that is unrestricted in quantity. The four-digit UN identification number for these mixtures of ammonium nitrate fertilizer is 2071.

Solid carbon dioxide, also known as dry ice, presents a danger in transport because carbon dioxide gas is produced as it warms. This warming is much like melting ice, although no liquid is formed in the case of dry ice. While carbon dioxide gas is nontoxic, it is an asphyxiant and can displace oxygen in the air or in a confined space. It is nonflammable; in fact, carbon dioxide gas is used as a fire-extinguishing agent. The four-digit UN identification number for solid carbon dioxide is 1845.

Box truck with over pack drum underneath, usually hauling hazardous waste.

Solutions of formaldehyde, 30 to 50%, such as those used in preservatives, are listed as miscellaneous hazardous materials. These solutions are nonflammable, and the toxicity is below the requirements for a poison liquid. However, the material may still be carcinogenic. Formaldehyde solutions usually contain up to 15% methanol to retard polymerization. The four-digit UN identification number for nonflammable solutions is 2209.

Polychlorinated biphenyls (PCBs) are composed of two benzene rings attached together with at least two chlorine atoms in the compound. PCBs were widely used in industry since 1930 because of their stability; however, it was this same stability that led to their downfall. They are highly toxic, colorless liquids with a specific gravity of 1.4 to 1.5, which is heavier than water. They are known carcinogens. In the human body, they tend to settle in the liver and fat cells, where they stay for a long period of time. They are not biodegraded and remain as an ecological hazard through water pollution. The only known way to remove PCBs from the environment is high-temperature incineration (at least 2200°F) for a proper length of time. The manufacture was discontinued in the United States in 1976. The material that remains is considered hazardous waste and is shipped as a miscellaneous hazardous material. The structure for PCB is shown in Figure 11.1.

Batteries containing lithium are listed as miscellaneous hazardous materials. The storage batteries are composed of lithium, sulfur, selenium, tellurium, and chlorine. These batteries have four-digit UN identification numbers assigned depending on the use and composition of the battery. Lithium batteries contained in some kind of equipment have the four-digit number 3091. Batteries with liquid or solid cathodes, not in any kind of equipment, are given the number 3090.

Cl — — Cl

Polychlorinated biphenyl
$(C_6H_5)_2(Cl)_2$

Figure 11.1

Other miscellaneous hazardous materials listed in CFR 49 Hazardous Material Tables include solid materials and fish meal or fish scrap that has been stabilized. Fish meal is subject to spontaneous heating. These materials are given the four-digit UN identification number of 2216. Castor beans, meal, or flakes may also undergo spontaneous heating. The four-digit UN identification number is 2969. Additional materials include cotton (wet) 1365, polystyrene beads 2211, lifesaving appliances, self-inflating 2990, and not self-inflating 3072, environmentally hazardous liquids or substances 3077, hazardous waste liquid 3082, and hazardous waste solid 3077. Additionally, self-propelled vehicles, including internal combustion engines or other apparatus containing internal combustion engines, or electric storage batteries are regulated. Self-propelled vehicles include electric wheelchairs with spillable or non-spillable batteries.

Highway tanker of molten sulfur.

Titanium dioxide, TiO_2, is a white powder and has the greatest hiding power of all white pigments. It is noncombustible; however, it is a powder and, when suspended in air, may cause a dust explosion if an ignition source is present. It is not listed in the DOT Hazardous Materials Table, and the DOT does not consider it hazardous in transportation. The primary uses are as a white pigment in paints, paper, rubber, and plastics, in cosmetics, welding rods, and in radioactive decontamination of the skin.

Sodium silicate, $2Na_2OSiO_2$, is the simplest form of glass. It is found as lumps of greenish glass soluble in steam under pressure, white powders of varying degrees of solubility, or liquids cloudy or clear. It is noncombustible; however, when the powdered form is suspended in air, it could cause a dust explosion if an ignition source is present. Breathing the dust may also cause health problems. The glass form could also create a hazard to responders in an accident. It is not listed as a hazardous material in the DOT Hazardous Materials Tables. The primary uses are as catalysts, soaps, adhesives, water treatment, bleaching, waterproofing, and as a flame retardant.

Bisphenol A, $(CH_3)_2C(C_6H_4OH)_2$, is made up of white flakes that have a mild phenolic odor. It is insoluble in water. Bisphenol A is combustible, with a flash point of 175°F. It is not listed in the DOT Hazardous Materials Tables. It is used in the manufacture of epoxy, polycarbonate, polysulfone, and polyester resins, as a flame retardant, and as a fungicide. The structure and molecular formula are shown in Figure 11.2.

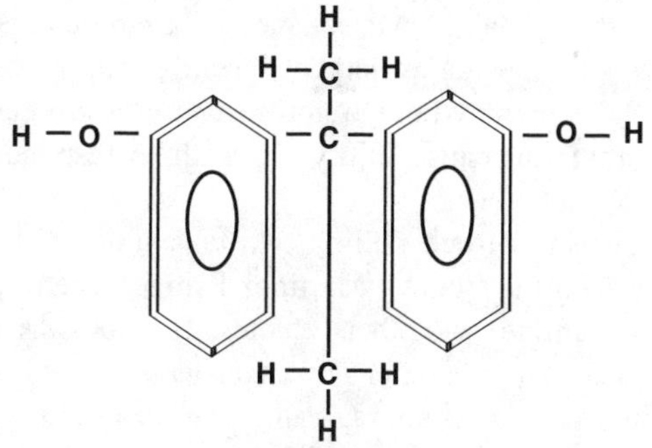

Bisphenol A
$(CH_3)_2C(C_6H_4OH)_2$

Figure 11.2

Urea (carbamide), $CO(NH_2)_2$, is composed of white crystals or powder, almost odorless, with a saline taste. It is soluble in water and decomposes before reaching its boiling point. Urea is noncombustible. The primary uses of urea are in fertilizers, animal feed, plastics, cosmetics, flame-proofing agents, pharmaceuticals, and as a stabilizer in explosives. Urea appears to be both a ketone and an amine by structure and molecular formula; however, it is neither, nor does it have any of the characteristics of either family. The structure and molecular formula are shown in Figure 11.3.

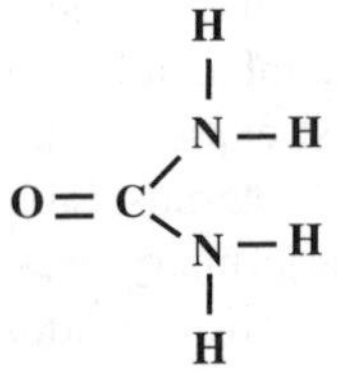

```
          H
          |
          N—H
         /
  O = C
         \
          N—H
          |
          H
```

Urea
$CO(NH_2)_2$

Figure 11.3

INCIDENTS

Hot materials, such as asphalt, can cause serious thermal burns if contacted with parts of the body. The television show *Rescue 911* highlighted a rescue operation that involved hot asphalt. A dump truck used to haul solid hot asphalt to patch holes in the road collided with a car. As a result of the collision, the load of hot asphalt was dumped into the car, covering the driver to the point that he could not escape. Before rescuers could remove the driver, he suffered severe second- and third-degree thermal burns from the hot asphalt.

This same hazard exists with all elevated-temperature materials. Responders should work carefully around transportation vehicles that have the HOT placard or the word "HOT" on the container. Miscellaneous hazardous materials can expose emergency responders to a wide variety of hazards. The placard itself does not indicate what those hazards may be. Do not treat this hazard class lightly, as there are some materials that can cause injury or death to responders if not handled properly.

In Benicia, California, a truck pulling two tank trailers of molten sulfur was involved in a collision on the Benicia-Martinez Bridge. One of the tanks ruptured, spilling the molten sulfur onto two other vehicles. The truck driver died, along with a passenger in one of the cars; another passenger was severely burned. At the time of the accident, molten sulfur was not regulated as a hazardous material. The molten sulfur produced sulfur dioxide vapors, which hampered visibility, along with fog.

Hazardous materials, regardless of class, almost always have multiple hazards. It is important for emergency responders to recognize that the hazard classes indicate only the most severe hazard of the materials as determined by the DOT. Research has to be conducted with all hazardous materials even if the hazard class is known. The correct chemical name must be identified, and all of the associated hazards evaluated before tactics are determined. Responders should have a thorough understanding of the physical and chemical characteristics of hazardous materials, including parameters of combustion, water and air reactivity, incompatibilities with other materials, and the effects of temperature and pressure on hazardous materials. Emergency responders should have the same level of understanding of hazardous materials as they do for firefighting, EMS protocols, and law enforcement procedures. All

Railcar with molten sulfur, a Class 9 miscellaneous hazardous material.

Highway tanker used to haul molten asphalt with "HOT" marking.

emergency response incidents have the potential to involve hazardous materials; your knowledge of the physical and chemical characteristics of hazardous materials will help ensure a safe outcome to incidents.

REVIEW QUESTIONS CHAPTER 11

1. What is the word that elevated-temperature materials may have on a placard on a vehicle?
   A. Dangerous When Wet
   B. Infectious Substance
   C. HOT
   D. Thermal Hazard
2. ORM-D materials are also known as which of the following?
   A. Consumer Commodities
   B. Household Substances
   C. Other Regulated Materials
   D. Not Otherwise Specified
3. A Class 9 hazardous material is best defined as which of the following?
   A. Smaller amounts of other classes
   B. Unregulated materials
   C. Shipped by highway only
   D. Do not fit in other categories

CHAPTER 12

Incompatible and Unstable Chemicals

Chemical processing and mixing under controlled circumstances safely creates many of the useful products we have come to depend on in our daily lives. However, when certain chemicals come in contact with each other during uncontrolled situations, such as accidents and spills, dangerous reactions can occur. This chapter will focus on incompatible chemicals, safe storage practices, water- and air-reactive materials, and some of the hazards to responders that may occur as a result of an uncontrolled chemical reaction.

Responders often ask what will happen when certain chemicals mix together. That question may not always have an answer. Many chemicals that are stored or shipped together may not normally be mixed together, therefore there may be little knowledge of the outcome of mixing. Other chemicals or families of chemicals are commonly used or stored together, and it is fairly easy to predict the consequences of mixing them.

Mixing of chemicals may cause a number of reactions, from mild to violent. Reactions can produce heat or cold; splattering of the material; spontaneous combustion; production of flammable, oxidizing, toxic, or corrosive vapors; and explosions. Some reactions occur when chemicals contact each other or when a chemical contacts air, oxidizers, or water. Reactions can also occur within a chemical without contact with other chemicals. Compounds that have double and triple bonds are highly reactive. When these bonds break, heat is produced and may result in spontaneous combustion, which may be slow or quite rapid. Resulting reactions, if they occur inside a container, may result in the container coming apart explosively. Figure 12.1 provides a brief listing of chemicals and some of the things they are incompatible with.

ACIDS AND BASES

One of the most common groups of chemicals that are violently incompatible are acids and bases. While grouped together in the same DOT hazard class of corrosives, these materials are generally not stored together because of the potential dangers.

Chemical Incompatibilities

Chemical	Incompatible with
Acetylene	Bromine, chlorine, fluorine, copper, silver, mercury
Ammonia	Mercury, hydrogen fluoride, calcium hypochlorite, chlorine
Flammable liquids	Ammonium nitrate, chromic acid, hydrogen peroxide, sodium peroxide, nitric acid, and the halogens
Oxygen	Oils, grease, hydrogen, flammable liquids, solids, gases
Nitric acid	Acetic acid, hydrogen sulfide, flammable liquids and gases
Phosphorus	Air, oxygen
Sulfuric acid	Potassium chlorate, perchlorate, permanganate
Alkali metals	Carbon tetrachloride, carbon dioxide, water, halogens
Alkaline Earth metals	Carbon tetrachloride, halogens, hydrocarbons

Figure 12.1

Compatibility of Chemical Families

	Chemical Family	1	2	3	4	5	6	7	8	9	10	11	12	13	14	15	16
1	Inorganic acids																
2	Organic acids	×															
3	Caustics	×	×														
4	Amines and alkoholamines	×	×														
5	Halogenated compounds	×		×	×												
6	Alcohols, glycols	×															
7	Aldehydes	×	×	×	×		×										
8	Ketones	×		×	×			×									
9	Saturated hydrocarbons																
10	Aromatic hydrocarbons	×															
11	Olefins	×				×											
12	Esters	×		×	×												
13	Halogens			×			×	×	×	×	×	×	×				
14	Ethers	×															
15	Acid anhydrides	×		×			×	×									
16	Oxidizers	×	×	×	×	×	×	×	×	×	×	×	×	×	×	×	

Figure 12.2

Under controlled circumstances, acids and bases can be used as neutralizing agents for each other. Under controlled circumstances, weak acids are used to neutralize strong bases, and weak bases are used to neutralize strong acids. During an accident

Chemical Compatibility Chart

Group	Name	Incompatible Goups
Group 1	Inorganic acids	2, 3, 4, 5, 6, 7, 8, 10, 13, 14, 16, 17, 18, 19, 21, 22, 23
Group 2	Organic acids	1, 2, 4, 7, 14, 16, 17, 18, 19, 22
Group 3	Caustics	1, 2, 6, 7, 8, 13, 14, 15, 16, 17, 18, 20, 23
Group 4	Amines	1, 2, 5, 7, 8, 13, 14, 15, 16, 17, 18, 23
Group 5	Halogenated compounds	1, 3, 4, 11, 14, 17
Group 6	Alcohols	1, 7, 14, 16, 20, 23
Group 7	Aldehydes	1, 2, 3, 4, 6, 8, 15, 16, 17, 19, 20, 23
Group 8	Ketones	1, 3, 4, 7, 19, 20
Group 9	Saturated hydrocarbons	20
Group 10	Aromatic hydrocarbons	1, 20
Group 11	Olefins	1, 5, 20
Group 12	Petroleum oils	20
Group 13	Esters	1, 2, 4, 19, 20
Group 14	Monomers/Esters	1, 2, 3, 4, 5, 6, 15, 16, 19, 20, 21, 23
Group 15	Phenols	3, 4, 7, 14, 16, 19, 20
Group 16	Alkylene oxides	1, 2, 3, 4, 6, 7, 14, 15, 17, 18, 19, 23
Group 17	Cyanohydrins	1, 2, 3, 4, 5, 7, 16, 19, 23
Group 18	Nitriles	1, 2, 3, 4, 16, 23
Group 19	Ammonia	1, 2, 7, 8, 13, 14, 15, 16, 17, 20, 23
Group 20	Halogens	3, 6, 7, 8, 9, 10, 11, 12, 13, 14, 15, 19, 21, 22
Group 21	Ethers	1, 14, 20
Group 22	Phosphorus	1, 2, 3, 20
Group 23	Acid anhydrides	1, 3, 4, 6, 7, 14, 16, 17, 18, 19

Figure 12.3

or spill, strong acids and strong bases could come in contact with each other, resulting in violent reactions and placing emergency responders in danger. Acids include sulfuric, hydrochloric, nitric, acetic, propionic, butyric, and others. Bases include sodium and potassium hydroxide, ammonium hydroxide, and others.

OXIDIZERS AND ORGANIC MATERIALS

As previously discussed in Chapter 5, in order for combustion to occur, there must be oxygen, fuel, heat, and a chemical chain reaction. When oxidizers (oxygen) encounter organic materials (fuel), the result of the reaction may produce enough heat for combustion to occur. Oxidizer sources include elemental oxygen compressed, cryogenic liquid oxygen, the halogen family of the Periodic Table, peroxide salts, oxysalts, and organic-peroxide hydrocarbon derivatives. Reactions may vary, from just dissolving as with the oxysalts, to violent contact explosions from liquid oxygen mixing with asphalt paving.

AGING CHEMICALS

Incompatibility can also result from chemicals that are stored too long and become unstable. Organic peroxides are formed in the containers when exposed to the oxygen

Ethyl ether will form explosive peroxides if the container is opened and it is not discarded within 6 months.

in the air. This forms an organic peroxide compound, which is highly explosive that can react from exposure to heat, shock, and friction. Compounds susceptible to peroxide formation include ether, formaldehyde, and potassium metal. All compounds subject to peroxide formation should be stored away from heat and light. Sunlight is a particularly good promoter of peroxidation. Protection from physical damage and ignition sources during storage is also important. Make sure enclosures of the chemicals are tightly closed. Loose or leaky closures may permit or enhance evaporation of the stored material, leaving a hazardous concentration of peroxides in the containers.

Some peroxide-forming compounds can be inhibited during storage using hydroquinone, alkyl phenols, aromatic amines, or similar materials. The selection of a proper inhibitor should be made to avoid possible conflicts with use or purity requirements of the compound. The more volatile the peroxidizable compound, the easier it is to concentrate the peroxides. Pure compounds are subject to peroxide accumulation because impurities may inhibit peroxide formation or catalize their slow decomposition. Compounds that are suspected of having high peroxide levels because of visual observation of unusual viscosity or crystal formation, or because of age, should be considered extremely dangerous. If crystals form in a peroxidizable liquid, or discoloration occurs in a peroxidizable solid, peroxidation may have occurred, and the product should be considered extremely dangerous and should be destroyed without opening the container.

Testing procedures are available to determine if peroxides have formed by sampling outsides of containers; however, they should only be conducted by persons with a chemical background and experience with the test procedures. The precautions taken for disposal of these materials should be the same as for any material that can be detonated by friction or shock. In general, the material should be carefully removed, using explosive handling procedures.

Peroxide-Forming Compounds

Severe Storage Hazard	Storage Time Limits
Butadiene	3 months
Chloroprene	
Divinyl acetylene	
Isopropyl ether	
Potassium amide	
Potassium metal	
Sodium amide	
Tetrafluoroethylene	
Vinylidene chloride	
Moderate Storage Hazard	1 year
Acetal	
Acetaldehyde	
Benzyl alcohol	
2-Butanol dioxanes	
Chlorofluoroethylene	
Cyclohexene	
2-Cyclohexen-1-o1	
Cyclopentene	
Decahydronaphthalene (decalin)	
Diacetylene (butaldiyne)	
Dicyclopentadiene	
Diethylene glycol dimethyl ether (diglyme)	
Diethyl ether	
Ethylene glycol ether acetates (cellosolves)	
Furan	
4-Heptanol	
2-Hexanol	
Methyl acetylene	
3-Methyl-1-butanol	
Methyl-isobutyl ketone	
4-Methyl-2-pentanol	
2-Pentanol	
4-Penten-1-o1	
1-Phenylethanol	
2-Phenylethanol	
Tetrahydrofuran	
Tetrahydronaphthalene	
Vinyl ethers	
Other secondary alcohols	
Subject to Autopolymerization	6 months
Butadiene	
Chlorobutadiene	
Chloroprene	
Chlorotrifluoroethylene	
Stryene	
Tetrafluoroethylene	
Vinyl acetate	
Vinyl acetylene	
Vinyl chloride	
Vinyl pyridine	
Vinyldiene chloride	

Figure 12.4

Generally, chemicals with the above functional groups are prone to instability

O—O	(peroxide)	—N	(imino)
$-NO_2$	(Nitro)	$-N_3$	(azide)
—N=N—	(azo)	—N=O	(nitroso)
—ONO	(nitrate ester)	$-NHNO_2$	(nitramine)

Figure 12.5

WATER- AND AIR-REACTIVE MATERIALS

Phosphorus is a material that spontaneously combusts when exposed to air. Thus, it is shipped under water to keep it from contacting air. Fires involving phosphorus should be fought using large quantities of water, as it will not react with the water. Care should be taken to ensure these types of materials are not stored in breakable containers. Alkali metals, such as lithium, sodium, and potassium, are considered water-reactive, and explosive reactions can occur when these metals contact water, releasing flammable hydrogen gas.

Calcium carbide is a water-reactive salt, which releases acetylene gas when in contact with water. Other materials may release toxic gases, oxygen, and flammable gas when contacting water. The *Emergency Response Guide Book* contains a section at the end of the Green pages that contains a listing of water-reactive materials and the toxic gases released when they contact water.

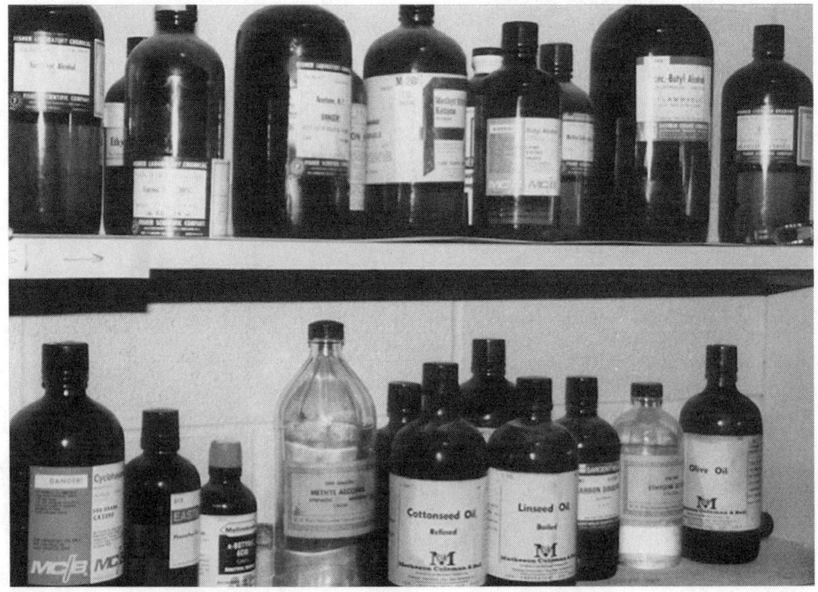

Chemicals not stored in proper hazard groups can be a hazard during a fire or accident.

When water-reactive materials contact flammable solids or oxidizers and fire occurs, water may be the only thing available in a large enough quantity to extinguish the fire. Magnesium is an element that is water-reactive, but only when it is on fire. Putting water on a magnesium fire can cause serious explosions, which may produce shock waves. Personnel should be at a safe distance using unmanned monitors and aerial devices to put water on a magnesium fire.

INCOMPATABLE CHEMICAL LISTING

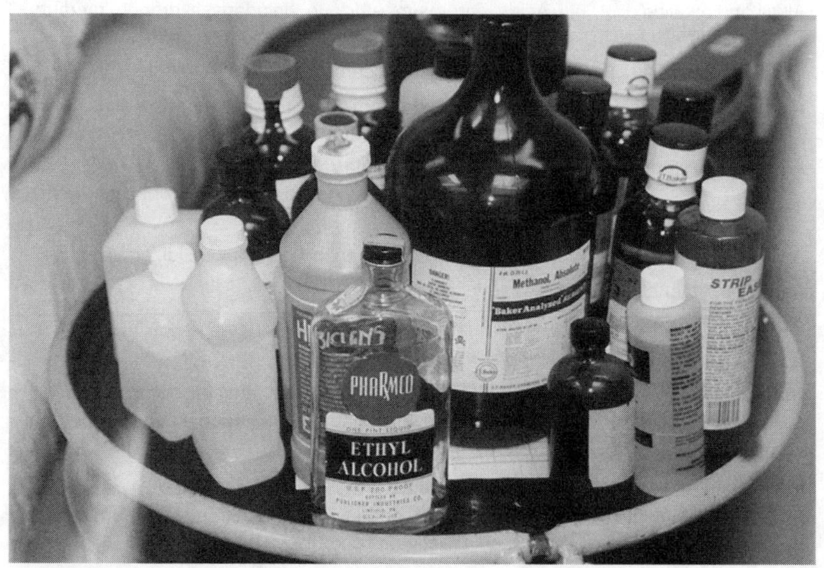

Corrosion of a flammable liquid container from an improperly stored acid.

Chemical	Incompatible Chemical(s)
Acetic acid	Aldehyde, bases, carbonates, hydroxides, metals, oxidizers, peroxides, phosphates, xylene
Acetic anhydride	Chromic acid, nitric acid, hydroxyl containing compounds, ethylene glycol, perchloric acid, peroxides and permanganates
Acetylene	Halogens (chlorine, fluorine, etc.), mercury, potassium, oxidizers, silver
Acetone	Acids, amines, oxidizers, plastics
Alkali and alkaline Earth metals	Acids, chromium, ethylene, halogens, hydrogen, mercury, nitrogen, oxidizers, plastics, sodium chloride, sulfur
Aluminum alkyls	Halogenated hydrocarbons, water
Ammonia	Acids, aldehydes, amides, halogens, heavy metals, oxidizers, plastics, sulfur
Ammonium nitrate	Acids, alkalis, chloride salts, combustible materials, metals, organic materials, phosphorous, reducing agents, urea
Aniline	Acids, aluminum, dibenzoyl peroxide, oxidizers, plastics
Arsenical materials	Any reducing agent
Azides	Acids, heavy metals, oxidizers
Benzoyl peroxide	Chloroform, organic materials
Bromine	Acetaldehyde, alcohols, alkalis, amines, combustible materials, ethylene, fluorine, hydrogen, ketones (acetone, carbonyls, etc.), metals, sulfur
Calcium carbide	Water
Calcium hypochlorite	Methyl carbitol, phenol, glycerol, nitromethane, iron oxide, ammonia, activated carbon
Calcium oxide	Acids, ethanol, fluorine, organic materials
Carbon (activated)	Alkali metals, calcium hypochlorite, halogens, oxidizers
Carbon tetrachloride	Benzoyl peroxide, ethylene, fluorine, metals, oxygen, plastics, silanes
Chlorates	Powdered metals, sulfur, finely divided organic or combustible materials
Chromic acid	Acetone, alcohols, alkalis, ammonia, bases
Chromium trioxide	Benzene, combustible materials, hydrocarbons, metals, organic materials, phosphorous, plastics
Chlorine	Alcohols, ammonia, benzene, combustible materials, flammable compounds (hydrazine), hydrocarbons (acetylene, ethylene, etc.), hydrogen peroxide, iodine, metals, nitrogen, oxygen, sodium hydroxide
Chlorine dioxide	Hydrogen, mercury, organic materials, phosphorous, potassium hydroxide, sulfur
Chlorosulfonic acid	Organic materials, water, powdered metals
Copper	Calcium, hydrocarbons, oxidizers
Cumene hydroperoxide	Acids, organic or mineral

Figure 12.6

Chemical	Incompatible Chemical(s)
Cyanide	Acids
Hydroperoxide	Reducing agents
Cyanides	Acids, alkaloids, aluminum, iodine, oxidizers, strong bases
Ethylene oxide	Acids, bases, copper, magnesium perchlorate
Flammable liquids	Ammonium nitrate, chromic acid, hydrogen peroxide, nitric acid, sodium peroxide, halogens
Fluorine	Alcohols, aldehydes, ammonia, combustible materials, halocarbons, halogens, hydrocarbons, ketones, metals, organic acids
Hydrocarbons (such as butane, propane, benzene, turpentine, etc.)	Acids, bases, oxidizers, plastics
Hydrocyanic acid	Nitric acid, alkalis
Hydrides	Water, air, carbon dioxide, chlorinated hydrocarbons
Hydrofluoric acid	Metals, organic materials, plastics, silica (glass), (anhydrous) sodium
Hydrogen peroxide	Acetyladehyde, acetic acid, acetone, alcohols, carboxylic acid, combustible materials, metals, nitric acid, organic compounds, phosphorous, sulfuric acid, sodium, aniline
Hydrogen sulfide	Acetylaldehyde, metals, oxidizers, sodium
Hypochlorites	Acids, activated carbon
Iodine	Acetylaldehyde, acetylene, ammonia, metals, sodium
Maleic anhydride	Sodium hydroxide, pyridine, and other tertiary amines
Mercury	Acetylene, aluminum, amines, ammonia, calcium, fulminic acid, lithium, oxidizers, sodium
Nitrates	Acids, nitrites, metals, sulfur, sulfuric acid
Nitric acid	Acetic acid, acetonitrile, alcohols, amines, (concentrated) ammonia, aniline, bases benzene, cumene, formic acid, ketones, metals, organic materials, plastics, sodium, toluene
Nitroparaffins	Inorganic bases, amines
Oxalic acid	Oxidizers, silver, sodium chlorite
Oxygen	Acetaldehyde, secondary alcohols, alkalis, and alkalines, ammonia, carbon monoxide, combustible materials, ethers, flammable materials, hydrocarbons, metals, phosphorous, polymers
Oxalic acid	Silver, mercury, organic peroxides
Perchlorates	Acids
Perchloric acid	Acetic acid, alcohols, aniline, combustible materials, dehydrating agents, ethyl benzene, hydriotic acid, hydrochloric acid, iodides, ketones, organic material, oxidizers, pyridine
Peroxides, organic	Acids (organic or mineral)
Phosphorus (white)	Oxygen (pure and in air), alkalis

Figure 12.6 (Continued)

Chemical	Incompatible Chemical(s)
Phosphorus pentoxide	Propargyl alcohol
Potassium	Acetylene, acids, alcohols, halogens, hydrazine, mercury, oxidizers, selenium, sulfur
Potassium chlorate	Acids, ammonia, combustible materials, fluorine, hydrocarbons, metals, organic materials, sugars
Potassium perchlorate	Alcohols, combustible materials, fluorine, hydrazine, metals, organic matter, reducing agents, sulfuric acid
Potassium permanganate	Benzaldehyde, ethylene glycol, glycerol, sulfuric acid
Selenides	Reducing agents
Silver	Acetylene, ammonia, oxidizers, ozonides, peroxyformic acid
Sodium	Acids, hydrazine, metals, oxidizers, water
Sodium amide	Air, water
Sodium nitrate	Ammonium nitrate and other ammonium salts
Sodium oxide	Water, any free acid
Sodium peroxide	Acitic acid, benzene, hydrogen sulfide metals, oxidizers, peroxyformic acid, phosphorous, reducers, sugars, water
Sulfides	Acids
Sulfuric acid	Potassium chlorates, potassium perchlorate, potassium permanganate
Tellurides	Reducing agents
UDMH (1,1-Dimethylhydrazine)	Oxidizing agents such as hydrogen peroxide and fuming nitric acid
Zirconium	Prohibit water, carbon tetrachloride, foam, and dry chemical on zirconium fires

Figure 12.6 (Continued)

Basic Chemical Storage Segregation

Class of Chemicals	Recommended Storage Method	Examples	Incompatibilities
Compressed gases flammable	Store in a cool, dry area, away from oxidizing gases. Securely strap or chain cylinders to prevent falling over.	Methane, acetylene, propane, hydrogen	Oxidizing and toxic compressed gases, oxidizing solids
Compressed gases oxidizing	Store in a cool, dry area, away from flammable gases and liquids. Securely strap or chain cylinders to prevent falling over.	Oxygen, fluorine, chlorine, bromine	Flammable gases
Compressed gases poisonous	Store in cool, dry area, away from flammable gases and liquids. Securely strap or chain cylinders to prevent falling over.	Carbon monoxide, hydrogen sulphide	Flammable and/or oxidizing gases
Corrosives-acids	Store in separate acid storage cabinet or area.	Mineral acids-hydrochloric, sulfuric, nitric, perchloric, chromic, chromerge	Flammable liquids, flammable solids, bases, oxidizers
Corrosives-bases	Store in separate storage area	Ammonium hydroxide, sodium hydroxide	Flammable liquids, oxidizers, poisons, and acids
Explosives	Store in secure location away from all other chemicals.	Ammonium nitrate, nitro urea, picric acid, trinitroaniline, trinitroanisole, trinitrobenzene, trinitrobenzenesulfonic acid	Flammable liquids, oxidizers, poisons, acids, and bases
Flammable liquids	In grounded flammable liquid storage cabinets or specially constructed rooms.	Acetone, benzene, diethyl ether, methanol, ethanol, toluene, glacial acetic acid	Acids, bases, oxidizers, and poisons
Flammable solids	Store in a separate dry, cool area away from oxidizers, corrosives, and flammable liquids.	Phosphorus	Acids, bases, oxidizers, and poisons
General chemicals non-reactive	Store on general laboratory benches or shelving, in chemical storage rooms or warehouse locations.	Agar, sodium chloride, sodium bicarbonate, and most non-reactive salts	See MSDS

Figure 12.7

Basic Chemical Storage Segregation

Class of Chemicals	Recommended Storage Method	Examples	Incompatibilities
Oxidizers	Store in a spill tray inside a noncombustible cabinet, separate from flammable and combustible materials.	Sodium hypochlorite, benzoyl peroxide, potassium permanganate, potassium chlorate, potassium dichromate. The following are generally considered oxidizing substances: peroxides, perchlorates, chlorates, nitrates, bromates, and superoxides.	Separate from reducing agents, flammables, and combustibles
Poisons	Store separately in vented, cool, dry, area, in unbreakable chemically resistant secondary containers.	Cyanides, heavy metals compounds, i.e. cadmium, mercury, osmium	Flammable liquids, acids, bases, and oxidizers
Water-reactive chemicals	Store in dry, cool, location; protect from water from fire sprinkler.	Sodium metal, potassium metal, lithium metal, lithium aluminum hydride	Separate from all aqueous solutions and oxidizers

Figure 12.7 (Continued)

Glossary

Absorption A route of exposure. It occurs when a toxic material contacts the skin and enters the bloodstream by passing through the skin.

Accidental explosion An unplanned or premature detonation/ignition of explosive/incendiary material or material possessing explosive properties. The activity leading to the detonation/ignition has no criminal intent and is primarily associated with legal, industrial, or commercial activities.

Acid 1. Any of a class of chemical compounds whose aqueous solutions turn litmus paper red (has a pH less than 7) or react with and dissolve certain metals or react with bases to form salts. 2. A compound capable of transferring a hydrogen ion in solution. 3. A molecule or ion that combines with another molecule or ion by forming a covalent bond with two electrons from other species.

Acetylcholine A chemical compound formed from an acid and an alcohol that causes muscle to contract (neurotransmitter). It is rapidly broken down by an enzyme, cholinesterase.

Acetylcholinesterase An enzyme present in nerve tissue, muscles, and red blood cells that catalyzes the hydrolysis of acetylcholine to choline and acetic acid, allowing neural transmission across synapses to occur; true cholinesterase.

Acid, corrosive A material that usually contains an H^+ ion and is capable of dehydrating other materials.

Acute exposure The adverse effects resulting from a single dose or exposure to a material. Ordinarily used to denote effects observed in experimental animals.

Acute toxicity Any harmful effect produced by a single short-term exposure that may result in severe biological harm or death.

Aerosol The dispersion of very fine particles of a solid or liquid in a gas, fog, foam, or mist.

Agent dosage The concentration of a toxic vapor in the air multiplied by the time that the concentration is present or the time that an individual is exposed (mg-min/m^3).

Alcohol foam A type of foam developed to suppress ignitable vapors on polar solvents (those miscible in water). Examples of polar flammable liquids are alcohols and ketones.

Alkaline Any compound having the qualities of a base. Simplified, a substance that readily ionizes in aqueous solution to yield hydroxyl (OH^-) anions. Alkalis have a pH greater than 7 and turn litmus paper blue.

Alpha particle A form of ionizing radiation that consists of two protons and neutrons.

Ambient temperature The normal temperature of the environment.

ANFO An ammonium nitrate and fuel oil mixture, commonly used as a blasting agent. The proportions are determined by the manufacturer or user. It is commonly mixed with the addition of an "enhancer," such as magnesium or aluminum, to increase the rate of burn.

Anhydrous Describes a material that contains no water (water-free).

Anion A negatively charged ion that moves toward the anode (+ terminal) during electrolysis. Oxidation occurs at the anode.

Anticholinergic An agent or chemical that blocks or impedes the action of acetylcholine, such as the (also cholinolytic) antidote atropine.

Anticholinesterase A substance that blocks the action of cholinesterase (acetylcholinesterase), such as nerve agents.

Antidote A material administered to an individual who has been exposed to a poison in order to counteract its toxic effects.

Arsenical Pertaining to or containing arsenic; a reference to the vesicant lewisite.

Asphyxia Lack of oxygen and interference with oxygenation of the blood. Can lead to unconsciousness.

Asphyxiant A vapor or gas that can cause unconsciousness or death by suffocation (lack of oxygen). Most simple asphyxiants are harmful to the body when they become so concentrated that they reduce (displace) the available oxygen in air (normally about 21%) to dangerous levels (18% or lower). Chemical asphyxiants, like carbon monoxide (CO), reduce the blood's ability to carry oxygen or, like cyanide, interfere with the body's utilization of oxygen.

Asphyxiation Asphyxia or suffocation. Asphyxiation is one of the principal potential hazards of working in confined spaces.

Atmospheric container A type of container that holds products at atmospheric pressure (760 mm).

Atom The smallest unit into which a material may be broken by chemical means. In order to be broken into any smaller units, a material must be subjected to a nuclear reaction.

Atomic weight (at. wt.) The relative mass of an atom. Basically, it equals the number of protons plus neutrons.

Autoignition A process in which a material ignites without any apparent outside ignition source. In the process, the temperature of the material is raised to its ignition temperature by heat transferred by radiation, convection, combustion, or some combination of all three.

Autoignition temperature *See* Ignition temperature.

Atropine An anticholinergic used as an antidote for nerve agents to counteract excessive amounts of acetylcholine. It also has other medical uses.

B-NICE The acronym developed by the National Fire Academy for identifying the five categories of terrorist incidents: **B**iological, **N**uclear, **I**ncendiary, **C**hemical, and **E**xplosive.

Bacteria Single-celled organisms that multiply by cell division and can cause disease in humans, plants, or animals. Examples include anthrax, cholera, plague, tularemia, and Q fever.

Base A chemical compound that reacts with an acid to form a salt. The term is applied to the hydroxides of the metals, to certain metallic oxides, and to groups of atoms containing one or more hydroxyl groups (OH^-) in which hydrogen is replaceable by an acid radical. *See* Alkaline.

Beta particle A form of ionizing radiation that consists of either electrons or positrons.

Biohazard Those organisms that have a pathogenic effect on life and the environment, and can exist in normal ambient environments. These hazards can represent themselves as disease germs and viruses.

Biological agent Living organism, or the materials derived from it, that cause disease in or harm humans, animals, or plants, or cause deterioration of material. Biological agents may be found as liquid droplets, aerosols, or dry powders. A biological agent can be adapted and used as a terrorist weapon, such as anthrax, tularemia, cholera, encephalitis, plague, and botulism. There are three different types of biological agents: bacteria, viruses, and toxins.

Blasting agent A material designed for blasting that has been tested in accordance with Sec. 173.114a (49 CFR). It must be so insensitive that there is little probability of accidental explosion or going from burning to detonation.

Blepharospasm A twitching or spasmodic contraction of the orbicular oculi muscle around the eye.

BLEVE *See* Boiling liquid, expanding vapor, explosion.

Blister agent A chemical agent, also called a vesicant, that causes severe blistering and burns to eyes, skin, and tissues of the respiratory tract. Exposure is through liquid or vapor contact. Also referred to as mustard agent. Examples include mustard and lewisite.

Blood agent A chemical agent that interferes with the ability of blood to transport oxygen and causes asphyxiation. These substances injure a person by interfering with cell respiration (the exchange of oxygen and carbon dioxide between blood and tissues). Common examples are hydrogen cyanide and cyanogens chloride.

Blood asphyxiant A chemical that is absorbed by the blood and changes or prevents the blood from flowing or carrying oxygen to cells. An example is carbon monoxide poisoning.

Boiling liquid, expanding vapor, explosion (BLEVE) The explosion and rupture of a container caused by the expanding vapor pressure as liquids in the container become overheated.

Boiling point At this temperature, vapor pressure of a liquid now equals the surrounding atmospheric pressure (14.7 psi at sea level).

BTU British Thermal Unit. Amount of heat required to raise 1 lb of H_2O –1°F at sea level.

Cation A positively charged ion that moves toward the cathode during eletrolysis. Reduction occurs at the cathode.

Carcinogen A material that either causes cancer in humans or, because it causes cancer in animals, is considered capable of causing cancer in humans.

Caustic 1. Burning or corrosive. 2. A hydroxide of a light metal. Broadly, any compound having highly basic properties. A compound that readily ionizes in aqueous solution to yields OH^- anions, with a pH above 7, and turns litmus paper blue. *See* Alkaline; Base.

Cellular asphyxiant A material that, upon entering the body, inhibits the normal function of cells. Examples are CO, hydrogen cyanide, or hydrogen sulfide poisoning.

Central nervous system (CNS) In humans, the brain and spinal cord, as opposed to the peripheral nerves found in the fingers, etc.

Chemical agent There are five classes of chemical agents, all of which produce incapacitation, serious injury, or death: nerve agents, blister agents, blood agents, choking agents, and irritating agents. A chemical substance used in military operations intended to kill, seriously injure, or incapacitate people through its physiological effects.

Chemical burn A burn that occurs when the skin comes into contact with strong acids, strong alkalis, or other corrosive materials. These agents literally eat through the skin and, in many cases, continue to do damage as long as they remain in contact with the skin.

Chemical properties A property of matter that describes how it reacts with other substances.

Chemical reaction A process that involves the bonding, unbonding, or rebonding of atoms. A chemical change takes place that actually changes substances into other substances.

Chemical reactivity The process whereby substances are changed into other substances by the rearrangement, or recombination, of atoms.

CHEMTREC Chemical Transportation Emergency Center operated by the Chemical Manufacturers Association. Provides information or assistance to emergency responders. CHEMTREC contacts the shipper or producer of the material for more detailed information, including on-scene assistance when feasible. Can be reached 24 hours a day by calling 1-800-424-9300.

CHLOREP Chlorine Emergency Plan operated by the Chlorine Institute. A 24-hour mutual-aid program. Response is activated by a CHEMTREC call to the designated CHLOREP's geographical-sector assignments for teams.

Cholinesterase (Ache) Acetylcholinesterase is the enzyme that breaks down the neurotransmitter acetylcholine after it has transmitted a signal from a nerve ending to another nerve, muscle, or gland. Organophosphate pesticides and military nerve agents block the normal activity of Ache, which results in the accumulation of excess acetylcholine at nerve endings.

Choking agents These agents exert their effects solely on the lungs and result in the irritation of the alveoli of the lungs. Agents cause the alveoli to constantly secrete watery fluid into the air sacs, which is called pulmonary edema. When a lethal amount of a choking agent is received, the air sacs become so flooded that the air cannot enter and the victim dies of anoxia (oxygen deficiency); also known as dry drowning.

Cholinergic Resembling acetylcholine, especially in physiological action. Cholinergic symptoms include nausea, vomiting, headache, and sweating.

Chronic Applies to long periods of action, such as weeks, months, or years.

Chronic effects An adverse health effect on a human or animal body with symptoms that develop slowly or that recur frequently due to the exposure of hazardous chemicals.

Chronic exposure Repeated doses or exposure to a material over a relatively prolonged period of time.

Closed-cup tester A device for determining flash points of flammable and combustible liquids, utilizing an enclosed cup or container for the liquid. Recognized types are the Tagliabue (Tag) Closed Tester, the Pensky-Martens Closed Tester, and the Setaflash Closed-Cup Tester.

CNS *See* Central nervous system.

Combustibility The ability of a substance to undergo rapid chemical combination with oxygen, with the evolution of heat.

Combustible dust Particulate material that, when mixed in air, will burn or explode.

Combustible liquid Term commonly used for liquids that emit burnable vapors or mists. Technically, a liquid whose vapors will ignite at a temperature of 100°F or above.

Compound A substance composed of two or more elements that have chemically reacted. The compound that results from the chemical reaction is unique in its chemical and physical properties.

Compressed gas Any material or mixture having in the container an absolute pressure exceeding 40 psi at 70°F or, regardless of the pressure at 70°F, having an absolute pressure exceeding 104 psi at 130°F; or any liquid flammable material having a vapor pressure exceeding 40 psi absolute at 100°F as determined by testing. Also includes cryogenic or "refrigerated liquids" (DOT) with boiling points lower than –130°F at 1 atmosphere.

Concentration The amount of a material that is mixed with another material.

Concentration (corrosives) In corrosives, the amount of acid or base compared to the amount of water present. Corrosives have "strength" and "concentration." *See* Strength.

Contaminant 1. A toxic substance that is potentially harmful to people, animals, and the environment. 2. A substance not in pure form.

Corrosive A chemical that causes visible destruction of or irreversible alterations in living tissue by chemical action at the site of contact; a liquid that causes a severe corrosion rate in steel. A corrosive is either an acid or a caustic (a material that reads at either end of the pH scale).

Corrosive material (DOT) A material that causes the destruction of living tissue and metals.

Covalent bond A chemical bond in which atoms share electrons in order to form a molecule.

Critical pressure The pressure required to liquefy a gas at its critical temperature.

Critical temperature The temperature above which a gas cannot be liquefied by pressure.

Cryogenic burn Frostbite; damage to tissues as the result of exposure to low temperatures. It may involve only the skin, extend to the tissue immediately beneath it, or lead to gangrene and loss of affected parts.

Cryogenic cylinder An insulated metal cylinder contained within an outer protective metal jacket. The area between the cylinder and the jacket is normally under vacuum. The cylinders range in size from a Dewier (similar to a small thermos) up to 24 in in diameter and 5 ft in length. Examples of materials found in these types of cylinders are argon, helium, nitrogen, and oxygen.

Cryogenic liquid A liquid with a boiling point below −130ºF.

Cylinder A container for liquids, gases, or solids under pressure. Ranges in size from aerosol containers found at home, such as spray deodorant, to the cryogenic (insulated) cylinders for nitrogen that can be approximately 24 in in diameter and 5 ft in length. Pressure ranges from a few pounds to 6000 pounds per square inch.

Dangerous When Wet Materials that when exposed to water allow a chemical reaction to take place and often produce flammable or poisonous gases, heat, and a caustic solution. An example is sodium.

Decomposition Separation of larger molecules into separate constituent and smaller parts.

Decomposition (chemical) A reaction in which the molecules of a chemical break down to its basic elements, such as carbon, hydrogen, or nitrogen, or to more simple compounds. This often occurs spontaneously, liberating considerable heat and often large volumes of gas.

Decontamination The physical or chemical process of reducing and preventing the spread of contamination from persons and equipment used at a hazardous materials incident.

Deflagration Explosion, with rapid combustion, up to 1250 feet per second.

Detonating cord A flexible cord containing a center cord of high explosives used to detonate other explosives with which it comes in contact.

Detonation An explosion at speeds above 1250 feet per second and many times over 3300 feet per second.

Detonator Any device containing a detonating charge that is used for initiating detonation in an explosive. This term includes, but is not limited to, electric and nonelectric detonators (either instantaneous or delayed) and detonating connectors.

Dewier container Small (less than 25 gallons) container used for temporary storage or handling of cryogenic liquids.

Dilution The application of water to water-miscible hazardous materials. The goal is to reduce the hazard of a material to safe levels by reducing its concentration.

Dose The accumulated amount of a chemical to which a person is exposed.

DOT U.S. Department of Transportation. Regulates transportation of materials to protect the public as well as fire, law, and other emergency response personnel.

Dry bulk A type of container used to carry large amounts of solid materials (more than 882 lbs, or 400 kg). It can either be placed on or in a transport vehicle or vessel constructed as an integral part of the transport vehicle.

Dyspnea Shortness of breath, a subjective difficulty or distress in breathing, usually associated with disease of the heart or lungs; occurs normally during intense physical exertion or at high altitudes.

Element A substance that cannot be broken down into any other substance by chemical means.

Empirical formula Describes the *ratio* of the number of each element in the molecule, but not the exact number of atoms in the molecule.

Emulsification The process of dispersing one liquid in a second immiscible liquid. The largest group of emulsifying agents are soaps, detergents, and other compounds whose basic structure is a paraffin chain terminating in a polar group.

Encephalitis, pl. Encephalitides Inflammation of the brain.

Endothermic A process or chemical reaction that is accompanied by absorption of heat.

Enterotoxin A cytotoxin specific for the cells of the intestinal mucosa.

Erythema Red area of the skin caused by heat or cold injury, trauma, or inflammation.

Etiologic agent Those living organisms or their toxins that contribute to the cause of infection, disease, or other abnormal condition.

Evaporation The process in which liquid becomes vapor as more molecules leave the vapor than return.

Exothermic reaction A chemical reaction that liberates heat during the reaction.

Expansion ratio The amount of gas produced from a given volume of liquid escaping from a container at a given temperature.

Explosion The sudden and rapid production of gas, heat, noise, and many times a shock wave, within a confined space.

Explosive (DOT) Any chemical compound or mixture whose primary function is to produce an explosion.

Explosives, high Explosive materials that can be used to detonate by means of a detonator when unconfined (e.g., dynamite).

Explosives, low Explosive materials that deflagrate rather than detonate (e.g., black powder, safety fuses, and "special fireworks" as defined by Class 1.3 Explosives).

Explosive limits *See* Flammable limits.

Febrile Denoting or relating to fever.

Fire point The lowest temperature at which the vapor above the liquid will ignite and continue to burn, usually a few degrees above the flash point.

Flammable gas A gas that at ambient temperature and pressure forms a flammable mixture with air at a concentration of 13% by volume or less; or a gas that at ambient temperature and pressure forms a range of flammable mixtures with air greater than 12% by volume, regardless of the lower explosive limit.

Flammable limits The range of the percentages of vapor mixed with air that are capable of ignition, as opposed to those mixtures that have too much or too little vapor to be ignited. Also called explosive limits.

Flammable liquid A liquid that gives off readily ignitable vapors. Defined by the NFPA and DOT as a liquid with a flash point below 100°F (38°C).

Flammable range The percentage of fuel vapors in air where ignition can occur. Flammable range has an upper and lower limit.

Flammable solid A solid (other than an explosive) that ignites readily and continues to burn. It is liable to cause fires under ordinary conditions or during transportation through friction or retained heat from manufacturing or processing. It burns so vigorously and persistently as to create a serious transportation hazard. Included in this class are spontaneously combustible and water-reactive materials. An example is white phosphorus.

Flash back The ignition of vapors and the travel of the flame back to the liquid/vapor-release source.

Flash point The minimum temperature at which a liquid gives off vapor within a test vessel in sufficient concentration to form an ignitable mixture with air near the surface of the liquid.

Foam Firefighting material consisting of small bubbles of air, water, and concentrating agents. Chemically, the air in the bubbles is suspended in the fluid. The foam clings to vertical and horizontal surfaces and flows freely over burning or vaporizing materials. Foam puts out a fire by blanketing it, excluding air, and blocking the escape of volatile vapor. Its flowing properties resist mechanical interruption and reseal the burning material.

Formula A combination of the symbols for atoms or ions that are held together chemically.

Freezing point The temperature at which a material changes its physical state from a liquid to a solid.

Frothing A foaming action caused when water, turning to steam in contact with a liquid at a temperature higher than the boiling point (212°F), picks up a part of a viscous liquid.

Gas A formless fluid that occupies the space of its enclosure. It can settle to the bottom or top of an enclosure when mixed with other chemicals. It can be changed to its liquid or solid state only by increased pressure and decreased temperature.

Gastrointestinal tract GI tract. The entire digestive canal from mouth to anus.

Halogens A chemical family that includes fluorine, chlorine, bromine, and iodine.

Halon Halogenated hydrocarbons (containing the elements F, Cl, Br, or I) used to suppress or prevent combustion.

Hazard class One of nine classes of hazardous materials as categorized and defined by the DOT in 49 CFR.

Hazmat foam A special vapor-suppressing mix that can be applied to liquids or solids to prevent off-gassing.

Hematemesis Vomiting of blood.

Hemolytic anemia Anemia caused by increased destruction of red blood cells where the bone marrow is not able to compensate for it.

Hemoptysis The spitting of blood derived from the lungs or bronchial tubes as a result of pulmonary or bronchial hemorrhage.

Hepatoxin A chemical that is injurious to the liver.

High-expansion foam A detergent-based foam (low water content) that expands at ratios of 1000 to 1.

High-order explosion Materials that require moderate heat and reducing agents to initiate combustion.

Hypergolic materials Materials that ignite upon contact with one another.

Hypergolic reaction The immediate spontaneous ignition when two or more materials are mixed.

Hypotension Subnormal arterial blood pressure.

Hypovolemia A decreased amount of blood in the body.

Hypoxemia Subnormal oxygenation of arterial blood, short of anoxia.

IDLH Immediately Dangerous to Life and Health. The maximum levels to which a healthy worker can be exposed for 30 min to a chemical and escape without suffering irreversible health effects or escape impairing symptoms.

Ignition temperature The minimum temperature at which a material will ignite without a spark or flame present. This is also the temperature the ignition source must be.

Immiscible Matter that cannot be mixed. For example, water and gasoline are immiscible.

Incapacitating agent An agent that produces physiological or mental effects or both that may persist for hours or days after exposure, rendering an individual incapable of performing his or her assigned duties.

Incompatibility The inability to function or exist in the presence of something else, such as when a chemical will destroy the container.

Inert A material that under normal temperatures and pressures does not react with other materials.

Inhibited A substance that has had another substance added to prevent or deter its reaction either with other materials or itself (polymerization). Usually used to deter polymerization.

Inhibitor A substance that is capable of stopping or retarding a chemical reaction. To be technically useful, it must be effective in low concentration (i.e., to stop polymerization).

Initiator The substance or molecule (other than reactant) that initiates a chain reaction, as in polymerization.

Inorganic Pertaining to or composed of chemical compounds that do not contain carbon as the principal element (except carbonates, cyanides, and cyanates). Matter other than plant or animal.

Inorganic peroxides Inorganic compounds containing an element at its highest state of oxidation (such as sodium peroxide), or having the peroxy group –O–O– (such as perchloric acid).

Ion An atom that possesses an electrical charge, either (+) positive or (–) negative.

Ionic bond A chemical bond in which atoms of different elements transfer (exchange) electrons. As the electrons are exchanged, charged particles known as ions are formed.

Ionizing radiation High-energy radiation, such as an x-ray, that causes the formation of ions in substances through which it passes (gamma rays). Excessive amounts of ionizing radiation will cause permanent genetic or bodily damage.

Irritant A noncorrosive material that causes a reversible inflammatory effect on living tissue by chemical action at the site of contact.

Latent period Specifically in the case of mustard, the period between exposure and onset of signs and symptoms; otherwise, an incubation period.

Lethal chemical agent An agent that may be used effectively in a field concentration to produce death.

Lacrimation Secretion and discharge of tears.

LC_{50} Lethal concentration 50, median lethal concentration. The concentration of a material in air that on the basis of laboratory tests (respiratory route) is expected to kill 50% of a group of test animals when administered as a single exposure in a specific time period.

LD_{50} Lethal dose 50. The single dose of a substance that causes death of 50% of an animal population from exposure to the substance by any route other than inhalation.

Liquid A substance that is neither a solid nor a gas; a substance that flows freely, like water.

Low-order explosion Materials that require excessive heat and reducing agents to initiate combustion.

Low-pressure container A container designed to withstand pressures from 5 to 100 psi.

Lower explosive limit The lowest concentration of gas or vapor (% by volume in air) that burns or explodes if an ignition source is present at ambient temperatures.

LOX Liquid oxygen.

Mechanical foam A substance introduced into the water line by various means at a 6% concentration. Air is then introduced to yield foam consisting generally of 90 volumes air, 9.4 volumes water, and 0.6 volumes foam liquid. It uses hydrolyzed soybean, fish scales, hoof and horn meal, and peanut or corn protein as a base.

Median incapacitating dosage (ICT_{50}) The volume of a chemical agent vapor or aerosol inhaled that is sufficient to disable 50% of exposed, unprotected people (expressed as mg-min/m^3).

Median lethal dosage (LCT_{50}) The dosage of a chemical-agent vapor or aerosol inhaled that is lethal to 50% of exposed, unprotected people (expressed as mg-min/m^3).

Median lethal dosage (LD_{50}) The amount of liquid chemical agent expected to kill 50% of a group of exposed, unprotected individuals.

Median incapacitating dosage (LD_{50}) The volume of a liquid chemical agent expected to incapacitate 50% of a group of exposed, unprotected individuals.

Melting point The degree of temperature at which a solid substance becomes a liquid, especially under a pressure of one atmosphere.

Miosis A condition where the pupil of the eye becomes contracted (pinpointed), impairing night vision.

Miscible Mixable in any and all proportions to form a uniform mixture. Water and alcohol are miscible; water and oil are not.

Molecular formula Shows the *exact number* of each atom in the molecule.

Molecular weight The sum of the atomic weights of all the atoms in a molecule.

Molecule The smallest possible particle of a chemical compound that can exist in the free state and still retain the characteristics of the substance. Molecules are made up of atoms of various elements that form the compound.

Monomer A simple molecule capable of combining with a number of like or unlike molecules to form a polymer. It is a repeating structure unit within a polymer.

Mutagen A material that induces genetic changes (mutations) in the DNA of chromosomes. Chromosomes are the "blueprints" of life within individual cells.

Myalgia Muscular pain.

Mydriasis Dilation of the pupil.

Narcosis General and nonspecific reversible depression of neuronal excitability, produced by a number of physical and chemical agents, usually resulting in stupor rather than in anesthesia.

Necrosis Cell or tissue death due to disease or injury.

Nerve agent A substance that interferes with the central nervous system. Exposure is primarily through contact with the liquid (skin and eyes) and secondarily through inhalation of the vapor.

Neutralization A chemical reaction used to remove H^+ ions from acidic solutions and OH^- ions from basic solutions. The reaction can be violent and usually produces water, a salt, heat, and many times, a gas.

Nonpersistent agent An agent that remains in the target areas for a relatively short period of time. The hazard, predominantly vapor, will exist for minutes or, in exceptional cases, hours after dissemination of the agent. As a general rule, nonpersistent agent duration will be less than 12 hours.

Normal A solution that contains one equivalent of solute per liter of solution.

NRC National Response Center, a communications center for activities related to response actions, is located at Coast Guard headquarters in Washington, D.C. The toll-free number, (800) 424-8802, can be reached 24 hours a day for reporting actual or potential pollution incidents.

NRT National Response Team, consisting of representatives of 14 government agencies (DOD, DOI, DOT/RSPA, DOT/USCG, EPA, DOC, FEMA, DOS, USDA, DOJ, HHS, DOL, NRC, and DOE), is the principal organization for implementing the NCP. When the NRT is not activated for a response action, it serves as a standing committee to develop and maintain preparedness, to evaluate methods of responding to discharges or releases, to recommend needed changes in the response organization, and to recommend revisions to the NCP.

Odor A quality of something that affects the sense of smell; fragrance.

Odor threshold The greatest dilution of a sample with odor-free water to yield the least definitely perceptible odor.

OHMTADS Oil and Hazardous Materials Technical Assistance Data System, a computerized database containing chemical, biological, and toxicological information about hazardous substances. OSCs use OHMTADS to identify unknown chemicals and to learn how to best handle known chemicals.

Open-cup tester A device for determining flash points of flammable and combustible liquids, utilizing an open cup or container for the liquid. Recognized types are the Tagliabue (Tag) Open-Cup Apparatus and the Cleveland Open-Cup Apparatus.

Organic A material that comes from living plants or animals, such as waste or decay products. Distinguished from mineral matter. Organic chemistry deals with materials that contain the element carbon (C).

Organic peroxides Any organic compound containing oxygen (O) in the bivalent –O–O– structure and that may be considered a derivative of hydrogen peroxide, where one or more of the hydrogen atoms have been replaced by organic radicals.

Organophosphate A compound with a specific phosphate group that inhibits acetylcholinesterase. Used in chemical warfare and as an insecticide.

Oxidizer A material that gives up oxygen easily, removes hydrogen from another compound, or attracts negative electrons (such as chlorine or fluorine), thus enhancing the combustion of other materials.

Oxidizing agent A material that gains electrons from the fuel during combustion.

Oxime A compound that blocks acetylcholinesterase from combining with organophosphates, formed by the action of hydroxylamine upon an aldehyde or a ketone.

Oxygen deficient Defined by OSHA as ambient air containing less than 19.5% oxygen concentration.

Oxygen enriched Defined by OSHA as ambient air containing above 24% oxygen concentration.

2-PAM CL Pralidoxime chloride, Protopam®, is an antidote to organophosphate poisoning such as might result from exposure to nerve agents or some insecticides. The drug, which helps restore an enzyme called acetylcholinesterase, must be used in conjunction with atropine to be effective. Restores normal control of skeletal muscle contraction (relieves twitching and paralysis).

Papule A small, circumscribed, solid elevation on the skin.

PEL (permissible exposure limit) Term used by OSHA for its health standards covering exposures to hazardous chemicals. PEL generally relates to legally enforceable TLV limits.

Persistency An expression of the duration of effectiveness of a chemical agent, dependent on physical and chemical properties of the agent, weather, method of dissemination, and terrain conditions.

Persistent agent An agent that remains in the target area for longer periods of time. Hazards from both vapor and liquids may exist for hours, days, or in exceptional cases, weeks or months after dissemination of the agent. As a general rule, persistent-agent duration will be greater than 12 hours.

Photophobia Morbid dread and avoidance of light. Photosensitivity, or pain in the eyes with exposure to light, can be a cause.

pH The "power of hydrogen." A measure of the acidity or basicity of a solution, that is, of the concentration of H^+ or OH^- ions in solution. Scale ranges from 0 to 14, where a reading of 7 is neutral.

Physical properties A property of matter that describes only its condition, not the way it reacts with other substances. Examples are size, density, color, and electrical conductivity.

Poison Any substance (solid, liquid, or gas) that, by reason of an inherent deleterious property, tends to destroy life or impair health.

Polar-solvent liquids Those liquids that mix (are miscible with water).

Polymer A long chain of molecules having extremely high molecular weights made up of many repeating smaller units called monomers or comonomers.

Polymerization A chemical reaction in which small molecules combine to form larger molecules. A hazardous polymerization is a reaction that takes place at a rate that releases large amounts of energy that can cause fires or explosions or burst containers. Materials that can polymerize usually contain inhibitors that can delay the reaction.

Powder A solid reduced to dust by pounding, crushing, or grinding.

Ppm (parts per million) Parts of vapor or gas per million parts of contaminated air by volume at 25ºC and 1 torr pressure.

Pressure vessel A tank or other container constructed so as to withstand interior pressure greater than that of the atmosphere.

Psi Pounds per square inch.

Presynaptic Pertaining to the area on the proximal side of a synaptic cleft.

Pruritus Syn: itching

Ptosis, pl. Ptoses In reference to the eyes, drooping of the eyelids.

Pustule A small, circumscribed elevation of the skin containing pus and having an inflamed base.

Pyrogenic Causing fever.

Pyrolysis A chemical decomposition or breaking apart of molecules produced by heating in the absence of air.

Pyrophoric Material that ignites spontaneously in air below 130ºF (54ºC). Occasionally caused by friction.

Pyrophoric gas Gaseous materials that spontaneously ignite when exposed to air under ambient conditions. An example is trimethyl aluminum.

Pyrophoric liquid Liquid materials that spontaneously ignite when exposed to air under ambient conditions.

Pyrophoric solid Solid materials that spontaneously ignite when exposed to air under ambient conditions. An example is phosphorus.

RAD Radiation-absorbed dose.

Radiation Ionizing energy, either particulate or wave, that is spontaneously emitted by a material or combination of materials.

Radioactive material (DOT) Materials that emit ionizing radiation.

Radioactivity Any process by which unstable nuclei increase their stability by emitting particles (alpha or beta) or gamma rays.

Rate of explosion Rate of decomposition measured in feet per second in relation to the speed of sound. If subsonic, the rate is described as a deflagration. If supersonic, the rate of decomposition is defined as a detonation.

Reducing agent A substance that gives electrons to (and thereby reduces) another substance.

Respiratory asphyxiant A material that prevents or reduces the available oxygen necessary for normal breathing. Divided into simple and chemical asphyxiants.

Respiratory dosage This is equal to the time in minutes an individual is unmasked in an agent cloud multiplied by the concentration of the cloud.

Retinitis Inflammation of the retina.

Rhinorrhea A runny nose.

RCRA Resource Conservation and Recovery Act (of 1976), which established a framework for the proper management and disposal of all wastes.

Roentgen The amount of ionization that occurs per cubic centimeter of air.

Routes of exposure Ways in which chemicals get in contact with or enter the body. These are inhalation, absorption, ingestion, or injection.

RRT Regional response teams composed of representatives of federal agencies and a representative from each state in the federal region.

SADT *See* Self-accelerating decomposition temperature.

Safety fuse A flexible cord containing an internal burning medium by which fire or flame is conveyed at a uniform rate from point of ignition to point of use, usually a detonator.

Safety-relief valve A safety-relief device containing an operating part that is held normally in a position closing a relief channel by spring force, and is intended to open and close at a predetermined pressure.

SARA The Superfund Amendments and Reauthorization Act of 1986. Title III of SARA includes detailed provisions for community planning.

Self-accelerating decomposition temperature (SADT) Organic peroxides or other synthetic chemicals that decompose at ambient temperature, or react to light or heat, resulting in a chemical breakdown. This releases oxygen, energy, and fuel in the form of rapid fire or explosion. To ensure stabilization, these materials must be kept in a dark or refrigerated environment.

Sensitizer A substance that on first exposure causes little or no reaction in humans or test animals, but that on repeated exposure may cause a marked response not necessarily limited to the contact site.

Septic shock Shock associated with septicemia caused by gram-negative bacteria.

Shigellosis Bacillary dysentery caused by bacteria of the genus *Shigella*, often occurring in epidemic patterns.

Simple asphyxiant A material that replaces the amount of oxygen admitted into the body without further damage to tissue or poisoning. Examples are nitrogen and carbon dioxide.

Slurry A pourable mixture of solid and liquid.

Solid The state of matter having definite volume and rigid shapes. Its atoms or molecules are restricted to vibration only.

Solubility The ability of a substance to form a solution with another substance.

Solution The even dispersion (mixing) of molecules of two or more substances. The most commonly encountered solutions involve mixing of liquids and liquids or solids and liquids.

Solvent A substance, usually a liquid, capable of absorbing another liquid, gas, or solid to form a homogeneous mixture.

Specific gravity The weight of a solid or liquid substance as compared to the weight of an equal volume of water; specific gravity of water equals 1.

Spontaneous combustion A process by which heat is generated within material by either a slow oxidation reaction or by microorganisms.

Spontaneous ignition Ignition that can occur when certain materials, such as tung oil, are stored in bulk, resulting from the generation of heat, which cannot be readily dissipated; often heat is generated by microbial action.

Spontaneous ignition temperature *See* Ignition temperature.

States of matter Any of three physical forms of matter: solid, liquid, or gas.

Strength (acid/base) The amount of ionization that occurs when an acid or a base is dissolved in a liquid.

Stridor A high-pitched, noisy respiration, like the blowing of the wind; a sign of respiratory obstruction, especially in the trachea or larynx.

Subacute exposure 1. Less than acute. 2. Of or pertaining to a disease or other abnormal condition present in a person who appears to be clinically well. The condition may be identified or discovered by means of a laboratory test or by radiologic examination.

Sublimation The direct change of state from solid to vapor.

Subscripts Identify the number of atomic weights of the element present in the molecule.

Symbol Letters used to identify each element. The symbol for an element represents a definite weight (1 atomic weight) of that element.

Systemic toxicity Poisoning of the whole system or organism, rather than poisoning that affects, for example, a single organ.

Target organ The primary organ to which specific chemicals cause harm. Examples are the lungs, liver, or kidneys.

Temperature Measure of the vibratory rate of a molecule.

Teratogen material that affects the offspring when a developing embryo or fetus is exposed to that material.

Thermal burn Pertaining to or characterized by heat.

TLV Threshold limit value, *estimated* exposure value, below which no ill health effects *should* occur to the individual.

Toxemia A condition in which toxins produced by cells at a local source of infection or derived from the growth of microorganisms are contained in the blood.

Toxic Harmful, poisonous.

Toxicity The ability of a substance to cause damage to living tissue, impairment of the central nervous system, severe illness, or death when ingested, inhaled, or absorbed by the skin.

Toxins Toxic substance of natural origin produced by an animal, plant, or microbe. They differ from chemical substances in that they are not manmade. Toxins may include botulism, ricin, and mycotoxins.

TTL Threshold toxic limit, *estimated* exposure value, below which no ill health effects *should* occur to the individual.

Upper explosive limit The maximum fuel-to-air mixture in which combustion can occur.

Uricant A chemical agent that produces irritation at the point of contact, resembling a stinging sensation, such as a bee sting. For example, the initial physiological effects of phosgene oxime (CX) upon contact with a person's skin.

Vapor density The weight of a vapor or gas compared to the weight of an equal volume of air; an expression of the density of the vapor or gas calculated as the ratio of the molecule weight of the gas to the average molecule weight of air, which is 29. Materials lighter than air have vapor densities less than 1.

Vapor pressure The pressure exerted by a saturated vapor above its own liquid in a closed container. Vapor pressure reported on MSDS are in millimeters of mercury at 68°F (20°C), unless stated otherwise.

Vapors Molecules of liquid in air; moisture, such as steam, fog, mist, etc., often forming a cloud suspended or floating in the air, usually due to the effect of heat upon a liquid.

Variola Syn: smallpox.

Vesicants Chemical agents also called blister agents, that cause severe burns to eyes, skin, and tissues of the respiratory tract. Also referred to as mustard agents, examples include mustard and lewisite.

Vesicles Blisters on the skin.

Virus The simplest type of microorganism, lacking a system for its own metabolism. It depends on living cells to multiply and cannot live long outside of a host. Types of viruses are smallpox, Ebola, Marburg, and Lassa fever.

Violent reaction The action by which a chemical changes its composition near or exceeding the speed of sound, often releasing heat and gases.

Viscosity The measurement of the flow properties of a material expressed as its resistance to flow. Unit of measurement and temperature are included.

Vomiting agent Compounds that cause irritation of the upper respiratory tract and involuntary vomiting.

Water-reactive material A material that will decompose or react when exposed to moisture or water.

Water solubility The ability of a substance to mix with water.

Zoonosis An infection or infestation shared in nature by humans and other animals that are the normal host; a disease of humans acquired from an animal source.

APPENDIX

LIST OF ACRONYMS AND RECOGNIZED ABBREVIATIONS

AAR/BOE: Association of American Railroads/Bureau of Explosives
AIChE: American Institute of Chemical Engineers
ACC: American Chemical Council (Formerly CMA: Chemical Manufacturers Association, parent organization for CHEMTREC)
ASCS: Agricultural Stabilization and Conservation Service
ASME: American Society of Mechanical Engineers
ASSE: American Society of Safety Engineers
ATSDR: Agency for Toxic Substances and Disease Registry (HHS)
CAER: Community Awareness and Emergency Response (ACC)
CDC: Centers for Disease Control (HHS)
CEPP: Chemical Emergency Preparedness Program (EPA)
CERCLA: Comprehensive Environmental Response, Compensation, and Liability Act of 1980 (PL 96–510)
CFR: Code of Federal Regulations
Chemnet: A mutual-aid network of chemical shippers and contractors
CHEMTREC: Chemical Transportation Emergency Center
CHLOREP: A mutual-aid group comprised of shippers and carriers of chlorine
CHRIS/HACS: Chemical Hazards Response Information System/Hazard Assessment Computer System
CSEPP: Chemical Stockpile Emergency Preparedness Program
CWA: Clean Water Act
DOC: U.S. Department of Commerce
DOD: U.S. Department of Defense
DOE: U.S. Department of Energy
DOI: U.S. Department of Interior
DOJ: U.S. Department of Justice
DOL: U.S. Department of Labor
DOS: U.S. Department of State
DOT: U.S. Department of Transportation
EENET: Emergency Education Network (FEMA)
EMA: Emergency Management Agency
EMI: Emergency Management Institute
EOC: Emergency Operating Center
EOP: Emergency Operations Plan

EPA: U.S. Environmental Protection Agency
ERD: Emergency Response Division (EPA)
FEMA: Federal Emergency Management Agency
FWPCA: Federal Water Pollution Control Act
HazMat: Hazardous material
HazOp: Hazard and Operability Study
HHS: U.S. Department of Health and Human Services
ICS: incident command system
IEMS: integrated emergency management system
LEPC: Local Emergency Planning Committee
MSDS: material safety data sheets
NACA: National Agricultural Chemical Association
NCP: national contingency plan
NCRIC: National Chemical Response and Information Center (CMA)
NETC: National Emergency Training Center
NFA: National Fire Academy
NFPA: National Fire Protection Association
NIOSH: National Institute of Occupational Safety and Health
NOAA: National Oceanic and Atmospheric Administration
NRC: National Response Center
NRT: National Response Team
OHMTADS: Oil and Hazardous Materials Technical Assistance Data System
OSC: on-scene coordinator
OSHA: Occupational Safety and Health Administration (DOL)
PSTM: Pesticide Safety Team Network
RCRA: Resource Conservation and Recovery Act
RQs: reportable quantities
RRT: Regional Response Team
RSPA: Research and Special Programs Administration (DOT)
SARA: Superfund Amendments and Reauthorization Act of 1986 (PL 99–499)
SCBA: self-contained breathing apparatus
SERC: State Emergency Response Commission
SPCC: spill prevention control and countermeasures
TSD: treatment, storage, and disposal facilities
USCG: U.S. Coast Guard (DOT)
USDA: U.S. Department of Agriculture
USGS: U.S. Geological Survey
USNRC: U.S. Nuclear Regulatory Commission

NUMBERS TO REMEMBER

Agency for Toxic Substance and Disease Registry	(404) 639–0615
American Chemical Council (Formerly CMA)	(703) 741–5000
American Petroleum Institute	(202) 682–8000
American Trucking Association	(800) 282–5463
Ashland Chemical Company	(614) 790–3333
Association of American Railroads	(202) 639–2100
Bureau of Explosives	(202) 639–2222
CAS Registry Service	(800) 848–6538
Centers for Disease Control (CDC)	(404) 633–5313
CHEMTREC	(800) 424–9300
Non-Emergency	(800) 262–8200
Chlorine Institute	(202) 775–2790
CHEMTREC International Emergency	(703) 527–3887
Compressed Gas Association	(703) 412–0900
Defense Threat Reduction Agency (DRTA)	(703) 325–2102
DOW Chemical Emergency	(517) 636–4400
DuPont Chemical Emergency	(302) 774–7500
FEMA 24-hour Emergency	(202) 646–2400
Fertilizer Institute	(202) 675–8250
Kerr-McGee Chemical Company	(405) 962–0490
Malline Krodf Chemical Company	(636) 530–2000
Monsanto Agricultural Company	(314) 694–1000
National Poison Control Center	(800) 222–1222
National Fire Protection Association	(781) 837–6012
National Foam "Red Alert" 24-hour	(610) 363–1400
National Pesticide Information Center	(800) 858–7378
National Response Center	(800) 424–8802
National Safety Council	(202) 293–2270
National Transportation Safety Board (NTSB)	(202) 314–6000
Nuclear Regulatory Commission	(301) 492–7000
Poison Control Center	(800) 922–1117
Hazardous Materials Information Center	(800) 467–4922
U.S. DOT Information	(202) 366–4000
U.S. DOT Hazardous Materials Registration	(202) 366–4109
U.S. Army Explosives Disposal	(301) 677–9770
Toxic Substance Control Act Hotline	(202) 554–1404

RESOURCE GUIDE

Chemical Hazard Information Response System	CHRIS
Condensed Chemical Dictionary	CCD
Cross-handling Guide for Potentially Hazardous Materials	CROSS
Dangerous Properties of Hazardous Materials	SAX
North American Emergency Response Guide Book	NAERG
Emergency Handling of Hazardous Materials in Surface Transportation	EHHM
NFPA Fire Protection Guide on Hazardous Materials	FPG
NIOSH/OSHA Pocket Guide to Chemical Hazards	NIOSH
TLV Guide (American Conference of Governmental Industrial Hygienists)	TLV

Sources of Specific Information

Chemical name to four-digit UN number	NAERG, EHHM
Four-digit UN number to chemical name	NAERG, EHHM
Chemical name to STCC	EHHM
STCC to chemical name	EHHM
STCC to four-digit UN number	EHHM
Four-digit UN number to STCC	EHHM
Chemical name to synonym	CHRIS(MAN 2), NIOSH, CROSS, CCD (Limited), Sax (Limited)
Synonym to chemical name	CHRIS(MAN 2), CCD, Sax, MNFC
(From trade name)	FPG (Flash Point Index), CCD
(From chemical name)	CHRIS(MAN 2)

Product Uses

(From trade name) (Flash Point Index)	FPG, CCD
(From chemical name)	CCD, CROSS
Product trade name to chemical composition	(contact manufacturer)
NFPA 704 designation	FPG (Sec. 325M, Sec. 49)
Chemical formula	NIOSH, CCDCHRIS(MAN 2), FPG (Sec. 325M, Sec. 49), Sax
Reactions	FPG (Sec. 491M), NIOSH, CHRIS, CROSS, EHHM
IDLH	CHRIS (MAN 2), NIOSH
TLV	TLV
PEL	NIOSH
LD_{50}	CHRIS(MAN 2), Sax
LC_{50}	CHRIS(MAN 2), Sax
Odor threshold	CHRIS(MAN 2)
Physical, chemical properties	FPG (Sec.49, Sec. 325), CHRIS(MAN 2), CROSS, EHHM, CCD, NIOSH, ERG
(BASIC) medical information	FPG (Sec. 49), CHRIS(MAN 2), CROSS, EHHM, CCD
NIOSH,	ERG
(BASIC) fire protection, personal protection	FPG (Sec. 49), CHRIS(MAN 2), CROSS, EHHM, CCD, NIOSH, ERG

SELECTED TECHNICAL REFERENCES

Computer-Aided Management of Emergency Operations (CAMEO), National Safety Council, 1019 19th Street N.W., Suite 401, Washington, D.C., 20036.

Chemical Hazard Response Information System (CHRIS), Superintendent of Documents, U.S. Government Printing Office, Washington, D.C., 20402–9328.

Condensed Chemical Dictionary, Van Nostrand and Reinhold Co., 115 Fifth Avenue, New York, 10003.

Dangerous Properties of Industrial Chemicals (Sax), Van Nostrand and Reinhold Co., 115 Fifth Avenue, New York, 10003.

Dictionary of Chemical Names and Synonyms, CRC/Lewis Publishers, 2000 Corporate Blvd. N.W., Boca Raton, FL, 33431.

North American Emergency Response Guide Book (NAERG), U.S. Department of Transportation, RSPA, 400 Seventh Street, S.W., Washington, D.C., 20590–0001.

Emergency Action Guides, Bureau of Explosives, Association of American Railroads, 50 F Street, N.W., Washington, D.C., 20001.

Emergency Handling of Hazardous Materials in Surface Transportation, Bureau of Explosives, Association of American Railroads, 50 F Street, N.W., Washington, D.C., 20001.

Farm Chemical Handbook, Meister Publishing, 37841 Euclid Ave., Willoughby, OH, 44094.

Fire Protection Guide on Hazardous Materials, National Fire Protection Association (NFPA), One Batterymarch Park, Quincy, MA, 02110.

Handbook of Chemistry and Physics, CRC/Lewis Publishers, 2000 Corporate Blvd. N.W., Boca Raton, FL, 33431.

Merck Index, Merck and Company, Rahway, N.J., 07065

NIOSH Pocket Guide to Chemical Hazards, Superintendent of Documents, U.S. Government Printing Office, Washington, D.C., 20402–9328.

IUPAC RULES OF NOMENCLATURE

Hydrocarbons with More Than 10 Carbons

$C_{11}H_{24}$	Undecane
$C_{20}H_{42}$	Eicosane
$C_{12}H_{26}$	Dodecane
$C_{21}H_{42}$	Heneicosane
$C_{13}H_{28}$	Tridecane
$C_{22}H_{46}$	Docosane
$C_{14}H_{30}$	Tetradecane
$C_{23}H_{48}$	Tricosane
$C_{15}H_{32}$	Pentadecane
$C_{26}H_{54}$	Hexacosane
$C_{16}H_{34}$	Hexadecane
$C_{30}H_{62}$	Triacontane
$C_{17}H_{36}$	Heptadecane
$C_{31}H_{64}$	Hentriacontane
$C_{18}H_{38}$	Octadecane
$C_{32}H_{66}$	Dotriacontane
$C_{19}H_{40}$	Nonadecane
$C_{33}H_{68}$	Tritriacontane
$C_{40}H_{82}$	Tetracontane
$C_{49}H_{100}$	Nonatetracontane
$C_{50}H_{102}$	Pentacontane
$C_{60}H_{122}$	Hexacontane
$C_{70}H_{142}$	Heptacontane
$C_{80}H_{162}$	Octacontane
$C_{90}H_{182}$	Nonacontane
$C_{100}H_{202}$	Hectane
$C_{132}H_{266}$	Dotriacontahectane

There are four types of structures:

Molecular:

C_2H_{10}

Structural:

```
     H   H   H   H   H
     |   |   |   |   |
H —  C — C — C — C — C — H
     |   |   |   |   |
     H   H   H   H   H
```

Condensed structural:

$CH_3CH_2CH_2CH_3$

Skeleton:

C–C–C–C

The naming of all the alkanes is based upon the number of carbon atoms in the longest continuous chain of carbon atoms. If, for example, the longest chain contains four carbon atoms, the compound would be called butane. If it has five carbon atoms, it is pentane, and so on.

Names of branched-chain hydrocarbons and hydrocarbon derivatives using the IUPAC system are based on the name of the longest continuous carbon chain in the molecule, with a number indicating the location of a branch or substituent.

In order to locate the position of a branch or substituent, the carbon chain is numbered consecutively from one end to the other, starting at that end that gives the lowest numbers to the substituents. For example:

```
     1    2    3    4    5
     H    H    H    H    H
     |    |    |    |    |
 H — C —  C —  C —  C —  C — H     Is 2-Methylpentane
     |    |    |    |    |
     H    |    H    H    H
      H — C — H
          |
          O
          |
          H
```

```
     1    2    3    4
     H    Cl   H    H
     |    |    |    |
 H — C —  C —  C —  C — H     Is 2,2-Dichlorobutane
     |    |    |    |
     H    Cl   H    H
```

The prefixes "di," "tri," "tetra," "penta," "hexa," etc. indicate how many of each substituent is in the molecule. A cyclic (ring) hydrocarbon is designated by the prefix "cyclo."

Double bonds in hydrocarbons are indicated by changing the suffix "ane" to "ene," and triple bonds by changing to "yne." The position of the multiple bond within the structure is indicated by the number of the first or lowest-numbered carbon atom attached to the multiple bond. For example:

```
     1    2    3    4    5
     H    H    H    H    H
     |    |    |    |    |
 H — C —  C —  C =  C —  C — H     Is 2-Pentene
     |    |              |
     H    H              H
```

```
     5    4    3     2    1
     H    H    H
     |    |    |
H—   C —  C —  C —  C ≡ C — H      is 1-Pentyne
     |    |    |
     H    H    H
```

```
     4    3    2    1
     H    H    H    H
     |    |    |    |
H—   C =  C —  C =  C — H          is 1,3,-Butadiene
```

Most of the hydrocarbon-derivative functional groups in organic compounds are designated by either a suffix or a prefix, as shown in Table B. Rules regarding whether a prefix or a suffix designation is used are as follows:

1. When one such group is present, the suffix will be used.
2. When more than one such group is present, only one will be designated by a suffix; the others are designated by prefixes.
3. The order of precedence for deciding which group takes the suffix designation is the same as the order in Table B.

Examples:

```
     H    H
     |    |
H—   C —  C — O — H          is ethanol, not Hydroxyethane
     |    |
     H    H
```

```
          3    2    1
          H    H    O
          |    |    ||
H — O —   C —  C —  C — O — H      is 3-Hydroxypropanoic acid
          |    |
          H    H
```

In numbering the carbon chain, the lowest numbers will be given preference to:

1. Groups in Table B named by suffixes
2. Double bonds
3. Triple bonds
4. Groups named by prefixes (groups named by prefixes are listed in alphabetical order)

Nomenclature of Aromatic Compounds

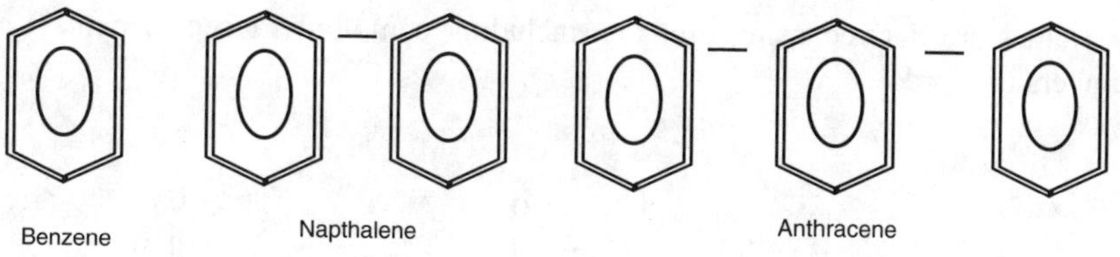

Monosubstituted Compounds

Names are derived using prefixes from Table B and Table C, followed by the name "benzene."

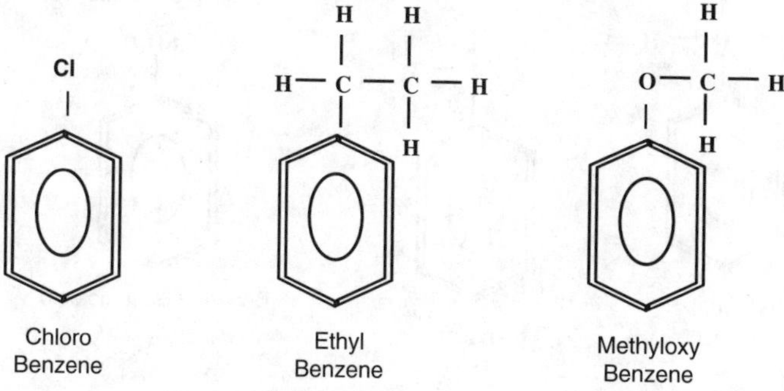

Names can be indicated by commonly accepted names.

Disubstituted Compounds

Names are derived using prefixes (including commonly accepted names) and numbers or words.

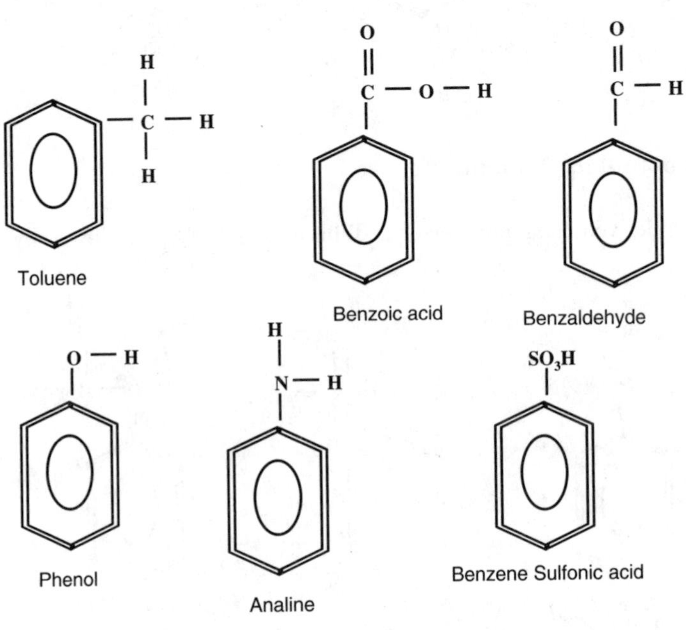

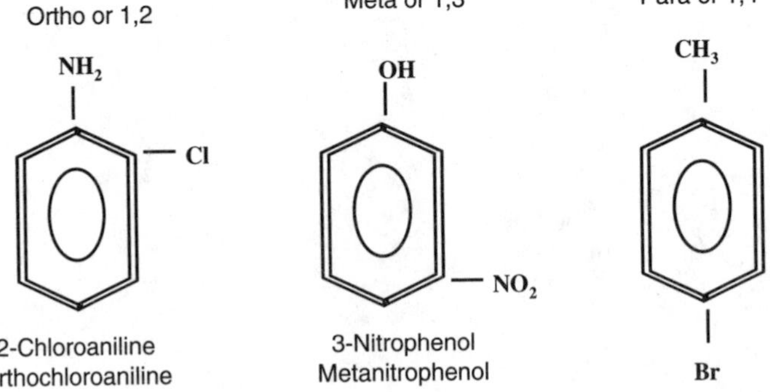

4-Bromotoluene
Parabromotoluene

STUDY GUIDE

CHEMISTRY OF HAZMAT STUDY SHEET

PERIODIC TABLE:

Family 1-Column 1-Alkali Metals
Family 2-Column-2Alkaline Earth Metals
Family 3-Between Towers-Transitional Metals
Family 4-Column 7-Halogens
Family 5-Column 8-Inert/Noble Gases

ATOMIC STRUCTURE:

Atomic Number=Number of Protons
Atomic Weight=Protons + Neutrons
Number of Protons = Number of Electrons
Octet-8 electrons outer shell
Duet-2 electrons outer shell

HAZMAT ELEMENTS:

Hydrogen	H
Lithium	Li
Sodium	Na
Potassium	K
Magnesium	Mg
Calcium	Ca
Chromium	Cr
Iron	Fe
Cobalt	Co
Copper	Cu
Mercury	Hg
Tin	Sn
Lead	Pb
Aluminum	Al
Carbon	C
Nitrogen	N
Phosphorus	P
Oxygen	O
Sulfur	S
Fluorine	F
Chlorine	Cl
Bromine	Br
Iodine	I
Boron	B
Krypton	Kr
Uranium	U
Xenon	Xe
Gold	Au
Arsenic	As
Plutonium	Pu
Zinc	Zn
Silicon	Si
Titanium	Ti
Radium	Ra
Silver	Ag
Manganese	Mn
Beryllium	Be
Helium	Hl
Argon	Ar

BINARY:

Metal + **Non Metal**
Not Oxygen
Ends in **ide**
General Hazard
Except ***NCHP (When Wet)**
***Nitride, Carbide, Hydride, Phosphide**

BINARY OXIDE:

Metal + **Oxygen**
Ends in **Oxide**
CL,RH

PEROXIDE:

Metal + O_2^{-2}
Ends in
Peroxide
CL,RH,RO

HYDROXIDE:

Metal + OH^{-1}
Ends in
Hydroxide
CL,RH

OXYSALTS:

Metal + **Oxy Radical**

-1	-2	-3
FO_3	CO_3	PO_4
ClO_3	SO_4	
BrO_3		
IO_3		
MnO_3		
NO_3		

+1 Oxygen Per___ate
Base State_____ate
-1 Oxygen_______ite
-2 Oxygen Hypo__ ite

(Cation=Positive Ion, Anion=Negative Ion)

H/C PREFIXES:

Meth	1
Eth	2
Prop	3
But	4
Pent	5
Hex	6
Hept	7
Oct	8
Non	9
Dec/Dek	10

ALKANES:

Single Bond
Saturated
Ends in ANE
CnH2n+2

ALKENES:

Double Bond
Unsaturated
Ends in ENE
CnH2n

ALKYNES:

Triple Bond
Unsaturated
Ends in YNE
CnH2n-2
Know (Acetylene)

AROMATICS:

Resonant Bond
Acts Saturated
"BTX" BENZENE
TOLUENE
XYLENE

(diene-2 double bonds, triene-3 double bonds)
Iso-Branched, **Neo**-Center Carbon surrounded by Carbons
Cyclo-All Carbons hooked together in a circle or end to end.
prefixes: Di = 2, Tri = 3, Tetra = 4

Transitional Metal Charges:

Mercury	I or II
Chromium	II or III
Iron	II or III
Copper	I or II
Manganese	II or III
Tin	II or IV

RADICAL PREFIXES:

SINGLE BOND:
Meth=Methyl/Form*
Eth=Ethyl/Acet*
Prop=Propyl
But=Butyl
Pent=Pentyl

DOUBLE BOND:
Ethene=**Vinyl=C_2H_3**
Propene=**Acryl (Allyl)= C_3H_5**

AROMATIC:
Benzene= **Phenyl=C_6H_5**
Toluene= Benzyl=$C_6H_5CH_2$

***Form and Acet prefixes used with Acids Aldehydes, and Esters. ©' RAB 4/1/93**

Hydrocarbon Derivatives

Func Grp	Gen Form	Structure	Hazard
Alkyl Halide	R-X	Cl F Br I	Toxic/Flam
Nitro	$R\text{-}NO_2$	$-N(=O)-O$ (N bonded to two O)	Explosive
Amine	$R\text{-}NH_2$, R_2NH R_3N	$-N(H)-H$; $-N(-)-H$; $-N(-)-$	Toxic/Flam
Ether	R-O-R	– O –	Anest/WFR**
Organic Peroxide	R-O-O-R	– O – O –	Explosive/Oxy
Alcohol	R-O-H	– O – H	Toxic/WFR
Ketone	R-C-O-R	$-C(=O)-$	Narc/Flammable
*Ester	R-C-O-O-R, $R\text{-}C\text{-}O_2\text{-}R$	$-C(=O)-O-$	Polymerize/Flam
*Aldehyde	R-C-H-O	$-C(=O)-H$	Toxic/WFR
*Organic Acid	R-C-O-O-H	$-C(=O)-O-H$	Toxic/Flam/Cor

* When naming the structure, count all carbon atoms hooked together and attached to the carbon atom in the functional group.

** WFR = Wide Flammable Range.

NAMING RULES FOR HYDROCARBON DERIVATIVES

ALKYL HALIDES: Name functional group first, such as chlorine, which would become **chloro**, fluorine would become **fluoro**, and so on. Then name the hydrocarbon backbone: one carbon would be methane, two carbons ethane, and so on. For exanple, **chloro methane**. Or name the hydrocarbon backbone first: one carbon would become **methyl**, two carbons **ethyl**, and so on. Then name the functional group and add "ide" for an ending. For example, chlorine would become **chloride**, fluorine would become **fluoride**, and so on. An example is **methyl chloride**.

NITROS: The word "nitro" comes first, followed by the hydrocarbon radicals. For example, **nitro methane**.

AMINES: The hydrocarbon radicals come first in the name, followed by the word "amine." When there is more than one radical, start with the smallest and go to the largest, and then end with the word "**amine**." For example, **vinyl amine**, **methyl vinyl amine**, or **methyl ethyl vinyl amine**.

ETHERS: Start with the smallest hydrocarbon radical and name it, then the other hydrocarbon radical and name it, then end with the word "**ether**." For example, **methyl ethyl ether**.

ORGANIC PEROXIDES: Start with the smallest hydrocarbon radical and name it, then find the other hydrocarbon radical and name it, then end with the word "**peroxide**." For example, **ethyl propyl peroxide**.

ALCOHOLS: Name the hydrocarbon radical, then end with the word "**alcohol**," or name the hydrocarbon radical and end with "**ol**." For example, **ethanol**.

KETONES: Start with the smallest hydrocarbon radical and name it, then find the other hydrocarbon radical and name it, then end with the word "**ketone**." For example, **methyl ethyl ketone**.

ESTERS: First of all, nothing is named "**ester**." If there is a **vinyl radical to the left** of the ester functional group, it is an **acrylate**; if there is a **methyl radical to the left** of the ester functional group, it is an **acetate**. (You can also count all the carbons.) Then name the radical on the right, and name the compound. Examples include **ethyl acetate** or **propyl acrylate**.

ALDEHYDES: Name the hydrocarbon radical first, counting the carbon atoms in the aldehyde functional group, and end with the word "**aldehyde**." When naming aldehydes, you use the prefix **"form" for one carbon** and **"acet" for two carbons**. For example, **formaldehyde** for a one-carbon aldehyde.

ORGANIC ACIDS: Name the hydrocarbon radical first, adding "**ic**" to the radical, and end with the word "**acid**." As with the aldehydes, when naming the organic acids, count the carbon atoms in the functional group and use "**form**" and "**acet**." For example, **acetic acid** or **formic acid**.

BRANCHING HYDROCARBON DERVATIVES

n-Normal: straight-chained

```
    H   H   H   H
    |   |   |   |
H - C - C - C - C - C - H
    |   |   |   |
    H   H   H   H
```

Iso: carbon atom attached to functional group is touching one other carbon atom.*

```
    H
    |
H - C     H   H
    | \   |   |
    H     C - C - C - H
    H   /     |
    |         H
H - C
    |
    H
```

Secondary: carbon atom attached to functional group is touching two other carbon atoms.

```
              H
              |
    H   H     O   H
    |   |     |   |
H - C - C  -  C - C - H
    |   |     |   |
    H   H     H   H
```

Tertiary: carbon ztom attached to functional group is touching three other carbon atoms.

```
        H
        |
    H   O   H
    |   |   |
H - C - C - C - H
    |   |   |
    HH- C -HH
        |
        H
```

*Propane has only one possible branch, which is iso.

POLARITY

ORGANIC ACID	SUPER-DUPER POLAR
ALCOHOL	SUPER POLAR
KETONE	POLAR
ALDEHYDE	POLAR
ESTER	POLAR
AMINE	SLIGHTLY POLAR
ETHER	NONPOLAR
NITRO	NONPOLAR
ALKYL HALIDE	NONPOLAR
ORGANIC PEROXIDE	NONPOLAR
HYDROCARBONS	NONPOLAR

THINGS THAT AFFECT BOILING POINT

1. WEIGHT
2. POLARITY
3. BRANCHING

BOILING POINT RELATIONSHIPS

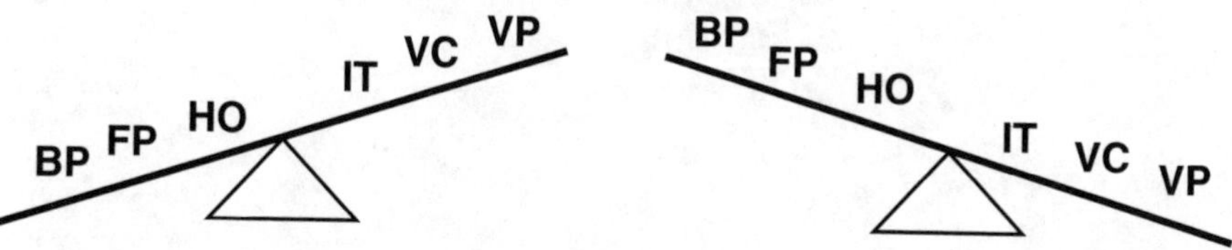

BP = Boiling Point
FP = Flash Point
HO = Heat Output
IT = Ignition Temperature
VC = Vapor Content
VP = Vapor Pressure

First determine boiling point, and then compare boiling point to the other components.

WEAK BONDS WITHIN A STRUCTURE

Carbon to Carbon Double Bond	Alkenes	C=C
Carbon to Carbon Triple Bond	Alkynes	C≡C
Oxygen to Oxygen Single Bond	Peroxide/Nitro	–O–O–
Carbonyl when on the last Carbon	Aldehyde	O ‖ –C–H

ANSWERS TO REVIEW QUESTIONS

Chapter 2

1. B
2. C
3. Noble Gases VIII, Bromine not a family, Alkaline Earth I, Alkali II, Halogens VII
4. B
5. A
6. C
7. B and C
8. B
9. NaCl, sodium chloride, binary salt, varying
   $Ca_3(PO_4)_2$, calcium phosphate, oxysalt oxidizer
   $Al_2(O_2)_3$, aluminum peroxide, peroxide RH, CL, RO
   $CuBr_2$, copper II, bromide, binary salt, varying
   KOH, potassium hydroxide, hydroxide RH, CL
   Li_2O, lithium oxide, metal oxide RH, CL
   $Mg(ClO)_2$, magnesium hypochlorite, oxysalt, oxidizer
   HgO_2, mercury II, peroxide, peroxide RH, CL, RO
   NaF, sodium fluoride, binary salt, varying
   $FeCO_3$, iron II, carbonate, oxysalt, oxidizer
10. $Ca(ClO)_2$, binary salt, varying
    $AlCl_3$, binary salt, varying
    LiOH, hydroxide RH, CL
    CuO_2, peroxide RH, CL, RO
    Na_2O, metal oxide RH, CL
    KI, binary salt, narying
    Mg_3P_2, binary salt, varying
    $HgClO_4$, oxysalt, oxidizer
    $Fe(FO_3)_3$, oxysalt oxidizer

Chapter 3

1. C
2. Mechanical overpressure, mechanical chemical, chemical reaction, dust, nuclear
3. Detonation and deflagration

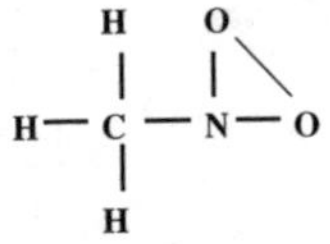

Nitro methane

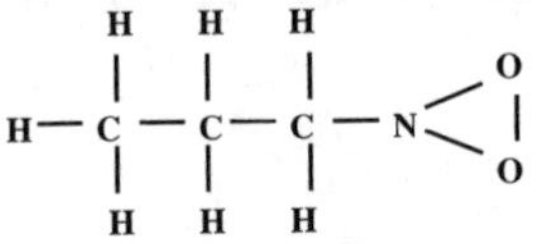

$C_3H_7NO_2$

4.

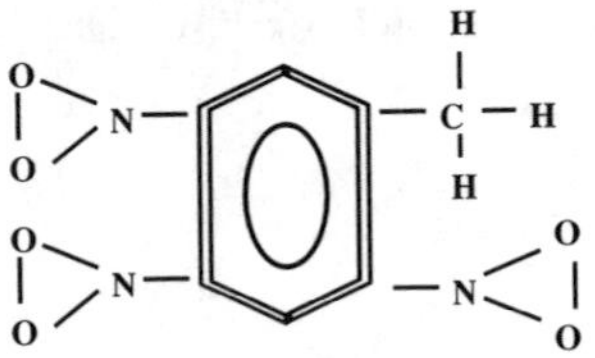

Tri-nitro toluene

$C_6H_2(NO_2)_3OH$

5. Confinement, fuel, chemical oxidizer, heat
6. B
7. A
8. C
9. C
10. B
11. 1.1, 1.2, 1.3, 1.4, 1.5, 1.6
12. B
13. C
14. C
15. A

Chapter 4

1. B
2. A
3. D
4. A
5. D
6. B
7. A
8. Isobutane C_4H_9, ethane C_2H_6, methane CH_4, ethene C_2H_4
9. Alkane, alkyne, alkene, alkane, alkene
10. Alkane saturated, alkene unsaturated, alkyne unsaturated

Chapter 5

1. B
2. B
3. A
4. C
5. Weight, polarity, branching
6. B, C, B, A, D, B
7. B

8. Alkane, aromatic, alkene, aromatic, alkene

9.

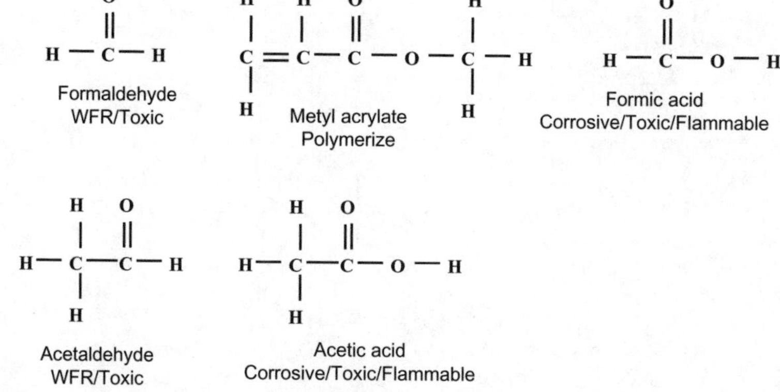

10. Acetaldehyde CH_3CHO WFR/toxic, butadiene C_4H_6 flammable, isopropyl alcohol C_3H_7OH WFR/toxic, methyl amine CH_3NH_2 toxic/flammable

Chapter 6

1. Flammable solid, spontaneously combustible, dangerous when wet
2. D
3. B
4. D
5. A, C, D
6. D
7. B
8. D
9. D
10. D

Chapter 7

1. Fluorine, chlorine, bromine, oxygen
2. Halogens, peroxide salts, oxysalts, organic peroxides
3. D
4. Aluminum persulfate $Al_2(SO_5)_3$
   Lithium chlorite, balanced
   Sodium fluorate, balanced
   Magnesium hypophosphite $Mg_3(PO_2)_2$
   Copper I chlorate, balanced

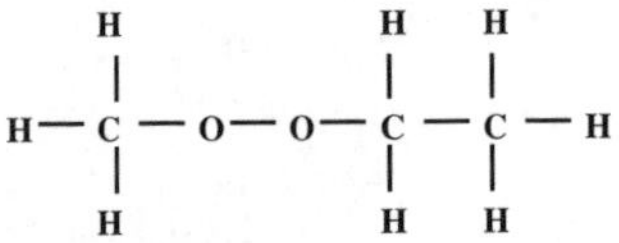

$CH_3OOC_2H_5$

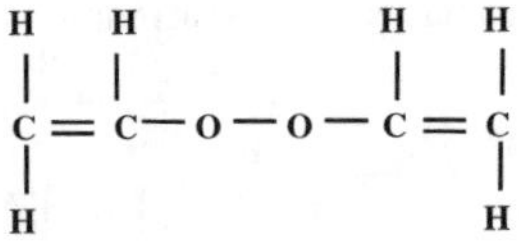

$C_2H_3OOC_2H_3$

5.

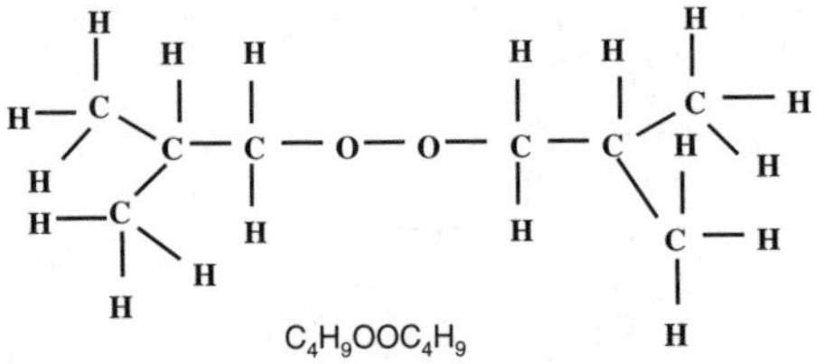

$C_4H_9OOC_4H_9$

6. C
7. D
8. B

Chapter 8

1. C
2. A
3. Inhalation, ingestion, absorption, injection
4. A
5. C
6. D
7. B

8.

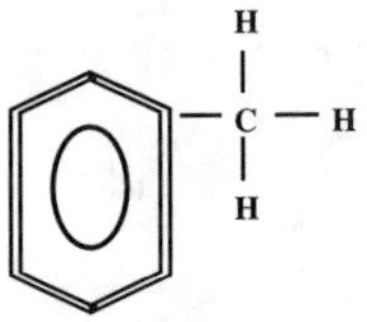

Toluene
Toxic/Flammable

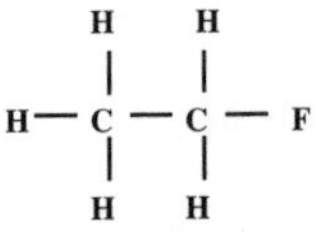

Ethyl fluoride
Toxic/Flammable

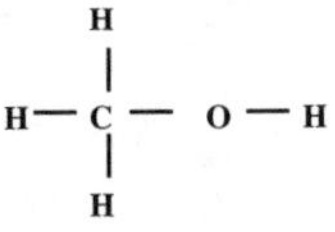

Methyl alcohol
WFR/Toxic

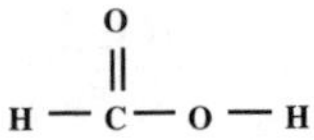

Formic acid
Corrosive/Toxic/Flammable

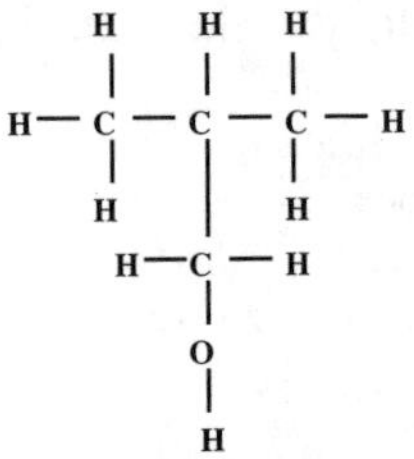

Tertiary butyl alcohol
WFR/Toxic

9. Caution, danger, warning
10. STEL, TLV-TWA, PEL, IDLH

Chapter 9

1. C
2. B
3. Alpha, beta, gamma
4. Alpha, beta
5. D
6. Ionizing, nonionizing
7. C, A, B
8. C
9. C
10. Time, distance, shielding

Chapter 10

1. Acids, bases
2. Inorganic, organic
3. C
4. B
5. D
6. A
7. C
8. Neutralization, dilution
9.

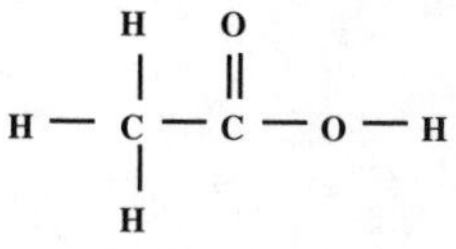

```
     H   O
     |   ||
H —  C — C — O — H
     |
     H
```

Acetic acid

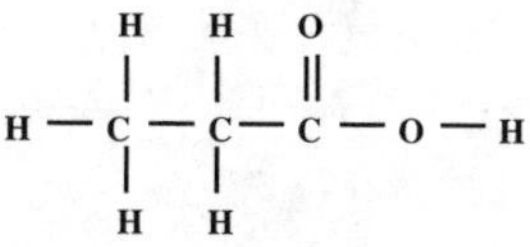

```
      H    H    O
      |    |    ||
H  —  C —  C —  C — O — H
      |    |
      H    H
```

Proprionic acid

```
     H    H    H    O
     |    |    |    ||
H  — C —  C —  C —  C — O — H
     |    |    |
     H    H    H
```

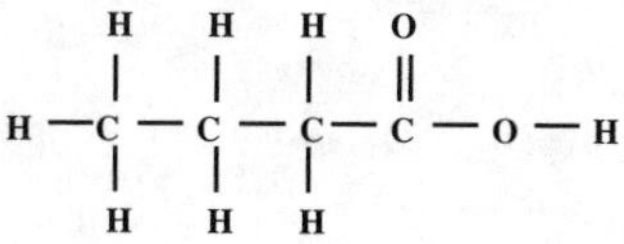

Butyric acid

10. Tertiary butric acid tC_4H_9COOH, acrylic acid C_2H_3COOH, benzoic acid C_6H_5COOH

Chapter 11

1. C
2. A
3. D

References

1. Killen, W.D., Eight Die in Chlorine Tanker Derailment, *Fire Command,* National Fire Protection Association, Quincy, MA.
2. *NFPA Quarterly,* National Fire Protection Association, Quincy, MA, October 1964.
3. Sullivan, B., Burns Illustrate Need for Bunker Gear, *Firehouse Magazine,* July 1994.
4. *Introduction to Chemistry,* Chisholm and Johnson, Usborn Publishing, London, England.
5. *Milwaukee Fire Department Basic Training Manual*, 1986.
6. Harte, J. et. al., *Toxics A-Z,* University of California Press, CA, 1991.
7. National Transportation Safety Board, Anhydrous hydrogen fluoride release from NATX 9408, Elkhart, IN, February 4, 1985.
8. Dektar, C., Peroxide Blast Shatters Chemical Plant, *Fire Engineering,* January 1979.
9. Klem, T.J., High-Rise Fire Claims Three Philadelphia Fire Fighters, *NFPA Journal,* September/October 1991, Quincy, MA: National Fire Protection Association.
10. *JEMS,* Proper Use of Highway Flares, August 1982.
11. Hill, I., *Crescent City Remembers,* Scheiwe's Print Shop, Crescent City, IL, 1995.
12. Ryczkowski, J.J., , Kingman Revisited, *American Fire Journal*, July 1993.
13. *North American Emergency Response Guide Book 1996,* U.S. Department of Transportation, U.S. Government Printing Office, 1996.
14. Code of Federal Regulations (CFR) 49, Parts 100–177, American Trucking Association, October 1, 1994.
15. *Hawley's Condensed Chemical Dictionary,* 12th ed., Van Nostrand Reinhold, New York, 1993.
16. *National Fire Academy Chemistry of Hazardous Materials Instructor Guide*, U.S. Government Printing Office, 1994.
17. *National Fire Academy Chemistry of Hazardous Materials, Student Manual*, U.S. Government Printing Office, 1994.
18. *National Fire Academy Initial Response to Hazardous Materials Incidents: Basic Concepts, Student Manual*, U.S. Government Printing Office, 1992.
19. *National Fire Academy Initial Response to Hazardous Materials Incidents: Concept Implementation*, U.S. Government Printing Office, 1992.
20. National Fire Protection Association, *NFPA Fire Protection Handbook,* 17th ed., Quincy, MA, 1992.
21. *NIOSH Pocket Guide to Chemical Hazards,* U.S. Government Printing Office, June 1997.
22. National Fire Protection Association, *Fire Protection Guide to Hazardous Materials,* 11th ed., Quincy, MA, 1994.
23. Kamrin, M.A., *Toxicology,* Lewis Publishers, 1988.
24. Brady, J.E. and Holm, J.R., *Fundamentals of Chemistry,* 3rd ed., John Wiley & Sons, 1988.

25. Cashman, J.R., *Hazardous Materials Emergencies Response and Control,* 1st ed., Technomic Publishing, 1983.
26. California Department of Justice Web page, Stop Drugs.org-Resources 8/17/01.
27. Burke, R., *Counter-Terrorism for Emergency Responders*, 1st ed., Lewis Publishers, Boca Raton, FL, 2000.
28. United States Chemical Safety and Hazard Investigation Board web site, *www.chem-safety.gov,* 3–26–02.
29. National Safety Council Data Sheet, I-655 Recognition and Handling of Peroxidizable Compounds, 1982.
30. Basic Chemical Segregation, *www.acs.ucalgary.ca*
31. Chemical Compatibility, *www.labsafety.org,* adapted from *CRC Press Laboratory Handbook,* CRC Press, Boca Raton, FL.
32. *2000 Emergency Response Guide Book*, United States Department of Transportation.
33. Los Alamos National Labs Periodic Table of Elements, http//:pearl1.lanl.gov/periodic/default.htm.

Index

A

B

C

D

E

F

G

I

K

L

M

N

O

P

Q

R

S

T